CW00548049

THE TRAVELS OF MARCO POLO

TRANSLATED BY

JOHN FRAMPTON

In the same series

A NEW VOYAGE ROUND THE WORLD

by WILLIAM DAMPIER

Introduction by Sir ALBERT GRAY, K.C.B., K.C.

The United States
THE MACMILLAN COMPANY, NEW YORK

Australia and New Zealand
THE OXFORD UNIVERSITY PRESS, MELBOURNE

Canada
THE MACMILLAN COMPANY OF CANADA, TORONTO

South Africa
THE OXFORD UNIVERSITY PRESS, CAPE TOWN

India and Burma
MACMILLAN AND COMPANY LIMITED
BOMBAY CALCUTTA MADRAS

THE MOST NOBLE AND FAMOUS TRAVELS OF MARCO POLO

TOGETHER WITH THE TRAVELS OF

NICOLÒ DE' CONTI

Edited
from the Elizabethan Translation of

JOHN FRAMPTON

With Introduction, Notes and Appendices by
N. M. PENZER, M.A.

SECOND EDITION

ADAM AND CHARLES BLACK
4, 5 & 6 SOHO SQUARE LONDON W.1
1937

370
F9
1937

FIRST PUBLISHED 1929 BY THE ARGONAUT PRESS
IN AN EDITION LIMITED TO 1050 COPIES
SECOND EDITION, ENLARGED, 1937
PUBLISHED BY ADAM AND CHARLES BLACK

MADE IN GREAT BRITAIN

PREFACE

APART from the general interest attaching itself to an Elizabethan translation of the Travels of Marco Polo, the present edition aims at supplying a long-felt want in Polian research—a series of maps embodying the latest work and discoveries of explorers and cartographers.

Owing to the kindness of Sir Aurel Stein in allowing me to use the maps illustrating his Third Journey in Innermost Asia, I have been able, with the help of my expert cartographer, Miss G. Heath, to construct eleven entirely new maps, which I trust will help to elucidate the itinerary of our great traveller.

With regard to the notes, the chief difficulty I have experienced is brevity, for Marco Polo offers unlimited resources to the student of research. I have, however, chiefly limited my notes to a consideration of Frampton's text, and to any fresh light that has been shed on the vexed question of Polo's itineraries.

I am greatly indebted to the Rev. A. C. Moule, who has read my proofs and given me valuable advice; to Dr C. O. Blagden, who has guided Polo's fleet safely through the Malay Archipelago; and to Miss Frances Welby, whose generous help in Mediaeval Spanish and Italian has been of the very greatest value. The exhaustive index is the work of the Cambridge University Press, whose help and care throughout the entire work has been beyond praise.

SAINT JOHN'S WOOD

April 1929

CONTENTS

MAPS AND ILLUSTRATIONS

THE TRAVELS OF MARCO POLO

TRANSLATED BY

JOHN FRAMPTON

INTRODUCTION

THE existence of an Elizabethan translation of the Travels of Marco Polo will probably come as a surprise to the majority of readers. This is not to be wondered at when we consider that only three copies of the work in question are known to exist, and that it has never been reprinted. The very rarity of the book would be of itself sufficient excuse for reprinting it, but in the present case there are other considerations which make its appearance little less than a necessity.

In the first place, its value to students of Elizabethan literature is self-evident. Bearing this in mind, I have made no attempt to alter the spelling in any way, nor have I marred the charm of the narrative as known to contemporary readers by the insertion of unsightly notes. These are relegated to the end of the volume. The original head- and tail-pieces have also been preserved, together with sixteenth-century capitals.

In the second place, the translation, made by John Frampton from the Castilian of Santaella, originates in a MS belonging to the Venetian recension, one of the most important of all the Polian recensions. Its editing, therefore, should be of considerable interest.

Then again, the recently issued work of Prof. Benedetto, to which we shall return later, has so largely helped to unravel the tangled skein of Polian texts, that it is now necessary to reconsider afresh many of our long-accepted theories.

Finally, thanks to the recent surveys carried out in Central Asia and Mongolia, we are able to trace the itineraries with a much greater degree of accuracy than before, and although many queries still remain, some of the blanks have been filled in, and a few of the old mistakes rectified.

§ i. JOHN FRAMPTON AND SANTAELLA

John Frampton

Apart from what Frampton tells us about himself in the Prefaces to one or two of his translations, we know nothing whatever about him. From these we learn that he was resident for many years in Spain, and that on his return to his native country about 1576, employed his leisure in translating several works from the Spanish. His knowledge of the language was very extensive as a comparison of the original with any of his translations will show. He must have worked hard during the first few years after his return to England, as between 1577 and 1581 six separate translations made their appearance.

His first work seems to have been an English rendering of Nicolas Monardes' *Primera Y Segunda Y Tercera Partes de la Historia Medicinal de las Cosas que se traen de nuestras Indias Occidentales que siruen en Medicina*, printed at London in 1577 by William Norton "in Poules Churche-yarde," under the title of *Joyfull Newes out of the Newe Founde Worlde wherein is declared the rare and ſingular vertues of diverſe and sundrie hearbes, trees, oyles, plantes and ſtones, with their aplications, aſwell for phiſicke as chirurgerie,...*

It was dedicated to Sir Edward Dyer (d. 1607), the Elizabethan courtier and poet, as was also *Marco Polo* and another of his translations, on China, to be mentioned later. A welcome reprint of *Joyfull Newes* has recently (1925) appeared in the Tudor Translations Series, edited by Stephen Gaselee. In his Introduction, the editor draws attention to a most interesting point: that it is by no means unlikely that to John Frampton is due the first interest taken in tobacco in England, leading shortly to the actual importation of the first smoking implements and the plant itself by Ralph Lane and Francis Drake.

To Monardes' description of tobacco, Frampton has added an account given him by Jean Nicot himself relating how, when French ambassador at Lisbon in 1559–61, he became acquainted with the new discovery and sent seeds to his Queen, Catherine de' Medici. An abstract of the actual report sent to France follows, in which we read of "the smoke of this Hearbe, the whiche thei receave at the mouth through certain coffins [paper cases of conical shape], ſuche as the Grocers do uſe to put in their Spices."

Thus nine years before Ralegh received the "herba santa" from Drake, a full description of it had been published in London by Frampton. A second edition, with some additions, came out in 1580, and a third edition in 1596.

His next work appears to be unrecorded, except in the *Registers of the Company of Stationers of London*. See Arber's *Transcript*, Vol. II. p. 325, where we find that Henry Bynneman obtained a licence on March 10th, 1578, for, *A brief Declaracon of all the portes. creekes. baies. and havens conteyned in the west India. the originall whereof was Dedicated to the mightie Kinge Charles the V Kinge of Castile.* I know of no copy in existence to-day. It was copied by Ames and Herbert, *Typographical Antiquities*, Vol. II. p. 982.

In January 1579 Frampton finished writing the Dedication of his *Marco Polo*, so we may assume that it appeared in the early spring of that year. We shall return to a full discussion of this work later.

On Oct. 1st he finished another Spanish translation which was published before the end of the year. It was Bernardino de Escalante's *Discurso de la Navegacion que los Portuguefes hazen à los Reinos y Prouincias del Oriente, y de la noticia q̄ fe tiene de las grandezas del Reino dela China*, Seuilla, 1577, which appeared as *A Difcourfe of the navigation which the Portugales doe make to the Realmes and Prouinces of the Eeft partes of the worlde, and of the Knowledge that growes by them of the great thinges, which are in the Dominions of China. Written by Barnardine of Efcalanta, of the Realme of Galifia Priest*, Imprinted in London at the three Cranes in the Vine-tree, by Thomas Dawson, 1579.

Two copies each of the original edition and Frampton's translation are in the British Museum. They are exceedingly rare books.

Most of the work deals with the customs, etc., of China. Thus, when in 1745 it was included in Vol. II. pp. 25–91 of *A Collection of Voyages and Travels...compiled from the curious and valuable library of the late Earl of Oxford*, we find the title-page altered as follows: *An Account of the Empire of China. ...to which is prefix'd A Difcourfe of the Navigation which the Portugueze do make....* As a matter of fact, the order of the chapters themselves are unaltered except that a few notes have been given to Chs. VI, IX, XI, XIV; Appendixes added to Chs. XI, XII–XIV, XV; and eleven "Reflections upon the Idolatry of the Jefuits, and other Affairs relating to Religion in China," inserted between Chs. XV and XVI [mis-numbered XIV].

Frampton's next work is of the utmost rarity, the only recorded copy being at the Lambeth Palace library. Its full title is as follows: *A Difcouerie of the countries of Tartaria, Scithia, & Cataya, by the North-Eaft: With the manners, fafhions, and orders which are ufed in thefe countries. Set foorth by Iohn Frampton merchaunt.* Imprinted at London at the three Cranes in the Vintree, by Thomas Dawfon. 1580.[1]

At first sight it would appear to be an original work by Frampton, but closer inspection shows it to consist of accounts of different parts of the East "collected and written by a certaine learned man called Francifco

[1] Owing to the excessive rarity of this work further bibliographical details will perhaps be welcome. The leaves are numbered on the recto only [1]–40, the actual number of pages being 6+80. Signatures are: ¶.3 +1 A. A.2 A.3. A.4+4 B. B2 B.3. B4+4 C. C.2. C.3. C.4. D. D.2. D.3. D.4.+4 E. E.2. E.3 E.4.+4. The dots before and after the figures are inserted or omitted as shown above. D.4 is marked in the middle of the page above the tail-piece. E. 3. is in a plain roman fount. The Colophon appears on the bottom of f 40 r°. The work forms No 6 in a volume of several similar items. It is numbered 30. 8.8, and bears the stamp and initials "R. B" on either side, showing it to have been the former property of Archbishop Bancroft.

Thamara of Cadiz. . . ." This Francisco Thamara, who flourished in the first half of the sixteenth century, was Professor of Belles Lettres at the University of Cadiz from 1550 to 1552, and made translations of selections from Cicero, and a collection of apophthegms. (See further *Diccionario Enciclopedico Hispano-Americano*, 1897, vol. v, p. 167.) In 1556 he compiled a book of travels taken mainly from Joannes Boemus' *Omnium gentium mores*, . . . , 1536. It was entitled *El libro de las Costumbres de todas las Gentes del mundo, y de las Indias*, Anvers, and is the work translated in part by Frampton. He merely selected those portions dealing with the East. Much of the information is taken direct from Marco Polo and Nicolo de' Conti. There are four distinct sections: (1) of the Region of Tartaria, and of the Lawes and power of the Tartars, ff. 1–13; (2) Of the Countrie of Scithia, and of the rude manners of the Scithians, ff. 13 v°–21; (3) Of the Countrie that is called, the other ſide of Ganges, and of Cataya, and the region of Sinas, which is a countrey of the great Cham; and of the meruailous things that haue bene ſeene in theſe countries, ff. 21 v°–28; and (4) of many notable things that are found in the land of Tartaria, and in the Eaſt India, ff. 28 v°–40 v.

On ff. 24 and 27 v, the same mistake is made as occurs in *Marco Polo* (see Appendix I. Note 53, p. 164) where Santaella translates "lingua per sì" as "lengua de persianos", thus making the natives of China and the Malay Archipelago speak Persian! On f. 27 Polo is referred to in connection with Ciampago, or Japan.

Frampton dedicated his translation "To the right worſhipfull ſyr Rowland Hayward Knight, and to maſter George Barne, Alderman of the citie of London, and gouernours of the worſhipfull company of merchaunts aduenturers for diſcouerie of newe trades, and to the aſiſtents & generalitie of all the ſayd worſhipfull fellowſhip, Iohn Frampton wiſheth all happye ſucceſſe in all their attempts."

His next publication was in 1851, when he made a translation of Pedro de Medina's *Arte de nauegar* . . . Valladolid. 1545, under the title of *The Arte of Nauigation, wherein is contained all the rules, declarations, ſecretes & aduiſes, which for good Nauigation are neceſſarie & ought to be knowen & practiſed: made by (maſter Peter de Medina) directed to the right excellent & renowned Lord don Philippe, prince of Spaine, & of both Siciles.* As with most of Frampton's books, it was dedicated to Sir Edward Dyer, and the date is given as Aug. 4, 1581. A second edition appeared in 1595.

Both are very rare. Dr Pollard, *Short-title Catalogue*, p. 402, No. 17771, records a copy of the first edition as being in the library of Sir R. L. Harmsworth, and the second edition in the H.E. Huntington library.

The British Museum contains the original 1545 work of de Medina, as well as six French, two Venetian, and one Dutch translation.

This comprises, as far as I can ascertain, all the translations made by Frampton. Although it is difficult to say for certain, it seems probable that he was alive in 1596, and personally re-edited the third edition of *Joyfull Newes*, which was, without doubt, his most successful work.

It remains to discuss the bibliographical difficulties of *Marco Polo*. It was published, as we know, by Ralph Newbery in Jan. 1579, but for some reason or other was not clearly entered in the registers of the Company of Stationers, and has become connected with another work with which it has been thought actually to coincide. It is duly entered in Ames and Herbert, *Typographical Antiquities*, Vol. II. 1786, p. 907, but a note is added as follows:

"He [Ralph Newbery] had licence about this time to print *A description of the East Indies, translated out of Italian*, Q. if this be the book intended?"

Now on reference to Arber's *Transcript* of the *Registers of the Company of Stationers of London 1554–1640 A.D.*, we find in the Index Volume (Vol. v. p. 113) that Frampton's *Marco Polo* is recorded as appearing in Vol. II. p. 342, but this reference is preceded by a query. On turning up the page in question we find no mention of Polo or Frampton at all. The entry is as follows:

"Raffe newbery. Receaued of him for his licence to printe
the description of the East Indies which was
translated out of Italian and lycencid
by master Tottell and master Cooke
vnder their handes in the tyme of their
being wardens."

The date of the licence was Dec. 3rd 1578. Arber has simply copied Ames. The trouble is that no such work on the East Indies has been traced, and the above bibliographers have come to the conclusion that as *Marco Polo* appeared in January 1579, its licence must have been obtained in December 1578. The only work on the "East" licensed to be published by Newbery about that time was "the description of the East Indies," and this has been taken to be intended for *Marco Polo*!

Although this may seem quite unjustifiable, I notice that Dr Pollard, in his *Short-title Catalogue of Books...1475–1640*, has accepted the connection, and gives *Marco Polo* as having received its licence on Dec. 3rd 1578. Personally I believe there is no connection whatever, and that the two works are quite distinct. It is possible that the *East Indies* was never published; so also *Marco Polo* may never have been entered in the register as it should have been.

Furthermore, all Frampton's works were translated from the Spanish, and we are distinctly told that the *East Indies* was from the Italian. Prolonged search has shed no light on the matter.

It is also rather strange that no subsequent edition of *Marco Polo* was issued. Here was a work that represented the first detailed information about the Far East, published at a time when English discovery and exploration was at its height, yet, as far as we can judge, its sale could not have warranted a reprint, and copies gradually got used up and lost. Thus to-day only three copies of the work are known to exist. Of these, two (979, f. 25 and G. 2755) are at the British Museum, while the third is in the Lambeth Palace library. For the present edition I have used 979, f. 25. In G. 2755 the title-page is missing, and has been copied out in ink.

The Lambeth copy is bound up in a volume of tracts, but is in very fine state with wide clean margins. The librarian tells me that as the volume bears the catalogue mark of Cambridge it certainly dates back to Archbishop Bancroft's time, when during the Commonwealth the library was transferred to Cambridge for twelve years. He considers, however, that it probably formed part of Archbishop Whitgift's collection.

Leaving John Frampton, we must pass on to a brief account of the life and writings of Santaella.

Santaella.

Rodrigo Fernández de Santaella y Córdoba was born in 1444 at Carmona, twenty-six miles north-east of Seville. Nothing is known of his early life, and we first hear of him in 1467 when he was presented with a fellowship of theology at the College of San Clemente de los Españoles at Bologna by the Archbishop and Chapter of Toledo. The fellowships lasted for eight years, so we may assume that Santaella remained at Bologna until 1475. After taking his degree as Doctor of Theology and Arts, he preached before Sixtus IV at Rome in 1477, in the presence of Innocent VIII.

Meanwhile Isabella had been recognized as heiress to Castile, and in 1469 had married Ferdinand of Aragon. The "Catholic Kings" were proclaimed in 1474, and soon after Santaella returned to Spain and embarked on his career of ecclesiastical preferment.

In 1499 his *magnum opus* appeared, the *Vocabulario Eclesiástico*, dedicated to the Illustrious Catholic Queen. It went through no less than thirty editions, which are duly recorded by D. Joaquin Hazañas y La Rua, whose work[1], *Maese Rodrigo, 1444–1509* (see pp. 155–196), is practically my sole authority for these few remarks on Santaella.

His *Sacerdotalis instructio circa missam* followed later in the same year, and the *Manual de Doctrina necesario al visitador y á los clérigos* in 1502.

In 1503 his Castilian translation of *Marco Polo* was published. In his Preface Santaella tells us that he was prompted to undertake the work since he realized its importance and no one had come forward to do it. It had already been printed in German, Latin, Venetian and Portuguese, and Santaella wished to see it in his native tongue. He also tells us that his library contained the treatise of Nicolò de' Conti, another Venetian, whose travels largely confirmed the narrative of Polo, and because of this fact he determined to include a translation in his work, "porque como nuestro señor dixo por boca de dos ó tres se confirma mas la verdad."

As is related on a later page (p. xxvi) the Polo MS used by Santaella is now preserved in the Biblioteca del Seminario at Seville. Subsequent editions appeared in 1507, 1518, 1520 and 1527, the last three being posthumous.

It is unnecessary here to enumerate the subsequent publications of Santaella. They consisted chiefly of sermons and other ecclesiastic writings of a similar nature, and are fully catalogued by La Rua.

On Sept. 12th 1502 Hurtado de Mendoza, Cardinal of Seville, had died, and Santaella was made "Visitador" for the whole of the see. On June 3rd 1503 the Chapter divided the Archbishopric into four sections, that including the city of Seville and Triana falling to Santaella. The vacancy was filled by Don Juan de Zúñiga, who made his entry into Seville on May 13th 1504, but he died on July 26th of the same year. The esteem in which Santaella was held is shown by the fact that at the death of Zúñiga, he was nominated "Provisor" during the interregnum, the next Archbishop, Fray Diego de Deza, not arriving at Seville till 1506.

For some years past Santaella had been deliberating on the founding of a university at Seville, and on June 13th 1503 the site was purchased for 4700 *maravedis*. A Bull, pointing out the necessity for a local university for the benefit of scholars and poor clergy studying in Seville, was approved by Julius III. Santaella's idea seems to have been to create a College for ecclesiastical studies, as well as a general university. In 1508 he obtained another Bull by which the College was united with three other

[1] It was published at Seville in 1909, being a greatly enlarged edition of a 46-page pamphlet issued in 1900, entitled *Maese Rodrigo Fernández de Santaella Fundador de la Universidad de Sevilla.*

benefices in order that medicine might be taught, and the whole establishment placed on the same footing as the university of Salamanca.

Santaella died on Jan. 20th 1509, and was buried in the chapel of his college. In 1771 the Colegio Mayor, as it was called, was separated from the university, and by 1847 hardly one stone remained upon another.

Thus the illustrious Archdeacon of the Realm, Maese Rodrigo Santaella, was almost completely forgotten, when the Rector of the university conceived the idea of erecting a statue to its founder.

This statue, more than life-size, was unveiled on Dec. 10th 1900, and stands in the great court of the university.

Having thus briefly given a short account both of Frampton and Santaella, we can pass on to a consideration of the extant texts of the Travels of Marco Polo.

§ ii. THE MANUSCRIPT TRADITION

PREVIOUS to 1928 it would have been practically impossible to have written anything new about the numerous Polian texts, unless it had been to have given more detailed accounts of the leading MSS already briefly described by Yule.

Early last year, however, the eagerly awaited work of Prof. L. F. Benedetto made its appearance in Florence[1], and for the first time the MSS were properly classified and arranged in the respective groups to which they belong.

But this is only a small portion of the work that Benedetto has accomplished. He has not only increased the Yule-Cordier list[2] of MSS from 78 to 138, but has discovered a copy of one that contains many of the passages used by Ramusio, the origin of which was not previously known. I shall return to this later.

All this forms the first part of Benedetto's work; the second half contains the text of the most famous MS of all, fr. 1116, correctly edited for the first time with textual notes and important passages from other MSS.

[1] *Marco Polo · Il Milione.* Prima edizione integrale, a cura di Luigi Foscolo Benedetto. Firenze, 1928. I have reviewed this great work at considerable length in *The Asiatic Review*, Oct 1928, Jan 1929, and April 1929. I have to thank my friend the editor, Mr F J. P. Richter, for allowing me to make what use I like of it in the present work.

[2] The Yule-Cordier list consists of 92 MSS (85 in the 1903 work and 7 in Cordier's *Notes and Addenda* of 1920), but, as Benedetto has shown, 14 are either duplicates, mistaken references, or are not MSS at all. Thus the total of 78 is obtained.

In order to derive the full benefit afforded for the elucidation of the complicated mass of MSS, it is necessary to study both parts in conjunction.

As is only to be expected in research of this nature, it is impossible to find proofs for every statement, and in the reconstruction of lost originals there is plenty of scope for what amounts to little less than pure guess-work.

I have never been able to understand exactly why Yule discarded fr. 1116, which he owned to be the best text, in preference for those used by Pauthier which were much inferior. His excuse that the awkwardnesses and tautologies in fr. 1116 prevented its use hardly seems sufficient to debar a scholar from attempting to overcome those difficulties.

But Yule was no paleographist; he was a commentator, and a very great commentator; just as Cordier was a bibliographer. Benedetto, on the other hand, is both a philologist and a paleographist, and only such a scholar can give us the thread that will guide us safely through the labyrinthine intricacies of Polian manuscript tradition.

As a close study of the works of these scholars is a *sine qua non* for every student of Marco Polo, it is to be regretted that Benedetto has not used Yule's chapter enumeration for facilitating reference, in addition to his own.

Owing to the fact that Benedetto's work is limited to only six hundred copies, that it is in Italian, and that its high price places it quite outside the reach of students, I make no excuse for giving here some account of the different groups of MSS as now first classified and described by him, together with such further information or comments as my own reading has suggested.

We will consider the MSS under the following headings:

1. The Geographic Text (fr. 1116).
2. The Grégoire Version.
3. The Tuscan Recension.
4. The Venetian Recension.
5. Ramusio's Version and the ante-*F* phase.

1. *The Geographic Text (fr. 1116).*

As is only natural, Benedetto first discusses the precious MS at the Bibliothèque Nationale, Paris, fr. 1116 (formerly 7367). It was published in 1824 by the French Geographical Society, since when it has been known as the Geographic Text. Benedetto refers to it as *F*. Although that letter also includes all French MSS (twenty in number) in this group, fr. 1116 is its only complete representative. We know little of its history, except that it is supposed to have come from the old library of the French kings at Blois. It is round this MS that scholastic controversy has chiefly centred, and since the appearance of Yule's *magnum opus* we have been perfectly content to accept the view that in fr. 1116 we have a direct representation of what Marco Polo dictated to his fellow-prisoner in Genoa.

In the light of Benedetto's new evidence we find that we have to reconsider the whole question. In the end we shall see all our pet theories destroyed, with little hope of settling points concerning the early history of the book until various new lines of research have been exhausted to their utmost.

At first sight this may seem a hopeless position, but one thing is certain, and that is that we can never hope to clear up the history of any important work until we know what data we have to work on, and are satisfied that such data are arranged in their correct order, each separate item in its proper place. This, then, is the achievement of Benedetto. He has brought order into chaos. We are now in a position to ascertain what the MS tradition can teach us, and once we are on the right path there is no telling what headway may be made in the future.

Our discussion opens in the prison at Genoa, where Polo's fellow-prisoner, a Pisan, is called in to help in the writing of the narrative. The name of this man is shown definitively to be Rustichello, instead of such forms as Rustician or Rusticiano[1]. It was natural to suppose that he had been chosen by the Genoese authorities because of his reputation as a writer of French Arthurian legends. Scholars have, therefore, been at pains to compare the style of fr. 1116 with that of his other works. They have considered (Yule especially) that the language of fr. 1116 is much more crude, inaccurate, and Italianized than that of Rustichello's other romances. This supported the theory of Polo's dictation, which, it was said, clearly betrayed itself in the halting style of the narrative.

Benedetto, however, after comparing numerous passages of fr. 1116 with portions of Rustichello's other works, has found practically identical phrases and idioms, some of which clearly betray the same hand. From this he argues that the same care and diligence that produced the romances also produced fr. 1116—in other words, that Rustichello did not copy down at Polo's dictation, but produced fr. 1116 (or rather a version of

[1] It should be noted that Yule fully realized that the form *Rustichello* was the correct one (see his Introduction, p. 63). He only used *Rusticiano* as being the nearest to the form given in his text. Certain reviewers have credited Benedetto with the sole discovery of this.

which that manuscript is a descendant) after a prolonged and detailed study of all the notes with which Polo supplied him. Polo was no trained writer, and, moreover, would not trust himself to present his story in a style acceptable to Western ears after his prolonged absence in the East. Here was a professional story-teller ready to hand! What more natural than to allow him to "write up" the work, after supplying him with all the necessary information! As Benedetto puts it:

"Compito espresso di Rustichello dev' essere stato quello di stendere in una lingua letteraria accettabile quelle note che Marco, vissuto così a lungo in oriente, non si sentiva di formulare con esattezza in nessuna parlata occidentale. Abbiamo intravisto abbastanza com' egli, assolvendo un tal compito, sia rimasto fedele allo stile ed alla visuale dei romanzi d'avventura. Ma non possiamo dire nulla di più."

Thus the style of fr. 1116, with all its "story-teller" mannerisms, does not necessarily betray dictation, but rather the usual style of a professional romance writer, who saw in Marco Polo a King Arthur come to life! Moreover, as regards the Italian words, we find quite a large percentage of them in fr. 1463, a MS which we *know* was not dictated. I may note in passing that Ramusio, in the Introduction to his version (to be discussed later), neither states that Polo *dictated* his work, nor that *a Pisan* had anything to do with it. He says that Polo was "assisted by a Genoese gentleman" who "used to spend many hours daily in prison with him," and helped him to write the book. It has always been taken for granted that facts had become muddled, and it was Rustichello the Pisan to whom reference was made. Now Benedetto argues (pp. xxxi *et seq.*) with considerable skill that fr. 1116 must represent only a later copy of the original Polo-Rustichello compilation. Might it not be possible that Ramusio, so correct and reliable in other points, is also correct here—and that one of the numerous Genoese, who without the slightest doubt *did* visit Polo, became very friendly with him, and helped in the editing of the work, *in addition to Rustichello*?[1]

However this may be, the fact remains that we must no longer regard *F* as the one and only direct and immediate descendant of the original Genoese text. Nor must we imagine that all subsequent recensions can be traced back to *F*. As will be seen later, they originate in lost prototypes dependent on lost MSS which we must regard as brothers of *F*. The Cottonian Codex Otho D. 5 at the British Museum, fragmentary though it be, is of importance in proving that the Franco-Italian recension was diffused, as well as all those MSS dependent on purer French texts.

[1] This suggestion was made to me by the Rev. A. C Moule.

2. *The Grégoire Version.*

A detailed study of this version has led Benedetto to believe in the existence of a lost version, F^1, very akin to *F*, but containing just those differences necessary to the production of an elaborated version (the lost *FG*) from which the Grégoire group is descended. In order to prove that *FG* is *not* a revision of *F*, as hitherto believed, it is necessary to determine the exact status of F^1 and to reconstruct it as far as possible.

This can be done chiefly by comparing the existing types of *FG* with *F*. This will show that *F* does not possess all the points necessary to produce *FG*—some of the *lacunae* should be different, and certain passages should be much more detailed. Thus the *FG* group must come from a MS similar to *F*, but certainly not *F* itself. This lost MS is Benedetto's F^1. *F* and F^1 can, therefore, be regarded as brother MSS.

We now examine *FG* as a separate group. Yule only knew of five MSS, while Benedetto has been able to add another ten. He divides *FG* into four sub-groups, A, B, C, and D. These again are subdivided into single MSS which are closely connected. Thus B has seven subgroups, of which B^1 and B^2 are closely related. So also B^4 and B^5. B^3 differs slightly from these two latter, while B^6 and B^7 form a more collateral branch. By arranging the MSS in this way a genealogical table can gradually be built up.

I might note in passing that Pauthier's "A" type, which formed the basis of his, and Yule's, translation, consisted of A^1; his "B" type of A^2; and his "C" of B^4. B^3 and B^4 (to which now must be added B^5) are especially interesting, as they bear the curious certificate of one Thibault de Cepoy, on which Pauthier placed such great importance. It appears that Thibault was a captain in the service of Philip the Fair. After beginning as valet and squire, he rose to the rank of Grand-Master of the Cross-bow men. He then entered the service of Charles de Valois, Philip's brother, who sent him to Constantinople to substantiate his claim to the throne on the grounds that his wife, Catherine de Courtenay, was the daughter of Philip de Courtenay, titular Emperor of Constantinople. Thibault left Paris on September 9th 1306, and proceeded to Venice, where he concluded a treaty of alliance in December, 1306. During his stay there he met Marco Polo, who in August, 1307, presented him with a copy of his book, inscribed as "the first copy of his said Book after he had made the same." After Thibault's death, his son Jean made a copy of the book, which he gave to Charles de Valois. He also made other copies for those of his friends who asked for them.

The three MSS mentioned above thus describe in the Note attached to them Polo's gift to Thibault, and how copies of it came to be distributed in France.

The great importance that Pauthier attached to these MSS on account of the Note has long since been proved quite unjustifiable. Although Yüle realized this, he still made Pauthier's MSS the basis of his own translation.

Benedetto has entirely discredited the Note and will not even allow Thibault to give his name to the group at all. He points out that it is impossible to believe that no copy of Polo's work should have been made until 1307. Certainly it is, but where is the evidence to prove it *was* made in 1307? Perhaps it had been written in 1299, and Polo had kept a copy by him for any important presentation such as this. Or, on the other hand, there may be something in Langlois' suggestion[1] when he says: "Mais, avant 1307, Ser Marco avait dû faire à bien des gens semblable politesse, peut-être avec des protestations analogues qu'il la faisait pour la première fois...."

Benedetto credits Grégoire with being the founder of this group because his name appears on two of the MSS (A^1 and A^3), while the date of the work is given as 1308 on the grounds that "this present year 1308" appears on another of the MSS (D). I cannot feel convinced, however, that Benedetto has proved his point in preference to accepting the original Thibault copy as the earliest extant MS of the group.

As I have already mentioned, *FG* is subdivided into four main groups. Among these, A^2 is the beautiful MS fr. 2810 at the Bib. Nat. containing 266 miniatures, of which 84 belong to the travels of Marco Polo, occupying the first 96 folios of the MS.

3. *The Tuscan Recension.*

At the commencement of the fourteenth century a Franco-Italian version of the original Genoese prototype was translated into Tuscan. It must have been very similar both to *F* and F^1, and can therefore be called F^2.

We possess five copies, which Benedetto has called TA^{1-5}. Of these TA^1 is the famous MS II. iv. 88 of the Bib. Naz. at Florence, better known as the *Codex della Crusca.*

The other copies are at the Bib. Naz. Florence ($TA^{2,5}$); the Bib. Nat. Paris (TA^3); and the Bib. Laurenziana (TA^4).

[1] *Histoire Littéraire de la France*, Tome xxxv, Paris, 1921, p. 255.

The Tuscan group contains two other versions which must be mentioned. The first is a Latin one (Bib. Nat. lat. 3195) in which the Tuscan translation is corrupted by Pipino's version (to be mentioned later). It was this text which formed the basis of H. Murray's English translation in 1844. It was published in 1824 by the French Geographical Society in the same volume as fr. 1116.

The second is a free *résumé* of *TA* found in the *Zibaldone* attributed to Antonio Pucci (d. 1388), the Florentine poet.

Owing to the differences found in the sub-groups of *TA*, it is necessary to utilize them all in attempting to restore the prototype of *TA*. Although TA^1 is the oldest codex, it is incomplete (as also TA^2) and less close to *F* than the others.

When we have restored *TA* as best we can with the help of all the sub-groups, we find that we have a complete text save for the omission of certain historic-military chapters and some minor details. It is of assistance in revising certain corruptions in *F*, as some of the *lacunae* in fr. 1116 could not have existed in F^2 from which *TA* is descended.

4. *The Venetian Recension.*

This group is of the utmost importance, and contains over eighty MSS. In order to fully appreciate the extensive ramifications of its sub- and sub-sub-groups, it is necessary to study the genealogical table given by Benedetto on p. cxxxii.

It is, moreover, of particular interest to us, as it contains the Spanish version of Santaella, the English translation of which is reprinted in the present volume. A glance at the table referred to above shows that the primitive Venetian codex is represented by five MSS (VA^{1-5}). Although VA^3 and VA^4 are the only complete ones, VA^1 is by far the most important, as it consists of the Casanatense fragment (Bib. Cas. 3999), which is a direct descendant from the prototype which served as the source of Fra Pipino's famous version. The great fame that this version achieved from its first appearance, and the eulogistic manner in which Pipino referred to his sources, led to the popular opinion that the Venetian version was nothing less than Polo's original! Consequently, the Pipino texts are more widely distributed than any others. To the previously known twenty-six MSS Benedetto has added another twenty-four. These fifty must be supplemented by seven more in the vulgar tongue, besides a very large number of printed versions. Nearly all the important European libraries possess one or more Pipino MSS. There are several copies in the British Museum, while others will be found at Oxford, Cambridge, Glasgow, and Dublin.

Of particular interest is the MS which once belonged to Baron Walckenaer. Benedetto describes it correctly as being in a volume containing other matter, including a version of the *Mirabilia* of Jordan de Sévérac. He regrets that its present locality is unknown, and conjectures that it has probably found its way to America. Both Yule and Cordier had previously made similar statements as regards the MS itself, yet only last year my friend, the Rev. A. C. Moule, "discovered" it properly catalogued and indexed at the British Museum![1]

When scholars and bibliographers[2] can pass over such fully recorded MSS, we can the more easily imagine that many unknown Polian treasures may still lie in European libraries *wrongly* catalogued, or not catalogued at all.

The fame of Pipino's version is well attested to by the numerous translations of it which exist—in French, Irish, Bohemian, Portuguese, and German. The French translation exists in two MSS, one at the British Museum (Egerton, 2176), and the other in the Royal Library at Stockholm. The Irish version is that in the famous "Book of Lismore," discovered in such a romantic manner[3] in 1814. The Bohemian version forms part of Cod. III, E. 42, in the Prague Museum, and dates from the middle of the fifteenth century. Benedetto considers, however, that the MS is copied from a still older Pipino text. The Portuguese translation was printed at Lisbon in 1502 (reprinted 1922).

The first printed Latin text appeared about 1485, while a second edition (1532) was included in the famous collection of travels known as the *Novus orbis regionum ac insularum veteribus incognitarum.* It was edited by Simon Grynaeus, but actually compiled by Jean Huttichius. The text is corrupt, and has been considered by many to be a retranslation from the Portuguese of 1502.

There were several editions of the *Novus orbis*—1535, 1537, and 1555[4], as well as translations—German (1534), French (1556), Castilian (1601), and Dutch (1664). Apart fiom this, Andreas Muller reprinted the Latin

[1] *Journal of the Royal Asiatic Society*, April, 1928, pp. 406 *et seq.* The MS is numbered Add. 19513

[2] Even when Cordier printed the entire Table of Contents of Walckenaer's volume in *Les Merveilles de l'Asie*, 1925, p. 44, he gave no indication that here was the long-lost Polo text.

[3] See Yule, Vol. I. pp. 102 *et seq* of his Introduction.

[4] Apparently the 1555 is the most complete edition There is a fine copy of this in the Grenville Library at the British Museum (G 7034), which contains the map that is so often missing.

in 1671 on which was based the French translation in Bergeron's *Voyages faits principalement en Asie* (1735).

The text of Ramusio (to be more fully discussed shortly) can be regarded as based on a version of Grynaeus, so that it is fundamentally a Pipinian text.

Apart from Pipino's version (*P*) and also that of an anonymous Latin writer (*LB*), a group of six Tuscan translations of the Venetian (TB^{1-6}) must be added. This Tuscan group in its turn gave rise to a German translation (*Ted.*) and another Latin one (*LA*).

We now turn to a group based on a MS similar to that which gave rise to the Tuscan group. It consists of two distinct sub-groups, the first of which comprises: (*a*) a fifteenth century Venetian MS at Lucca (Bib. Governativa, No. 1296[1]), and (*b*) a Spanish version from a Venetian codex, translated into English by John Frampton in 1579.

The second is also of importance as it consists of a mass of MSS and printed texts based on the early Venetian edition of 1496.

The Lucca MS is a paper codex of seventy-five pages, containing a brief epitome of Odoric besides the text of Polo. On the verso of the last page we are informed that it was completed on March 12th 1465 by one Daniele da Verona. The Spanish (Castilian) version of Santaella was taken from a MS of 78 folios, without pagination, which once belonged to the Biblioteca del Colegio Mayor de Santa Maria de Jesus at Seville. After the separation of the College and University in 1771 it entirely disappeared, and was given up as lost. Years later it was discovered with a number of papers in the garret of an old building belonging to the College, and is now preserved in the Biblioteca del Seminario of Seville. The manuscript is described by La Rua[2] as a quarto volume, written in two inks, in contemporary binding, somewhat deteriorated by the action of the weather. It contains 135 chapters (as in the present translation), and was completed on Aug. 20th 1493. All Santaella's editions are of extreme rarity, and it is hard to say for certain how many there were, or even to be sure of the date of the first edition.

As far as I can ascertain, the first edition was that described by Salvá (*Catálogo de la Biblioteca de Salvá*, Vol. II. No. 3278), and published at Seville on May 28th 1503. There is a fine copy at the British Museum (C. 32. m. 4.) which has been fully described by Yule (Vol. II. p. 566). An edition of 1502 is mentioned in some detail by Don Fernando Colon, but

[1] Yule wrongly refers to this MS as No 296 (Vol II. p. 544).

[2] *Maese Rodrigo*, p. 52.

as he gives the same printers and exactly the same date for the completion of the work (May 28th) as in the 1503 edition, it would seem that an error has been made.

The work was reprinted at Toledo in 1507, and, after the author's death, at Seville in 1518 and 1520, and again at Logroño in 1529[1]. This latter edition is also at the British Museum (G. 6788), and for all we know may be the actual copy used by Frampton. The excessive rarity of the work fully justifies such a possibility.

Turning, now, to the second sub-group, we find a large number of Venetian MSS and printed texts all based on the edition printed by Sessa in 1496. This edition was derived from a MS which, like the Lucca, began with an epitome of Odoric. Owing, however, to a large *lacuna* after the first folio, it has not only been sadly reduced, but the first chapters of Marco Polo itself have also suffered heavily.

Apart from these mutilations, and the fact that in places the text is abbreviated and somewhat corrupt, the early Venetian printed edition is identical with both the Lucca text and that of Santaella.

Without going further into the relationships of the various branches of the Venetian recension, we will pass on to Ramusio and the earlier connected MSS.

5. *Ramusio's Version and ante-F phase.*

In 1550 the first volume of a collection of travels appeared under the editorship of one Gian Battista Ramusio, an illustrious member of a noble Italian family of Rimini. In 1556, another volume (Vol. III) was issued, while Vol. II, containing Ramusio's account of Polo's travels, did not appear until 1559—two years after the editor's death.

Other editions of the *Navigationi et Viaggi*, as the collection was called, soon followed, and the "Ramusian Recension" of Marco Polo took a unique place of honour in Polian tradition.

Ramusio was a good scholar, and enjoyed a great reputation for learning and critical research. His chief pursuit was geography, and he is believed to have opened a school for its study in his own house at Venice. In fact, everything we know about him compels us to treat his work with the utmost consideration and credence, as he fully justifies his title of "the Italian Hakluyt" Bearing this in mind, we can more readily appreciate the disappointment with which Yule had to record the absence of those MSS from which Ramusio had obtained certain parts of his information. Turning to the volume itself, we find that in a letter to his friend Jerome

[1] For details of all these editions see La Rua, *op. cit.* pp. 198–201.

Fracastoro, Ramusio speaks of his sources, clearly indicating Pipino's text as well as another *di maravigliosa antichità*. Although Ramusio's text was at first ignored, its great importance has been gradually established, until, with Benedetto's discovery of *Z*, it is a *sine qua non* in helping to trace the earlier stages of the history of the book. At the same time, he admits that it is a composite text—*sbocco a tradizioni già sicuramente corrotte*—and therefore cannot be used as a basic text, especially when compared with *F*. Benedetto would analyse the Ramusio text as containing: (*a*) Pipino as the original and principal base; (*b*) three other MSS, *V*, *L*, and *VB*; (*c*) the newly discovered MS, *Z*, which corresponds to the Ghisi codex mentioned by Ramusio himself.

The history of the Milan copy of *Z*, so far as it is known, is very interesting. It is taken from an old lost Latin Codex Zeladiano, copied in 1795 by the Abate Toaldo to complete his collection of Polian documents. The original of this copy must be identified with the MS *cartaceo in*-8º, *del sec. xv.*, mentioned by Baldelli-Boni, who says it was left by the will of Cardinal Zelada to the Biblioteca Capitolare of Toledo. A close inspection of *Z* shows it to be a Latin version of a Franco-Italian codex, distinctly better than *F*. But, as we shall see later, *Z*, as represented in the Milan MS, is by no means complete.

The first three-quarters of *Z* seem like an epitome of a much fuller text, but after Chap. 147 *F* is faithfully followed, while the additional passages point to a pre-*F* codex, which must have been considerably more detailed than *F*. Benedetto suggests that the copyist of *Z* began with the idea of a limited selection of passages, but gradually became so interested in his work that he eventually found himself unable to sacrifice a single word.

A point of prime importance with regard to *Z* is that it clearly betrays Polo's mode of thought, showing that, as far as it goes, it is a literal translation of an early text now lost. This is also supported by the fact that the names of peoples and places appear in *Z* in less corrupted forms than in *F* or subsequent texts—*e.g.*, Mogdasio, Silingi, etc.

The various indications of *Z*'s anteriority to *F* suggest a subsequent suppression of certain passages by a copyist or by the cumulative work of several copyists. A large percentage of these passages occur in Ramusio, while some are found in *Z*. In those cases where *Z* only resembles an epitome, we must conclude that Ramusio had access to a text closer to the archetype of *Z* than *Z* itself. We can call this text Z^1. We can, therefore, agree that if *Z*, as represented by the Milan text (Y. 160 P.S.), can account for unique passages only in the latter part of Ramusio, it is not unreasonable to conclude that he had a complete *Z* text before him (Z^1), and took all the unidentified chapters in the first half of his book from it. The discovery of the archetype of both *Z* and *Z*[1] would doubtless help to settle the question.

We now come to *V*, *L*, and *VB*. They can be looked upon as coming somewhere between *F* and *Z*. They are of value because they occasionally contain passages neither in *F* nor in *Z*.

V is a curious Venetian recension (Staatsbib. Berlin, Hamilton 424[a]) which has undeniable echoes both of a Franco-Italian and a Latin text. It contains about thirty unique passages, and was undoubtedly used by Ramusio. *L* is an interesting Latin compendium represented in the four following codices: Ferrara, Bib. Pubb. 336NB 5; Venice, Mus. Corr 2408; Wolfenbuttel, Bib. Com. Weiss. 4I; and Antwerp, Mus. Plantin-Mor. 60. They are practically identical, and represent the best compendium of Marco Polo extant. Its Franco-Italian origin is proved by the survival of certain expressions which, not being understood, have been retained unaltered. It was probably used by Ramusio, though this cannot be said for certain.

Taken together, *V* and *L* must be regarded as closely related to, but distinctly a sub-group of *Z*[1] and *Z*.

VB is a Venetian version (Donà della Rose 224 Civ. Mus. Corr.) differing from any of the Venetian recensions we have already discussed. Two copies exist: one in Rome (Bib. Vat. Barb Lat. 536I) and the other in London (Brit. Mus. Slo. 25I). *VB* shows signs of a Franco-Italian origin, and in two cases contains details ignored by *F*, but preserved by *Z*. On the whole, however, this is the worst of all Polian texts, and it is a pity that Ramusio used it at all.

To sum up, we must not blind ourselves to the undoubted defects of Ramusio. Here is a man who has selected a distinctly ragged garment (*P*), with the intent to make it look new by the addition of various patches (*Z*, *V*, *L*, *VB*). Some of the patches are of very good material, but others are frayed and badly put on, and, moreover, not always in the best places. They do not harmonize well with the cloth to which they are sewn. In some cases they have been trimmed a little, but then again we find in other cases that our repairer has added extra pieces of his own.

Thus altogether, while the finished article contains much material, it does not approximate in any way to a complete and original garment.

In spite, however, of all this, Ramusio remains an essential source in the reconstruction of the richer text by which *F* was preceded. It has continually been assumed that from time to time additions were made to the original work of Polo. The researches of Benedetto clearly show that, on the contrary, as time went on, impoverishments have occurred.

Z gives occasional bits of folk-lore and details of intimate social customs; so also does the *Imago Mundi* of Jacopo d'Acqui (D. 526 Bib. Ambros.) called *I* by Benedetto. It may be that the church censored some of this material, for in the *Z* passages we have caught a glimpse of Marco Polo as the careful anthropologist, and how can we determine what curious and esoteric information was originally supplied to Rustichello? We do not find it hard to believe that there may well be some genuineness in the passage of Jacopo d'Acqui when he says in Polo's defence: "And because there are many great and strange things in that book, which are reckoned past all credence, he was asked by his friends on his death-bed to correct the book by removing everything that was not actual fact. To which he replied that he had not told *one-half* of what he really had seen."

The gradual decadence of the original text as proved in the cases of *FG*, *TA*, and *VA* must also have occurred in the stage anterior to *F*. The discovery of *Z*, the study of *V* and *L*, the analysis of Ramusio, and the reference of certain elements to the lost Ghisi codex all seem to point to the fact that *F* was preceded by more conservative and more exact copies. *Z*, *V*, and *L* not only help to bridge the distance from *F* back to the original Genoese archetype, but also prove the richness of the latter and its gradual impoverishment. They show as well, that each of the three phases (*Z*, *V* and *L*, *F*) is dependent on the same original Franco-Italian text. Thus, apart from restoring the lost passages of *F*, they also bear witness to its unique importance and authenticity.

Having thus briefly surveyed the five main groups into which, thanks to Benedetto's labours, we can now divide the Polian texts, it will be as well to summarize the conclusions:

(1) Fr. 1116 of the Bibliothèque Nationale is the best Polo MS that has come down to us.

(2) It does not represent a direct copy of the Genoese original, but is a later version, which, together with its three brother manuscripts, $F^{1,2,3}$, is described from a common Franco-Italian MS of earlier date, now lost.

(3) From $F^{1,2,3}$ were derived respectively the lost prototypes of the Grégoire, Tuscan, and Venetian recensions (*FG*, *TA*, *VA*).

(4) Of these *VA* is the largest and most important, Santaella's Castilian version being made from a MS in one of its sub-groups.

(5) There was an ante-*F* phase, as yet only represented by *Z*, *L*, *V* and *VB*.

(6) Ramusio based his version on Pipino, with additional help from all the MSS of the ante-*F* phase, as mentioned above. He also used one or more other MSS, at present undiscovered.

(7) The most complete account of Polo's travels, therefore, consists of fr. 1116 as a base, supplemented by Ramusio, together with a few unique passages from other MSS.

§ iii. THE ITINERARIES

OF all the Polian problems which still remain unsolved, or at any rate not entirely solved, the most important, and at the same time the most difficult, is that of the itineraries. We may well wonder what the Elizabethan readers made of Frampton's book. They read of places and customs of which they knew nothing, and of which, in many cases, nothing more was known until after 1860!

The marvel of Polo's achievement lies not only in the fact that he was the man who first drew aside for Western eyes the curtain veiling the "mysterious East," but that so many of the places visited and localities described remained unvisited again for over 600 years.

The curtain, pulled aside for a moment to reveal a world as unknown and amazing as it appeared unreal and fantastic, was soon to fall again. The audience had been charmed and amused, but that was all. For the majority it had been but a clever story, one which in later years the admirers of Galland might have enjoyed.

There were a few, however, to whom the real value of the work was at once manifest. Foremost among these was Christopher Columbus, whose copy (Pipino's Latin version) is copiously annotated in his own hand, and now lies in the Biblioteca Colombia at Seville (see Yule, Vol. II. p. 558).

When Frampton introduced it to England we like to think that the wish expressed in his Dedication—"that it mighte giue greate lighte to our Seamen, if euer this nation chaunced to find a paſſage out of the frozen Zone to the South Seas"—was not made in vain, and that Drake, Ralegh and Frobisher eagerly devoured the work of their great predecessor.

The clouds of scepticism and incredibility took a long time to disperse, but with the increasing light of subsequent discoveries the claims of the "Father of Geography" became accepted and his exaggerations explained. The identification of some of the places mentioned or visited by him still remains uncertain, while that of others has only been determined within the last few years.

In some instances, when the identification of a town is almost certain, we find that Polo's itinerary places it on the wrong side of a river or double the distance that it really is from a previously mentioned locality.

In order, therefore, to appreciate the difficulties in attempting to trace the itineraries, several points must be taken into consideration. In the first place, we must remember that we are not dealing with a single journey occupying a fixed time, but with many journeys spread over more than thirty years.

In the second place, we know that while in the service of the Khan, Polo was sent on various missions. In some cases the outward route appears to have coincided with that taken on the return, but details are sadly wanting. In other cases we suspect a slightly varying return route, necessitated, it may be, by differences of natural conditions. A fordable stream in the summer may become a raging torrent in the winter. As we shall see later, some such explanation may account for the difficulties in deciding the route at the crossing of the Hwang ho.

In the third place, we cannot always be certain that Polo is describing places on an itinerary at all. Two distinct possibilities at once suggest themselves. At times he may be speaking of places and peoples visited by the elder Polos alone, which information merely served as a supplement to that dependent on the actual itinerary being followed. Such seems to have been the case with the Turfān—Camul—Sachiu section. On the other hand, he may be quoting from local reports and the gossip of native traders. Thus we are still uncertain whether he personally visited either Karakorum or Baghdad. Unfortunately, Polo does not carry out his promise made in the Prologue that he will clearly differentiate between things actually seen, and those only heard about.

Finally, we must not forget the circumstances under which the book was written. The Polos had arrived back in Venice in 1295, and in 1298 Marco was taken to Genoa as a prisoner of war. Thus over two years had elapsed in which he must have related his travels constantly, and, we imagine, gone through his notes, continually adding and altering as his memory, or those of his father and uncle, dictated.

After Genoa had become fully acquainted with the oral relation of his travels, as Venice had done previously, we can well appreciate Marco's wish to have his notes with him if his experiences were to be put into writing. Accordingly, these precious notes were sent for. Ramusio tells us about them in his famous Preface to Vol. II of *Della Navigationi et Viaggi*. It was considered by some editors that Ramusio had invented this passage himself, but we may well ask how Polo could possibly have remembered such details as are found in his Book. In view of this, it is both interesting, and at the same time reassuring, to read the remarks of Sir Aurel Stein on the subject (*Journ. Roy. Geog. Soc.* Aug. 1919, Vol. LIV. p. 103):

"We have seen how accurately it reproduces information about territories difficult of access at all times, and far away from his own route. It appears to me quite impossible to believe that such exact data, learned at the very beginning of the great traveller's long wanderings, could have been reproduced by him from memory alone close on thirty years later when dictating his wonderful story to Rusticiano during his captivity at Genoa. Here, anyhow, we have definite proof of the use of those 'notes and memoranda which he had brought with him,' and which, as Ramusio's 'Preface' of 1553 tells us (see Yule, *Marco Polo*, Vol. I, Introduction, p. 6), Messer Marco, while prisoner of war, was believed to have had sent to him by his father from Venice. How grateful must geographer and historical student alike feel for these precious materials having reached the illustrious prisoner safely!"

In returning to the Book itself, we find that the first nineteen chapters of the best MS extant (fr. 1116) form a kind of general introduction to the whole. Most editors, accordingly, have given it the name of "Prologue," and divided it into what we may call an Invocation and eighteen chapters.

Frampton, however, while also having a Prologue, does not divide it up into chapters, nor does he end it in the customary and obviously correct place. He stops short after the presentation of the Polos to the Khan. Thus the first four chapters of Frampton appear in other editions as still part of the Prologue.

The first half of the Prologue deals in the briefest possible manner with the journey of the elder Polos, performed in the years 1260–1269 while Marco was but a boy in Venice.

In the second half of the Prologue no itinerary of the outward journey is given at all. We are merely told of the double start from Acre, the enthusiastic reception by the Khan, the reluctance with which he let them depart after seventeen years' employment in his service, and their being chosen as escort of the princess Cocachin (Kūkāchin), bride-elect of Arghun, Khan of Persia.

As the journey overland would be too strenuous for a lady, they decided to travel by sea, and started, apparently from Zayton (? Chüan-chau), in 1292. Their route to Persia lay *via* Little Java (Sumatra) and the Sea of

India. Depositing their charge, they continued to Trebizond, Constantinople, and Negroponte, finally reaching Venice in 1295.

Thus the Prologue ends, and we can now begin our attempt to trace the route from Acre to K'ai-p'ing fu, as described by Polo in Chapters xx–LXXV of fr. 1116; i.e. Book I of Yule, and Chs. 5–52 of Frampton.

Leaving Acre for the second time in Nov. 1271, on reaching Ayas (Frampton's "Gloza") once again, the Polos found the Egyptian invasion of Syria an obstacle to their taking the usual eastern caravan route (as well as to the enthusiasm of the two friars who were going to convert the East to Christianity!), and so they were forced to turn north-eastward. Thus Polo starts by giving a brief description of Lesser Hermenia (the classical Cilicia), Turconomia (Anatolia) and Greater Hermenia (Armenia). His route apparently was Laias, called Gloza by Frampton (the modern Ayas)—Casserie or Casaria (Kaisariya)—Savast or Sevasta (Sivas)—Arzingal or Arzinga (Erzingan)—Argiron (Erzerum)—Arzizi or Darzizi (Ardjish, near lake Van)—Toris or Tauris (Tabriz). This agrees with Yule as far as Argiron, but after that he prefers to include in the itinerary every place mentioned in the text, whereas I regard them as mere annotations to the main route. Thus from Argiron Yule makes Polo go to Mus (Mush)—Meridin (Mardin)—Mausul (Mosul)—Baudas or Baudac (Baghdad)—Bastra or Bascra (Basra)—the Persian Gulf—Kisi or Chisi (Kish or Kais)—Curmosa or Ormus (Hormuz). Arzizi is left out of the itinerary altogether, while Toris and all places between it and Kirmān are taken to refer to the return of the Polos[1]. I fail entirely in finding sufficient evidence to justify Yule in his preference.

First of all let us consult the actual passages in the best text extant (fr. 1116) Here we read: "The most noble city is Arçingal which is the See of an archbishop. Others are Argiron and Darçiçi...It is bounded on the south by a Kingdom which is called Mosul...on the north it is bounded by the Jorgiens of whom I shall tell you more later." This corresponds practically verbatim to the texts used by Yule. In Ch. XXIV of fr. 1116 (Ch. v of Yule) the kingdom of Mosul is described briefly. We are then told that Baudas is a great city. This is all. There is not a word about Mus or Meridin in any of the French MSS. They are merely mentioned in Ramusio as producing cotton. Yet Yule considered this sufficient evidence to include them all in his itinerary, at the same time completely ignoring Arzizi which occurs in all the best texts. But quite apart from this, a traveller at Erzerum having Mosul as his objective

[1] See Yule's Map I in Vol. I p. I of his edition of *Marco Polo*.

would certainly find the Tigris his best medium of progress. He would reach it from Mush either directly by the tributary the Batman Su, or else *via* Bitlis and the Bitlis Su. Mardin, forty miles to the west of the Tigris, would be quite out of the line of march.

But to continue, Polo next tells us that a river flows through Baudas, and that as you descend it you pass through Bastra and reach the Sea of India (Persian Gulf) at Kisi. Now surely if Polo had visited Baghdad personally and sailed through the Gulf of Hormuz he could never have placed Kisi (Kish) on the Tigris, when it is only about 165 miles from the mouth of the gulf. Furthermore, it seems very strange that he entirely omits to mention the buildings of the city, nor does he refer to the Tigris by name, or describe it at all, as he usually does when meeting with a large and important river. Finally, the five chapters devoted to the taking of Baghdad and the legend of the blind cobbler strike one as mere repetition, and in no way support the theory of a personal visit to the city. I imagine that these details were picked up by Polo on his return home, and, being fresh in his memory, found a place in the narrative when speaking of the locality in question.

So also I would account for the mention by Ramusio of the castle of Paipurth (Baiburt) between Trebizond and Tabriz, as well as the convent of St Leonard.

Turning to the alternative route which I have suggested above, if Polo *did* go to Tabriz from Erzerum, he would naturally skirt the northern shores of Lake Van, and mention some place near the lake. And this is exactly what he does; for Arzizi is close to the lake and in a direct line between Tabriz and Erzerum.

THE ROUTE THROUGH PERSIA AND AFGHANISTAN INTO EASTERN TURKISTAN

After Toris (Tabriz) I would give the itinerary as: Saba (Saveh)—Kashan—Yasdi (Yezd)—Bafk—Kirmān—Hormuz, where, finding the boats unseaworthy, they decided to continue their travels by land. So retracing their steps to Kirmān by a different route, they crossed Persia in a north-easterly direction. Polo describes the journey from Kirmān to Hormuz in detail, but as Yule has brought his travellers down the Persian Gulf to Hormuz he is forced to make Polo describe his itinerary backwards at this point. This is so highly improbable that unless some very good reason is given, its acceptance is quite impossible. Thus, when Sir Percy Sykes challenged the statement in 1905 (*Journ. Roy. Geog. Soc.* Vol. XXVI. pp. 462 *et seq.*), we expected an adequate explanation by Cordier, but this was not forthcoming. Instead of answering the points at issue, he contented himself by saying that Baghdad was not off the main route for some years after its fall.

In view, therefore, of all the above facts, it would seem certain that we must entirely abandon any attempt to trace the itinerary, *either on the outward or on the return route*, *via* Baghdad and the Persian Gulf. We shall require much more convincing evidence than we have at present before we can accept the position that Yule has given it.

The itinerary from Kirmān to Hormuz has always been a puzzle, and even now it is impossible to be absolutely sure of the route. Yule has devoted a long note to it (Vol. I. pp. 110–115), so that it is unnecessary to go into any great detail here.

Polo tells us that from Kirmān you ride seven days over a plain country when you come to a great mountain, and having reached the top of the pass, you find a great descent which continues for a good two days. The intense cold experienced after leaving Kirmān is especially noted. At the end of the descent you reach a vast plain with the city of Camadi at the beginning of it. The name of the province now entered is Reobarles. After crossing the plain in five days, you find another descent of twenty miles at the foot of which lies the Plain of Formosa. Two more days bring you to the sea at Curmosa, i.e. Old Hormuz, to the east of Bandar Abbas.

One of the various routes with which the above description had been thought to coincide, is that followed by Abbott and Smith running S.S.E. from Kirmān through the Deh Bakri pass and across the plain of Jirupt and Rudbar, over the pass of Nevergun to the coast; the Plain of Formosa being the plain of Harmuza between Nevergun and the present site of Old Hormuz near Mīnāb.

The latter part of this suggested route seems to be practically certain, but Gen. Houtum-Schindler has pointed out that the more westerly Sārdū route from Kirmān *via* Jupar, Bahramjird, over the Sarvistan pass, down a two-days' descent to the ruins now called Shehr-i-Daqīānūs (Camadi), and so to the plain of Jirupt and Rudbar, fits Polo's description much better.

This, then, is the route marked on the map opposite. On comparison with that facing page 112 of Yule's first volume, it will be seen to lie in its northern portion, between that given by Yule himself and that followed by Smith. As we have already mentioned, on arriving at Hormuz, Polo found the ships unseaworthy and accordingly returned to Kirmān. This seems to me to be so clearly stated that with this fact added to all the previous evidence against the Persian Gulf route, I am unable to discover

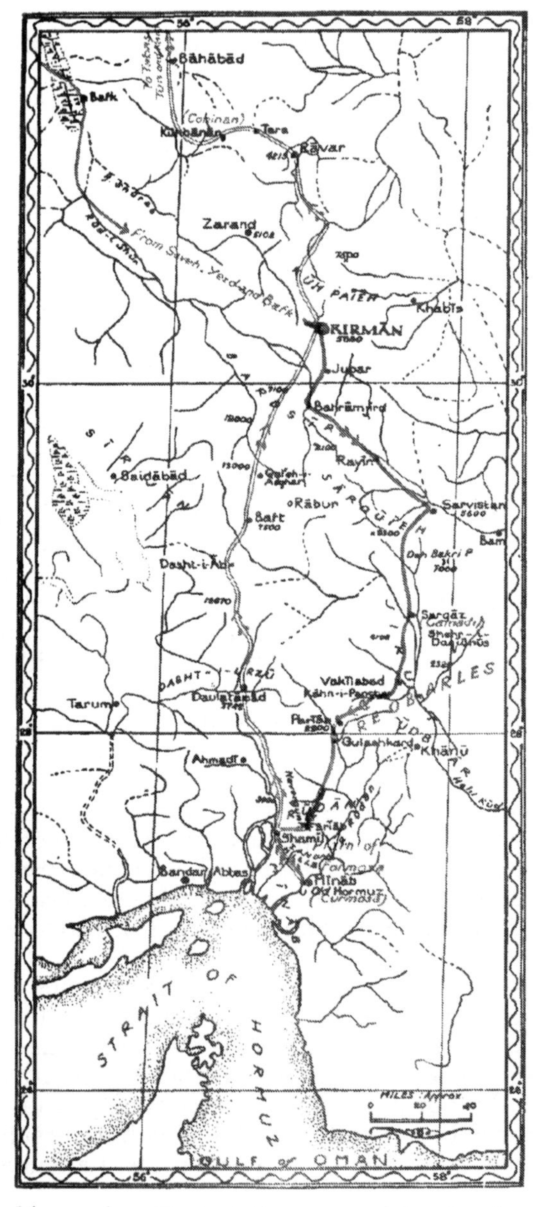

The Itinerary from Kirmān to the coast and the return Journey

any evidence whatsoever in support of Yule's theory. Polo tells us that the return route to Kirmān led through some very fine plains, and that you pass natural hot springs which cure skin diseases, and that there are plenty of partridges as well as dates and bitter bread. All these details fit the Urzū—Bāft route rather than the Tārum—Sīrjān route, which latter was that suggested by Yule. The medicinal springs occur both at Qal'eh-i-Asghar and Dashtāb, and the bitter bread is found only at Bāft and in Bardshir.

On departing from Kirmān, the route continues through a desert for seven days to Cobinan (Kuh-Banān). As the direct line of march *via* Zerend is only ninety-five miles, and Polo especially speaks of waterless deserts, it has been suggested that he went *via* Kūhpāyeh and the desert lying to the north of Khabīs. This seems to me too far east and quite unnecessary to account for the seven days' march. I suggest the route to the west of Kūhpāyeh running through desert and hilly country to Rāvar. From here he would take the westerly road to Cobinan *via* Tara. This would give an average daily march of a little over twenty miles, which is not at all unreasonable in this sort of country.

Eight days more, also through a desert, brings Polo to the province of Tunocain (Tūn-o-Kâin), but his exact itinerary at this point is not easy to determine. It has been discussed by many people, including Yule, Sykes, and Sven Hedin. Yule supposes that he travelled to Tebbes (or Tabas), while Sykes favoured the eastern route *via* Naibend. The evidence of Sven Hedin (*Overland to India*, Vol. II. Ch. XL; reprinted in Cordier, *Ser Marco Polo*, p. 27) in support of the Tabas route is so convincing that we need have no hesitation in accepting it. As will be seen from the accounts of the respective routes given at the above references, that from Kuh-Banān to Tabas agrees both with Polo's description and also with his distances. On the other hand, that from Kuh-Banān *via* Naibend to Tun does neither. In fact Sven Hedin quotes Sykes[1] to show how his description of the stage after Duhuk disagrees with Polo. It seems probable that Polo went to Tabas *via* Bāhābād, which is not only on the route, but is the branching-off place for caravans going from Tabas to Yezd. Sven Hedin

[1] It was very disappointing to find nothing of any value in a recent article by Sykes in *The Nineteenth Century*, May 1928, p 682. In fact, he skips all the difficulties: he ignores Cobinan, and even gives a wrong impression of the itinerary, for after informing us that Tonocain "represents Tun and Cain," he mentions the Assassins and continues: "Upon resuming the account of their journey, we find ourselves at Balkh...." Thus, not only are the various suggestions neither stated nor discussed, but the six days' journey through fair plains and the city of Sapurgan are entirely omitted.

actually met such a caravan at Tabas which had arrived from Sebsevar (Sabzawār), north of Tun.

When Polo arrived at Tabas he was in one of the "many towns and villages" which he mentions as being in the province of "Tonocain." From Tabas he undoubtedly proceeded to Tun, probably *via* Bushrūieh, but whether he also called at Kain we have no means of ascertaining. After Tun our difficulties in no way decrease, because Polo interrupts his itinerary to tell us about the Arbre Sec and the Old Man of the Mountain. When he returns to it again he is no longer at Tun, but at the "Castle" of the Old Man. It is almost impossible to decide where this "Castle" was. It may have been some ruined fortress which merely served as an excuse for telling Polo a romantic story which was the common property of the East[1]. It should be remembered that the number of "Castles" in Persia, as well as in Syria, had been steadily growing since the founder of the Assassins, Hasan Shabāh, had seized the fortress of Alamūt ("Eagle's teaching"[2]), near Kazvin in 1019.

I consider, therefore, that in attempting to assign to the "Castle" of Polo a definite locality such as Alamūt, Yule has deflected the itinerary much too far north. Thus he has been led to include Sebsevar (Sabzawār), Nishāpūr and Meshed in the route, giving it a most improbable right-angled turn just south of Shāhrūd and Bustām. No wonder he was surprised that none of the cities were mentioned in the narrative. Moreover, had Polo been going on to Sebsevar, etc., he would never refer to Tūn-o-Kâin as "the extremity of Persia towards the north." But what finally disposes of this suggested itinerary is the further evidence of Sven Hedin, who shows that the Sebsevar—Meshed road does not agree with Polo's description of "fine plains and beautiful valleys, and pretty hillsides producing excellent grass pasture...." This kind of country extended for six days, and then he arrived at Sapurgan.

Now this place has been identified with the modern Shībarghān, about seventy miles west of Balk, so that six days of fine plains could not possibly be anywhere near Meshed, as they must directly link up with Sapurgan. Thus we can have no hesitation in accepting Sven Hedin's suggestion that the six days must have been passed, after crossing the nemek-sār (salt

[1] See Prof. D S. Margoliouth's article "Assassins" in Vol. II. of Hastings' *Ency. Rel. Eth.*; the anonymous *ditto* in *Ency Islām*; and E. G. Browne, *Literary History of Persia*, Vol II pp. 190–211 (especially pp. 206–8).

[2] This appears to be the correct meaning in spite of the article in *Ency. Islām*, p. 249. See Browne, *op. cit. sup.*

desert) east of Tun and Kain, in the ranges Paropamisus, Firuz-Kuh, and Band-i-Turkistān.

To sum up, then, I would give the itinerary north from Kirmān as: Cobinan (Kuh-Banān)—Bāhābād (?)—Tabas—Tun—Kain (?)—nemeksār—Herāt, or some other place near or in the Paropamisus range—Firuz-Kuh (Firozkohi)—Band-i-Turkistān—Maimana(?)—Sapurgan (Shībarghān).

From Sapurgan Polo went to Balc (Balk), the "Baldach" of Frampton; thence to Dogana, the identification of which is still uncertain. Neither Ramusio nor Frampton mention it at all, but proceed straight to Taican or Thaychan (Talikan) which they both give as being two days' journey from Balk. This is obviously wrong, as the distance between these two places is at least 140 miles (Yule gives it as 170, but this is excessive).

On turning to the French texts, we find that after speaking of Balk, they say: "Now we will leave this city, and I will tell you of another called Dogana." We are not told the distance between Balk and Dogana, nor are we given any details of the latter place, for the texts immediately continue: "When one leaves this city that I have been telling you about, one goes a good twelve days (bien xii jornee) between north-east and east.. and...when one has gone this twelve days (doçe jornee) one finds a fortified town (caustiaus) called Taican."

The passage contains several difficulties. In the first place, what was the city that he has been telling us about? Does he mean Balk or Dogana? If he means the latter, as being the last mentioned, it seems obvious that there is a *lacuna* in the texts; but if he is referring to Balk we are still unable to adjust the distances, for the journey between Balk and Talikan would be easily accomplished in seven days.

Yule suggests that the "XII" is a mistake for "VII." If we accept this, it follows that "this city that I have been telling you about" was Balk, not Dogana. From the passage quoted above it will be noticed that the second mention of the distance is written in full, "Doçe" or "Doze," but this need not upset Yule's contention of the mistake in copying, for the scribe having once written "XII" a few lines above in mistake for "VII" would certainly have written it in full as "doçe."

The next point we must try and decide is the identification of Dogana. Yule is unable to make any satisfactory suggestion, and the interpretation given by Parker (see Cordier, *op. cit.* p. 34) with such assurance is quite unacceptable. He would connect it with the Chinese T'u-ho-lo or Tokhara. The limits of Tokharistan as given both in the Chinese annals and works of the Arabic geographers included a large part of

Badakhshan as well as Chitral, Kafiristan and Kabul. Thus it in no way fits into Polo's itinerary. Moreover, there is no evidence whatsoever to show that the name was used in Polo's day at all.

As we have already seen, Dogana must lie somewhere between Balk and Talikan, and from its method of introduction cannot possibly be a district of the size of Badakhshan, still less of the classical Tokharistan. Correspondence with Prof. Sten Konow of Oslo has entirely convinced me of this. In one of his letters he makes a most interesting suggestion after consulting Dr Morgenstierne, the well-known authority on Iranian languages. It is possible that Dogana is directly connected with the Persian *dogāna*, "double," because it was a "double district," embracing two main cities or two rivers. A glance at a large scale map will show that such a district exists to the west of Talikan, where the Āk Sarai branches into two main streams, on each of which is a city—Kunduz on the westerly and Khānābād on the easterly branch. It seems, therefore, that we are probably right in accepting Yule's reading of "VII" for "XII," and must look upon Dogana as being introduced *en passant* simply because it was the local name Polo heard applied to the Kunduz-Khānābād district as he was travelling to Talikan.

From Talikan he reaches Casem (Kishm) in three days, and the same distance again brings him to Badashan (Badakhshan). The line of march is quite clear, but we must not insist on a too definite determination of Badashan, because the district now bearing that name stretches south-westwards past Talikan and north-eastwards to the great bend of the Oxus at Kala Khum. It is hard to say if it extended so far west in Polo's time, but the approximate localization is identical.

After telling us of Badashan, Polo adds that the province of Pashai lies ten days to the south, and, later, that Keshimur is another seven days from Pashai. This is a digression, and really belongs to the information Polo acquired about the incursion of Nogodar into Kashmir, as mentioned by him previously in the chapter on Camadi (see Appendix I. Note 102).

As this route is of considerable interest, we too will make a digression in order to discuss some of the queries raised. Polo is talking of the Caraonas and explaining the methods of attack employed by them. Their king, he tells us, is one Nogodar, nephew of Chaghatai; and he goes on to relate how this man made an expedition through Badashan, Pashai-Dir and Ariora-Keshemur, and how that after subduing all these provinces, he entered India at the extreme point of a province named Dalivar, and seized the government from Asedin Soldan.

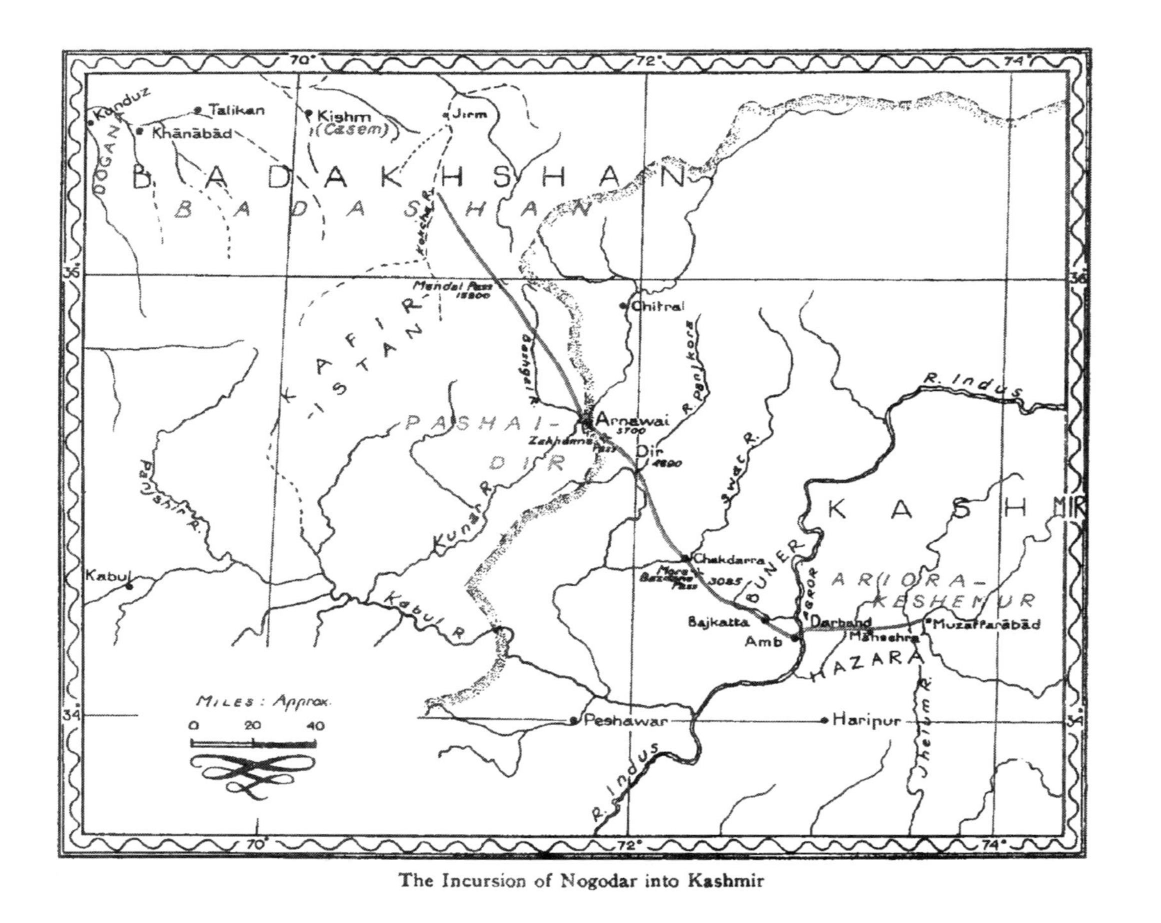

The Incursion of Nogodar into Kashmir

Yule has given us most interesting notes on many of the difficulties in ascertaining the actual route taken by Nogodar, but the identification of some of the localities, such as Ariora, has only recently been made possible owing to the explorations and research of Sir Aurel Stein. He has dealt fully with the whole itinerary in *Journ. Roy. Geog. Soc.* Aug. 1919, pp. 92–103 It will suffice, therefore, to give here his conclusions, with a brief note on "Ariora."

Badashan is, of course, Badakhshan, the province lying to the north of the Hindukush, and which Polo describes in some detail. Pashai-Dir is a copulate name: Pashai being a tribal designation applied to an area in Kafiristan which stretches to the south-east as far as the Kunar river and the tracts lying to the west of Dir. This wider application of the term Pashai has only been known since the results of Sir George Grierson's research on the Dardic languages have been published in the *Linguistic Survey*[1].

It is now possible to appreciate "Keshimur" as being seven days' journey from Pashai. Dir, which, by the way, does not appear in fr. 1116, has long since been recognized at the head of the western branch of the Panjkora river. The next locality is also a copulate name: Ariora-Keshemur, but here again I would point out that "Ariora" does not occur in fr. 1116, or in Ramusio.

Stein has now identified it with the modern Agror[2], the hill tract on the Hazara border which faces Buner on the east from across the left bank of the Indus. Keshemur is, of course, Kashmir.

With regard to Dalivar it would appear that Marsden's original suggestion was quite correct, and the name is a misunderstanding of "Città di Livar," for Lahawar or Lahore.

The name of the ruler, Asedin Soldan, has been definitely identified with Ghiasuddin, Sultan of Delhi (1266–1286).

The complete itinerary from Badashan would be, then, in all probability as follows: Badakshan—across the Mandal pass—down the Kafir valley of Bashgal—Arnawai on the Kunar river—across the Zakhanna pass—Dir—down the Panjkora river—Chakdarra—across the Mora-Bazdana pass in Lower Swat—Bajkatta in the Buner district—the Indus at Amb,

[1] Sir George tells me that since he wrote his account of Pashai our knowledge of both country and language has been greatly extended by Morgenstierne in his *Report on a Linguistic Mission to Afghanistan.*

[2] In his edition of *Marco Polo*, Charignon tells us that it corresponds to Haripur on the left bank of the Indus. His authority for this curious statement is entirely lacking. Incidentally, I was not aware that Haripur was on the Indus.

or Darband—over the Hazara district at about the latitude of Mānsehra —the Jhelum near Muzaffarabab.

We now return to Badakhshan whence the itinerary continues E.N.E. for twelve days to Vokhan (Wakhān), and thence another three days north-east to the plain of the Great Pāmir. The French texts merely tell us that Polo found a fine river running through a plain, but Ramusio also mentions "a great lake." This could be either Lake Victoria or Lake Chakmak, but is probably the former (see Stein, *Innermost Asia*, Vol. II. pp. 858 *et seq.*, also *Ancient Khotan*, Vol. I. pp. 30 *et seq.*; and *Serindia*, Vol. I. p. 65). He now describes a twelve days' desert ride across the plain, followed by another forty days of continuous desert tracts without any green thing to relieve the dreariness and monotony. To this country he gives the name of Bolor.

It is clear, then, that the itinerary could not have passed through the cultivated valleys of Tāsh-Kurghān or Tagharma, and Sir Aurel Stein reasonably suggests (*Ancient Khotan*, pp. 40–42) that, after visiting either the Great or Little Pāmir, he travelled down the Ak-su river for some distance, and then, crossing the watershed eastwards by one of the numerous passes, struck the route which leads past Muztāgh-Ata (24,000 ft.) and on towards the Gez defile. Previous suggestions of the itinerary will be found in Yule, Vol. I. pp. 173 *et seq.*

The route continues to Kashgar and then turns south-east past Yangi-Hissar to Yarcan (Yarkand) and Cotan (Khotān).

He mentions Samarkand *en passant*, but was obviously never there himself, and obtained his information from his father and uncle. From Khotān the route runs to Pein and Charchan, and thence, in five days, to Lop.

The position of Pein, the "Pimo" of Hiuen-Tsiang (Hsuan-tsang), has led to much speculation. Sir Aurel Stein would identify it with Uzun-tati, now forming part of a débris-covered area, thirty-five miles to the west of Keriya river. But Huntington (*Pulse of Asia*, pp. 387–8) shows that both distances and descriptions fit Keriya itself much better. (See further Charignon, *Marco Polo*, Vol. I. pp. 105 *et seq.*) From Pein the itinerary continues east through Niya to Charchan and along the Charchan river to Lop (Charkhlik); thence in a north-easterly direction past Mīrān and Kum-Kuduk, along the Su-lo-ho to the Khara-Nōr, thirty miles S.E. of which lies Tun-huang, or Sha-chau (Polo's "Sachiu") on the Tang-ho, a tributary of the Su-lo-ho.

At Tun-huang an important digression occurs, and Polo tells us of the Turfān-Hami district to the north-west. Its interest lies not only in the

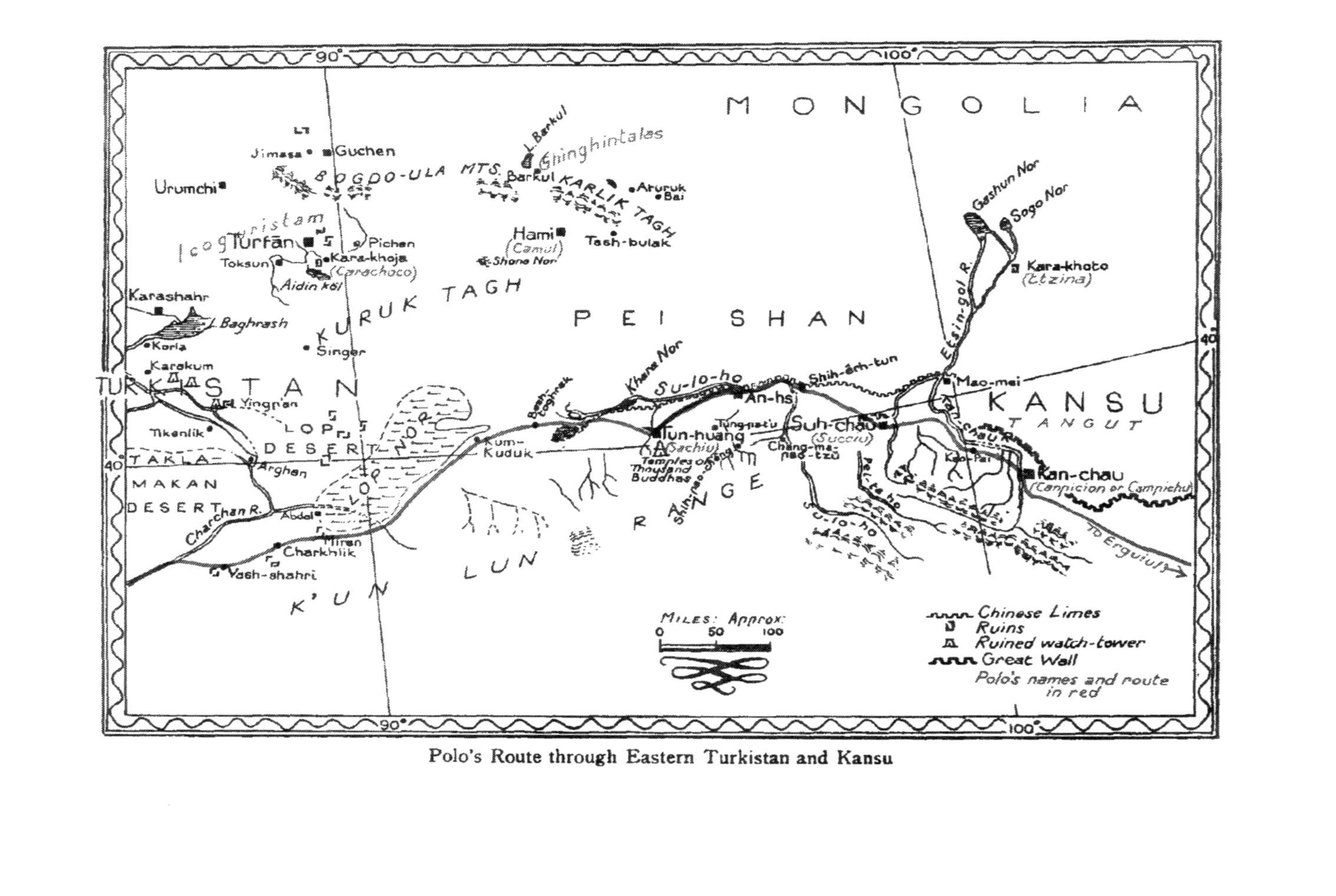

Polo's Route through Eastern Turkistan and Kansu

fact that it presents unsolved identification difficulties, but also that in all probability it formed part of the route of the elder Polos, and joined the itinerary we are following at Tun-huang.

In the leading MSS the localities mentioned are Camul (Hami) and Ghinghintalas, to which must now be added Carachoco (Kara-Khoja) which is mentioned in the newly found *Z* manuscript.

As Benedetto would identify this later place with Kara-Khoto[1], it is necessary to show how it agrees with Kara-Khoja in every detail, and is also a means of helping us to locate Ghinghintalas. From the map opposite it will be seen that Camul (Hami) lies about 200 miles N.N.W. of Sachiu (Tun-huang). After describing Camul, Polo says: "Or laison de Camul et vos conteron des autres que sunt entre tramontaine et maistre. Et sachiés que ceste provence est au grant can." He then continues: "Ghinghintalas est une provence que encor est juste le desert entre tramontaine et maistre. Elle est grant XVI jornee. Elle est au grant can." Thus it is clear that Ghinghintalas lies N.N.W. from Tun-huang past Hami.

Turning for a moment to the *Z* text (Benedetto, p. 46 note c) we find that in the very middle of the Camul chapter is inserted a passage which may help us in identifying Ghinghintalas.

It begins as follows: "Icoguristam quedam provincia magna est et subiacet magno can. In ea sunt civitates et castra multa sed principalior civitas Carachoco apellatur. Civitas ista sub se multos alias civitates et castra distringit." Now there is only one place that fits in with all the data, and that is Kara-Khoja, the ancient capital of Turfān, the ruins of which, dating down to Uigur times, have been proved so rich in archaeological spoil (see Stein, *Innermost Asia*, Vol. I. pp. 587–609).

The Turfān territory was known in medieval times as Uighuristan, being the chief seat of the Uigur domination. Thus we have no difficulty in identifying both the "Icoguristam" and "Carachoco" of the *Z* passage. We have seen from the above quotation that Carachoco "subiacet magno can," and it is this statement that should help us to locate Ghinghintalas. Yule, thinking that Kara-Khoja lay to the N.W. of Turfān, instead of

[1] Kara-Khoto is a ruined town on the Etsin-gol, 135 miles N.E of Mao-mei, and is to be identified with Polo's "Etzina," as we shall see later. Benedetto appears to have been misled partly by the similarity of name. Seeing the statement supported by Cav. de Filippi in his review in *Journ Roy. Geog Soc.* March 1928, I challenged the identification, and a correspondence ensued Filippi wrote to Sir Aurel Stein for his opinion, and the results were published in *Journ. Roy. Geog. Soc* Sept. 1928, pp. 300–302. The identification of Carachoco with Kara-Khoja was definitely established.

to the S.E., pointed out (Vol. I. p. 214) that it would be outside the Khan's boundary, yet Rashīd-ud-dīn, the famous Persian historian, a contemporary of Polo, distinctly says it was a neutral town on the border-line. He also tells us that a point near Chagan-Nōr (lat. 48° 10′; long. 99° 45′) was also on the boundary. Now if we take this boundary-line to run in a semi-circle from Kara-Khoja to Chagan-Nōr, we can surely place Ghinghintalas in the neighbourhood of Barkul.

Thus we would be "au grant can," as the *Z* text tells us it is, and also N N.W. of Hami.

After coming to this conclusion, I found that in his edition of *Marco Polo*, Vol. I. pp. 141 *et seq.*, Charignon had reached the same identification, but by entirely different means—etymological grounds and evidence of Chinese tradition. Thus there would seem to be little doubt that in Ghinghintalas we must recognize the Barkul district lying to the N.N.W. of Hami, and to the N.W. of the Karlik-Tagh.

Returning to Tun-huang (Sachiu), we continue our main itinerary eastwards. Ten days' ride E.N.E. brings the travellers to the province of Succiu whose chief city bears the same name. This is according to the reading in fr. 1116; but Yule speaks, not without considerable hesitation, of the province of Sukchur and the town of Sukchu.

Whatever reading we adopt, it is obvious that the province and town of Suhchau (Su-chow) are meant.

Polo tells that during these ten days' journey you find practically no dwellings, and that there is nothing of interest to report. It is hard to decide for certain the exact route followed, but after leaving Tun-huang the most natural route (according to Stein's maps of the district) would be north-east to the Su-lo-ho at An-hsi, along the line of the ancient *limes* (the fortified border constructed by the Chinese Emperor Wu-ti in the latter half of the second century B.C., recently discovered by Stein, *Serindia*, Ch. xv, sec. ii–v) to a point marked Shih-êrh-tun, where it would drop south-west to Suhchau. The other alternative would be nearly due east from Tun-huang. After visiting the cave temples of Thousand Buddhas to the south-east of the city, he would continue due east to the small oasis of Tung-pa-t'u, visit the cave temples on the T'a-shih river, follow the river to Shih-pao-ch'êng and reach Ch'ang-ma-pao-tzŭ on the So-lo-ho *via* its tributary the Sha-ho, whence Suhchau would be reached by a weary ride through gravel steppe and stony scrub slopes[1].

[1] It is impossible to follow these alternative routes without reference to Maps 41 and 42 of Stein's *Chinese Turkistan and Kansu* (or *ditto*, *Innermost Asia*, Vol IV).

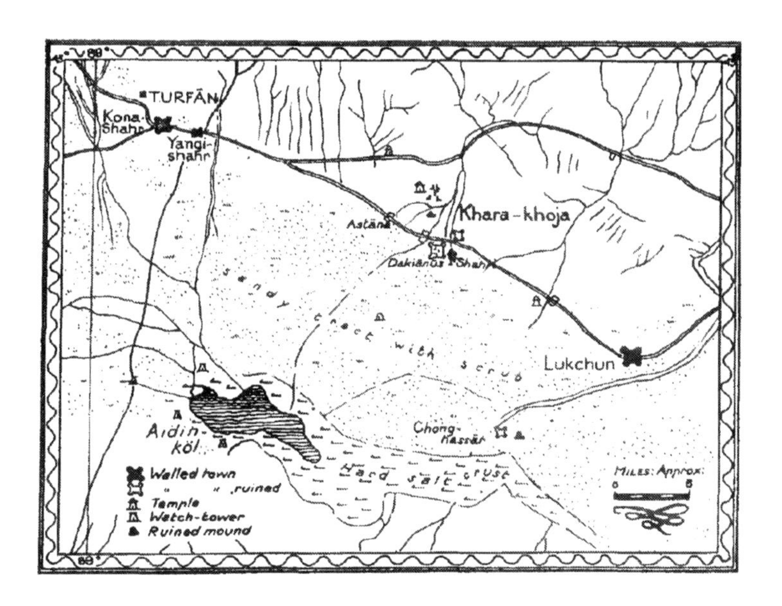

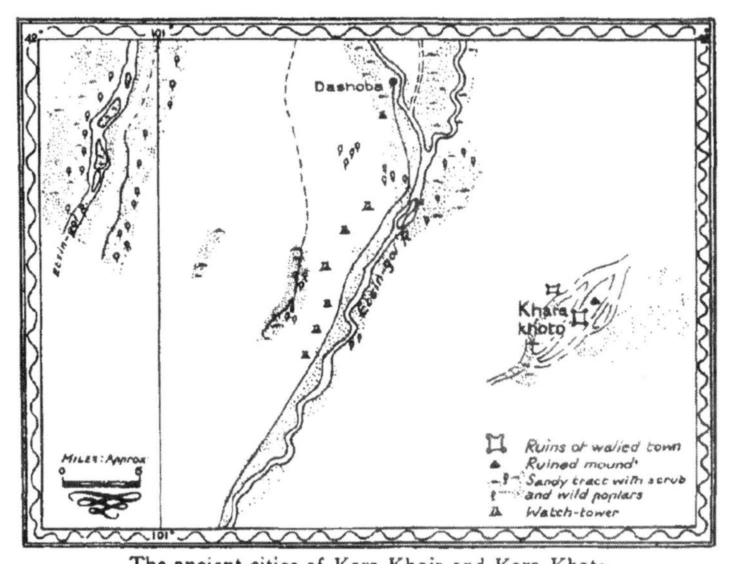

The ancient cities of Kara Khoja and Kara Khoto
(Reproduced by special permission from the recent surveys of Sir Aurel Stein)

The former route seems much preferable, and if he took it we need not be surprised that he makes no mention of the *limes*, because in the first place, unless it lay directly in his path, he might never have noticed it[1], and in the second place he even omits all mention of the Great Wall which commences at Suhchau, or rather at Chia-yu-kuan fourteen miles to the west of the city.

Proceeding in a south-easterly direction, Polo reaches Canpicion, or Campichu, the capital of Tangut. In this we have little difficulty in recognizing Kan-chau, the chief city of Kansu. Before continuing to Erguiul we have another digression—this time to Karakorum *via* the Etsin-gol.

As we are quite ignorant as to whether Polo visited Karakorum personally, it would be waste of time to try to ascertain his point of departure from the main route. If, however, it was Kan-chau his route would lie in a large semi-circle to the north-east along the Kan-chau river to Mao-mei, where the name changes to Etsin-gol after its confluence with the Pei-ta-ho.

Turning to the text, we find in fr. 1116 the following words as a kind of introduction to the digression:

"Et por ce nos partiron de ci et aleron seisante jornee ver tramontaine."

After twelve days' ride you reach Eçina, or Etzina, described as "chief dou desert do sablon, ver tramontaine."

As Cordier originally suggested, this town has proved to be on the Etsin-gol, and has now been identified with Kara-Khoto, the "Black Town," first visited by Col. Kozloff in 1908–9, and again by Stein in his Third Journey of Exploration (see *Innermost Asia*, Vol. I. pp. 435–506).

As has already been mentioned, it was this town that Benedetto and Filippi identified with the "Carachoco" of the *Z* text, instead of Kara-Khoja. De Filippi objected to the site of Kara-Khoto for Etzina because Polo describes the latter as a pastoral and agricultural community, whereas Kara-Khoto has proved rich in cultural relics. As I pointed out in *Journ Roy. Geog. Soc.* Sept. 1928, p. 302, de Filippi appears to forget the fate of Kara-Khoto under the ruthless hand of Chinghiz Khan in 1226. Polo knew of it only as an agricultural community, and perhaps only a small one at that (see Lattimore, *Journ. Roy. Geog. Soc.* Vol. LXXII. Dec. 1928, p. 510).

On the other hand, the large remains and important yields of Kara-Khoja, the old capital of Turfān, fully justify Polo's description of it already quoted (p. xliii).

[1] See, for instance, Plate 191 facing p 344 of *Innermost Asia*, Vol I.

Another forty days north lies Caracoron, of which Polo gives us practically no description at all, but appears to introduce it merely to include a long account of the wars of Prester John and Chinghiz Khan, together with the "customs of the Tartars." It is, of course, quite possible that he may have visited it during his long service with the Khan.

After speaking briefly of the Plain of Bargu and the provinces as far north as the Ocean Sea, Polo returns once more to Kan-chau, whence five days' ride brings him to Erginul, or Erguiul, which has been identified with Liang-chau fu. Eight days more takes him to Egrigaia, a province whose capital is Calacian, or Calachan. It appears that here Polo takes the route running north-west through Alashan, instead of following the Great Wall to the south-east of Liang-chau fu; and that Ning-sia fu and Ting-yüan-ying represent Egrigaia and Calacian respectively.

The province of Tanduc, which is mentioned next, is not easy to identify. It must include the district lying in the neighbourhood of the great northern bend of the Hwang ho, while what evidence we have leads us to favour the modern Tokto as its chief city.

After riding seven days eastwards through the province, Polo comes to a city called Sindachu, or Sindaciu, which Yule and others would identify with Siuen-hwa fu (Hsuan-hua fu), the Siuen-te-chau (Hsuan-tê chou) of the Chin dynasty. This may possibly be correct, but it seems curious that Polo should turn south-east towards Cambaluc (Peking) when he was going north-east to the Khan's summer residence at Chandu. We must not forget that in the Prologue Polo tells us that the Khan sent people a full forty days' journey to meet them, so at this point the Polos were already accompanied, and would *not* have gone to the capital by mistake thinking the Khan was in residence at the time.

Yule's text continues: "Now we will quit that province and go three days' journey forward. At the end of those three days you find a city called Chagan-Nōr." This Yule would place to the west of the Anguli-Nōr where there are many small lakes taking their names from neighbouring towns. It will thus be seen that if we accept Siuen-hwa fu, the itinerary bends back to Chagan-Nōr before continuing to Chandu. This seems most improbable.

Thus I would doubt Yule's suggested identifications. Charignon (*op. cit.* Vol. I. pp. 255 *et seq.*) considers "Syndatui" to be the ancient Hing-Houo (Chang-pei), fifty kilometres north-west of Kalgan. He also produces evidence to show that Chagan-Nōr (Ciagannor of fr. 1116) lay some considerable way to the *east* of Anguli-Nōr.

These identifications certainly deserve our close consideration, for apart

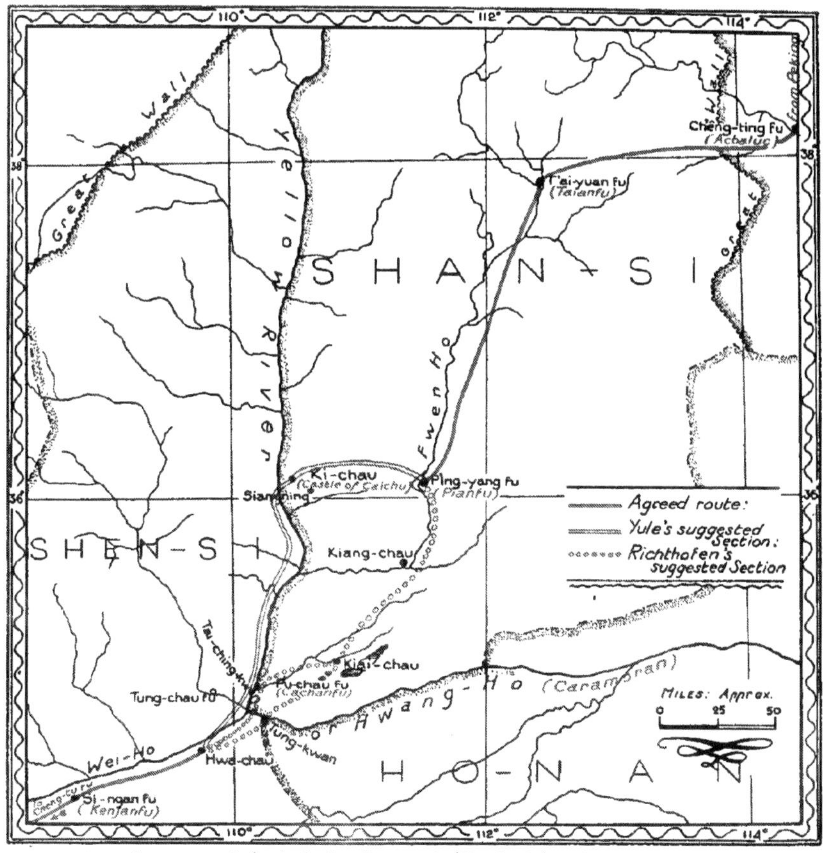

The Route from Chêng-ting fu (Acbaluc) to Si-ngan fu (Kenjanfu) and the crossing of the Hwang ho

from historical evidence they correspond with Polo's distances and enable us to trace the itinerary in a north-easterly sweep *via* Hing-Houo and Chagan-Nōr to Chandu, or K'ai-p'ing fu, twenty-six miles to the north-west of Dolon-Nōr.

Having safely conducted our worthy travellers to the presence of the Khan, we will proceed to the Chinese itineraries.

Polo gives us no details either as to the nature of the missions on which he was sent during his long service with the Khan, or as to the number of such missions. We have already suggested that he may have visited Karakorum when in the Khan's service. In his Preface, Polo speaks of continually going on missions, and when preparations were being made for the escorting of the fair Cocachin to Persia, we find he suddenly turns up from some mission to India. It will thus be seen that we must not attempt to include in the itinerary *every* place mentioned in the text, as the information given may have been picked up on a previous journey to be included here simply because we are somewhere in the same locality. As we shall see later, this is apparently what has happened in the case of Java.

THE MISSION TO YUNNAN AND BURMA

The first long journey given in detail is that through South-Western China to the Province of Mien, or Burma.

In attempting to trace this itinerary we shall continually be faced with such difficulties as have been enumerated at the beginning of this section of the Introduction.

Alternative routes present themselves at places, while on other occasions, unless slight errors in Polo's distances and geographical details be allowed, we find it impossible to complete the itinerary satisfactorily. At first the route is clear. Starting at Cambaluc, on the site of which is the modern Peking, it goes ten miles to the Pulisanghin river (Hun-ho), then thirty miles to Goygu or Juju (Cho-chau). One mile further Polo reaches branch roads, the western one leading through Cathay, and the southern one through Manzi. He takes the former, and, according to Ramusio, reaches Acbaluc (Ch'êng-ting fu) in five days. In another five days he arrives at Taianfu (T'ai-yuan fu), whence he reaches Pianfu (P'ing-yang fu) in seven days more. It is after leaving this place that our troubles begin. In two days he is at the "Castle of Caichu," which is described as being twenty miles from the Caramoran (Hwang ho) river. Two days later he is at Casiomphur, Cachanfu (P'u-chau fu), and at Bengomphu, Kenjanfu (Si-ngan fu) in another eight days.

The identity of the "Castle of Caichu" is unknown. Other forms of the word appear as Caicui, Caiciu, Caicin, Caytui, etc.; while Ramusio alone gives it as Thaigin or Taigin.

Yule suggests its identification with Ki-chau, a place lying about fifty miles from P'ing-yang fu, but not more than eight or nine miles from the Hwang ho (although Yule describes it as "just about 20 miles" from it). He would then trace the route either down the west bank of the river, or else on the river itself, to a point opposite P'u-chau fu, and then on to Si-ngan fu. Now Polo tells us that the "castle," or perhaps "fortress," is two days' ride *westward* of Pianfu, and that after crossing the river at a point twenty miles west of the "castle," he reaches Cachanfu (P'u-chau fu) in two days, which place, he says, is *on the west of the river*. This latter statement is a mistake, as P'u-chau fu is on the east of the river. Whether it is due to a lapse in Polo's memory, an error in his notes, or merely a slip on the part of the copyist, seems to be immaterial.

An alternative route has been suggested by Baron von Richthofen. He points out that Caicui or Caichu may be, as Marsden originally conjectured, Kiai-chau, or Chieh-chau, near the salt marsh half way between P'ing-yang fu and the fortress T'ung-kwan on the Hwang ho; and that Ramusio's Taigin may be Tai-ching-kwan (locally pronounced Taigin-kwan) close to P'u-chau fu. Thus as both forms can be separately identified, he would suggest that Polo passed one of these places on his outward, and the other on his return journey. From P'ing-yang fu he would go to Kiai-chau, on to T'ung-kwan, and so to Si-ngan fu; while on his return he would re-cross the river at Tai-ching-kwan and reach Kiai-chau again *via* P'u-chau fu.

The sketch-map opposite clearly indicates these alternative routes. It will be seen that either might be correct, but we have to make certain concessions whichever we accept. If Polo is to be taken literally when he says that the "castle" is two days *west* of Pianfu, Richthofen's theory cannot be correct. Added to this is the fact that the distance to the lake is rather too long (eighty miles) for a two-days' march. Kiai-chau is twenty-six miles from the river, which, according to Polo, is too far, just as the distance of Ki-chau was too short. Finally, it is only one day's march from Kiai-chau to the river, so that there seem to be a number of difficulties to overcome before we can accept this itinerary. The only objections I can see to Yule's route is that Ki-chau is W.N.W. of P'ing-yang fu, and is too near the river. The only other suggestion I can offer is that Siang-ning, which is forty-three miles from P'ing-yang fu and eighteen from the river, is the "castle." In either case I think Polo must

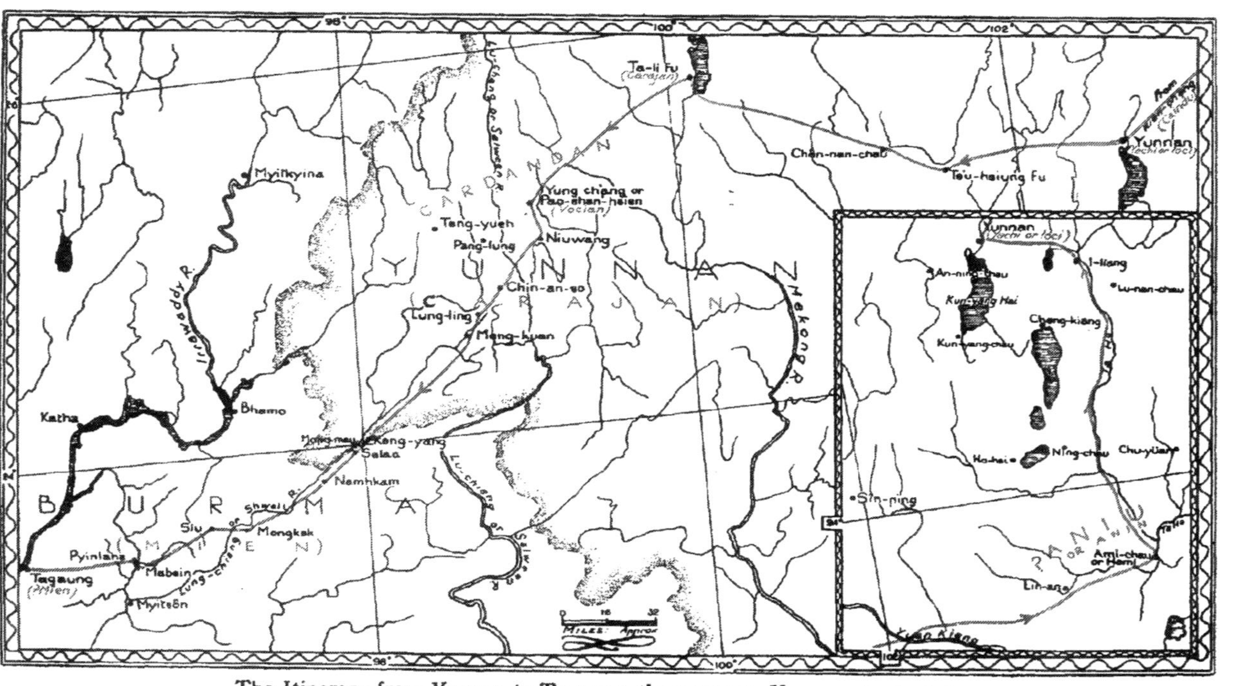

The Itinerary from Yunnan to Tagaung, the return to Yunnan via Ami-chau

have descended the river by boat if he was to reach P'u-chau fu in two days.

After leaving Kenjanfu (Si-ngan fu), the itinerary runs over difficult country across the Tsin-ling-shan, through the Han kiang valley, across the Ta-pa-shan, and then through the fertile regions of the province of Sze-ch'wan to the capital, Sindafu (Ch'êng-tu fu). The only trouble here is that Polo took forty-five days to do the journey, while recent travellers have shown that it can be done quite easily in six or seven days.

It is possible, of course, that some of his figures are wrong, or that certain delays occurred which he has counted in his reckoning. From Sindafu five days' journey brings the travellers to "Tebet," which must be taken as commencing at the mountainous region near Ya-chau. Proceeding in a S.S.W. direction through uninhabited country for twenty days, Polo arrives at the town of Caindu (Kien-ch'ang, usually called Ning-yuen fu in modern maps). Continuing now due south through the beautiful valley of Kien-ch'ang, with the mountainous and inhospitable country of the Lolos on his left and Menia on his right, Polo crosses the Kin-sha kiang at its great bend due north of Yunnan fu. He now enters the province of Yunnan (Carajan), and after five days' journey reaches the capital, Yachi, or Iaci (Yunnan fu)

Ten days' travelling westwards (really W.N.W.) through the province of Yunnan brings Polo to the *town* of Carajan (Ta-li fu). Five days more to the west takes him to Cardandan, or Zardandan (the "Nocteam" of Frampton), the land of the "Gold-teeth" people. Although its exact locality cannot be stated with absolute certainty, it can be taken as being a district near the present Yunnan-Burma boundary. Polo gives its capital as Vocian, Vochan, the "Nociam" of Frampton, in which we recognize Yung-ch'ang, half-way between the Mekong and the Salween. In the map of Burma issued by the Indian Survey (1918, corrected 1925) it appears as Pao-shan-hsien. The itinerary now becomes very confusing and it is not possible to trace it with certainty across the frontier, through Burma, and back again to Sindafu. We must, I think, agree with Yule in concluding that Polo never went personally further south than the city of "Mien." But let us see what he tells us himself in the French texts. On leaving Yung-ch'ang (Pao-shan-hsien), there is a great descent which lasts for two days and a half, at the end of which lies the province of Mien, or Amien, the "Machay" of Frampton. After travelling for fifteen days through an unfrequented and wooded country the *city* of Mien is reached.

As Yule has pointed out, the real capital of Burma at the time was Pagan in lat. 21° 13′, but fifteen days of overland travel would never be sufficient to reach so far. If, however, we take "Mien" to be Old Pagan, i.e. Tagaung on the Upper Irrawaddy in lat. 23° 28′, the distances would be reasonable.

In the first place we must try and determine the locale of the "great descent." Dr Anderson's suggestion that it is the descent into the plains near Bhamo need not detain us, nor need we consider further the route W.S.W. to Teng-yueh. After a study of altitudes and roads shown in Sheet No. 92 (3rd Provisional issue, 1926) of the "India and Adjacent Countries" series, I find it impossible to accept Yule's suggestion that the route lies direct to the Shweli valley. Yung-ch'ang is 5500 feet, and although the descent is continuous to the Salween, the altitudes in the vicinity of Pang-lung vary between 8000 and 10,000 feet.

The communication of Mr H. A. Ottewill published by Cordier (*Ser Marco Polo*, p. 89) contains a much more likely suggestion: that from Yung-ch'ang Polo went south to Niuwang, gradually dropped down to the Salween, and after crossing it, proceeded to Lung-ling and so to Keng-yang. I would suggest the full itinerary here as follows: Yung-ch'ang—Takwanshih—Niuwang—Hsiang-tou-shan—the Salween—Hochia Chai—Chin-an-so—Pawan Chai—Lung-ling—along the Nam Hkawn, tributary of the Shweli (Lung-chiang)—Keng-yang. After this, we are practically reduced to guesswork, but if his objective was Tagaung he would surely have followed either the land route from Keng-yang to Möng Mau or Selan—Namhkam—Siu—Mabein—Pyinlaha (just across the river)—Tagaung; or else he would have continued along the Shweli to Myitsôn and then turned N.W. to Tagaung.

This is, with but little doubt, the end of the present itinerary in its south-westerly direction. It is, however, practically impossible to say at what point we are on the return route to Sindafu (Ch'êng-tu fu). Polo speaks vaguely of Bangala (Bengal); Cangigu (? Upper Laos or Tonking); and Aniu, or Anin (? the district S.E. of Yunnan near Ami-chau, or Homi). We are now approaching Yunnan fu again, after which Polo seems to be following an itinerary once more. He mentions the province of Toloman, or Coloman (? N.E. of Yunnan fu to about Wei-ning), from which place he travels twelve days in an easterly direction through the province of Ciugiu (Yule reads "Cuiju") to the *city* of Ciugiu (Yule here reads "Fungul"). It is very doubtful which localities are meant, but if we have placed Toloman correctly, the distance and description would lead us to Sui fu in the province of Sze-ch'wan, at the point where the two chief branches of the Yangtze meet. If, however, Toloman stretched farther north, the town of Lu-chau opposite Nachi may be meant. Another

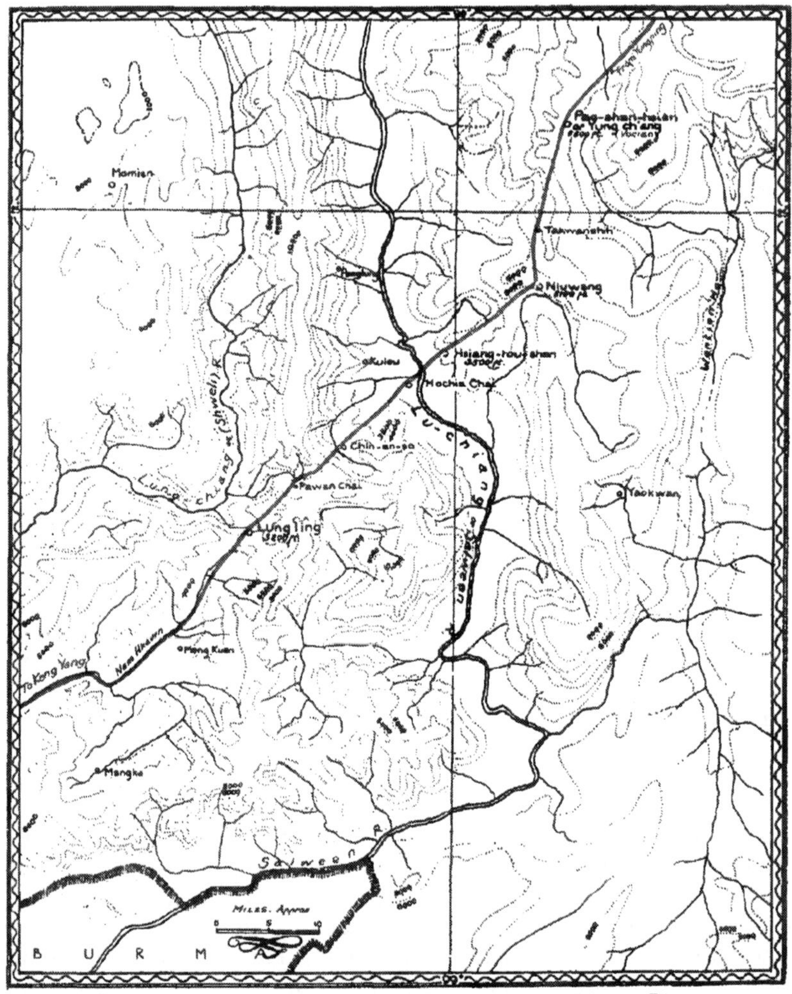

Contour map showing the crossing of the Lu-chiang or Salween River

twelve days brings Polo to Sindafu once again, so that unless the march was exceedingly slow, the city of Kiating fu would be too close to it to be the city of Ciugiu. In spite of the efforts of Yule and others to trace this portion of the route, it still remains very uncertain and at present our identifications are little more than guesswork.

It should be pointed out that Frampton has avoided the whole difficulty by entirely ignoring all places between Mien and the road-bifurcation near Cho-chau, which he calls "Cinguy." (See Appendix I. Note 317, p. 219.)

After leaving Sindafu (Ch'êng-tu fu), Polo travels seventy days back to Juju (Cho-chau). Apparently this is the end of the journey, although we are not told if he went on to Cambaluc before starting on his next mission to Manzi and south-eastern China. It seems highly probable that he did so, but he writes as if the itinerary was continuous, for having arrived at Juju, he says that four days south brings him to Cacianfu. In the next chapter, however, being still in Cacianfu, he writes: "We will now set out again, and travel three days to the south when you come to another city by name Cianglu."

In studying the itineraries as described in his missions, we can definitely say that the Yunnan-Burma route ends at Cho-chau.

THE MISSION TO THE EASTERN PROVINCES

Starting at Cacianfu or Cacanfu (Ho-kien fu), the itinerary runs three days to Cianglu (Tsang-chau) and thence five days to Ciangli (Tsi-nan fu), which Yule writes Chinangli. In another five days Polo reaches Tandinfu (Yen-chau), whence three days brings him to Singiumatu (Tsi-ning-chau). All is clear so far except that the *account* of Yen-chau fits T'si-nan much better. But after Singiumatu our troubles begin once again. Frampton, as well as Ramusio, speaks next of the Caramoran (Hwang ho), but the French texts make Polo go in turn to Linju, Piju, and Siju. The two latter have been fairly satisfactorily identified with Pei-chau and Su-t'sien, but "Linju" remains a mystery.

Yule, taking a hint from Murray, would identify it with Lin-ch'ing just under the 35th degree of latitude (not to be confounded with Lin-ch'ing chau on the canal, in lat. 36°. 51′). I have consulted all the old maps, and no two spell it exactly alike or put it in exactly the same place. It has, moreover, entirely disappeared from modern maps, unless it has become the Liuchuan in practically the same locality, north of Sü-chau fu. But it is a very small and insignificant place neither on the Grand Canal nor on the Hwang ho. The various forms of the name tell us nothing, being due merely to the mixing up of "n" and "u" in the MSS. We may, perhaps, gain some useful information by studying Friar Odoric's itinerary in the same district.

He is travelling north from a city called Menzu "towards the mouth of that great river Talay (Yangtze)." Menzu has been identified with Chin-kiang, but in Polo the name remains unchanged. After eight days' travelling from this place, Odoric arrives at a city called Lenzin, "which standeth on a river called Caramoran." Continuing "by that river towards the east," he comes to a city called Sunzumatu (Polo's Singiumatu). The next place he mentions is Cambaleck (Peking). Thus we see that (1) Odoric is on the same route as Polo, only going north instead of south; (2) he is only mentioning places of importance; (3) his "Lenzin" would seem to correspond to Polo's "Linju."

Now if we look at another section of Odoric's itinerary in order to ascertain his rate of travelling, we find that he goes from Cansay to Chilenfu (Hang-chau to Nan-king) in six days. Travelling from Menzu (taking it to be the modern Chin-kiang) at the same rate, eight days would bring him very near to Sü-chau fu, on the old course of the Yellow River (Hwang ho), or, if we go more east, to Han-chwang. Returning to Polo, we see that he takes three days to go from "Siju" to "Coiganju" (i.e. from Su-t'sien to Hwai-ngan-chau). Taking this distance as the radius of a circle whose centre is at "Piju" (Pei-chau), we find that it will actually pass through Sü-chau fu, as well as Han-chwang. As a matter of fact, it also cuts I-chau fu, but this is right out of the itinerary. Now as the distance between "Piju" and "Linju" is the same—three days—it will be seen that as far as *distances* are concerned, as given both by Polo and Odoric, we are fully justified in making "Linju" the modern Sü-chau fu (34° 12', 117° 20').

We now turn to *descriptions*. Odoric gives none, but Polo draws attention to the fact that although the inhabitants are great traders, they are also good soldiers. He adds that the necessities of life are found in abundance, and that the vessels transport much merchandise. In a certain secret report issued by the Admiralty during the war, I find these very points mentioned: the inhabitants are great traders, the town being the *entrepôt* for merchandise from East Honan, South Shan-tung, and North Anhwei. At the same time they are described as having a military disposition, while Su-chau fu has a great reputation as a recruiting centre. It lies on the great road from Peking to Nan-king, which is also the trade route for cart traffic. In Polo's time the shipping up the Hwang ho (dried up since 1851) must have been on a very large scale.

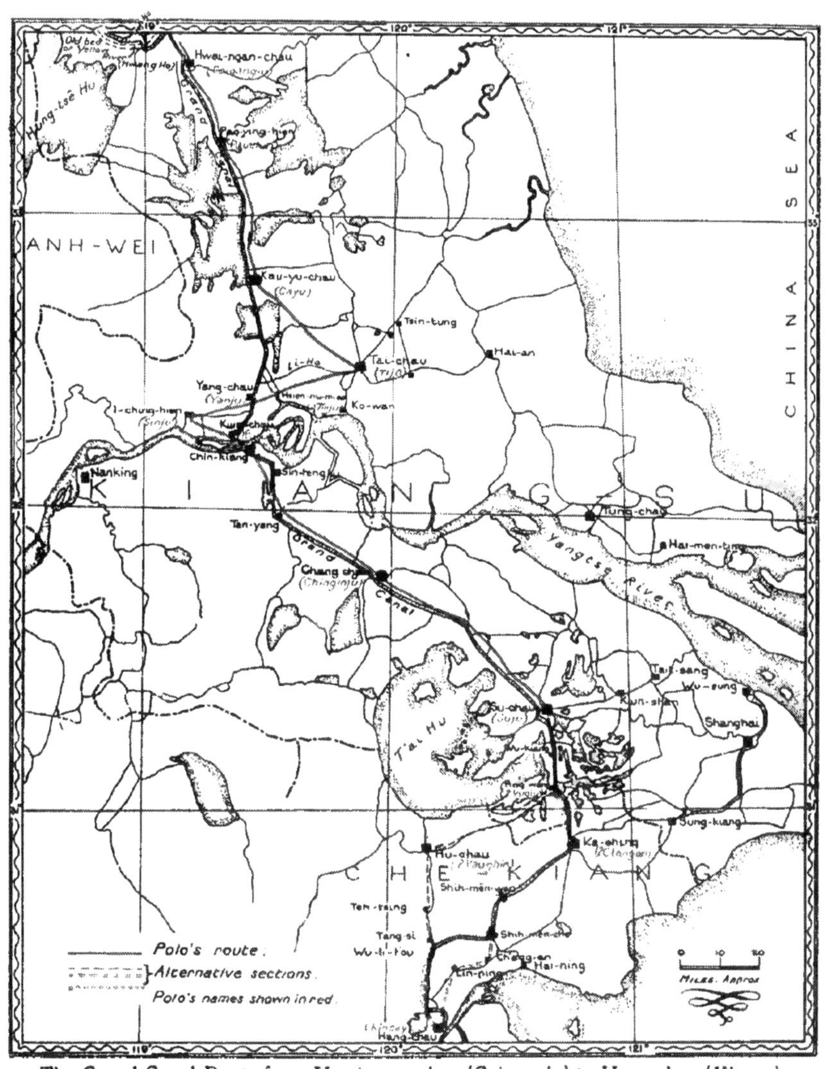

The Grand Canal Route from Hwai-ngan-chau (Coigangiu) to Hang-chau (Kinsay)

(Based, with permission from H. M. Stationery Office, on the Province of Kiangsu" map to semi-official Admiralty publication)

All this supports my suggestion; but if "Linju" is to be identified with Sü-chau fu, the itinerary will have to go slightly north to Pei-chau before continuing south to Su-t'sien and Hwai-ngan-chau. If, however, Han-chwang were "Linju," the *direction* would certainly be more in accordance with the text—that is, taking it absolutely literally. But in view of the fact that it is only a large village with apparently no past history, and that it is on the canal and not on the river, I have no hesitation in accepting Sü-chau fu in preference either to Han-chwang or any other place that can claim a possible agreement with the text.

After leaving Siju (Su-t'sien), Polo travels three days to Coiganju (Hwai-ngan-chau), on the east bank of the Grand Canal, opposite which, on the other side of the river, lay the small town of Caigiu. No trace of this latter place exists and we must conclude that it has been claimed by floods many centuries since. One day's march takes our traveller to Pauchin (Pao-ying-hien), whence the same distance brings him in turn to Cayu (Kao-yu-chau), Tiju (Tai-chau) and Yanju (Yang-chau). It was at Yanju that Polo was "Governor" for three years, for which see Appendix I. Note 351, p. 227). Tinju is also mentioned, being described as a great salt centre lying between Tiju and the sea. In this we should in all probability recognize the modern Hsien-nü-miao (see *Ser Marco Polo*, p. 94). At this point Polo leaves the itinerary to tell us of "Nanchin" (Ngan-king on the Kiang, not to be confused with the famous Nan-king near the mouth of the river), and the siege of Saianfu (Siang-yang fu), for which see Appendix I. Note 353, p. 228.

Returning to Yanju, the route runs fifteen miles to Sinju (I-ching-hien) and then on to Caiju (Kwa-chau) on the Kiang opposite the Golden Island and Chin-kiang fu, to which latter place he now proceeds. Three days more bring him to Chinginju (Chang-chau) and so to Suju (Su-chau), eighty miles west of Shang-hai.

Between Suju and Kinsay (Hang-chau) the route is uncertain. Two alternatives present themselves: he either (1) continued to follow the canal, in which case the itinerary would be: Wu-kiang—P'ing-wang—Ka-shing—Shih-mên-wan—Shih-mên-che—Wu-li-t'ou—Hang-chau (or from Shih-mên-che a shorter way would be *via* Ch'ang-an and Lin-ping) or else he (2) went from Ping-wang to Hu-chau, just south of the T'ai Hu, and then due south to Hang-chau *via* Teh-tsing, Tang-si and Wu-li-t'ou. The route across the lake is obviously not the one taken by Polo. Frampton jumps from Su-chau to Hang-chau without giving any details of the intermediate part. Fr. 1116 mentions three distinct places: "Vugiu," "Vughin," and "Ciangan," which Yule calls respectively "Vuju,"

"Vughin," and "Changan." The question to be answered is—which of the above mentioned routes do these places fit the best?

I can see no need to go as far east as Sung-kiang to look for our route as Pauthier did. Although fr. 1116 is alone in giving *three* distinct localities, we find that all the best MSS agree in stating that Kinsay was reached from some place or other (the names vary) *three days' journey away*, and that during these three days a number of towns and villages were passed through. Surely, therefore, it is useless to look for any of the three named places *within* a three days' area from Kinsay. Thus I fail to see how we can expect to find them in the neighbourhood of Shih-mên-che or Ch'ang-an, as Moule (*T'oung Pao*, July 1915, pp. 393 *et seq.*) rather hesitatingly suggests.

Now Hu-chau is forty-three miles, and Ka-shing sixty-three miles from Hang-chau (Kinsay). Thus either of these could be described as three days' journey away, but Ka-shing would seem preferable, being on the more direct route from Su-chau, and giving a fair average of over twenty miles a day.

Returning to Su-chau we read that Polo goes one day's journey to "Vugiu." It has been suggested (Yule, II. 184) that Wu-kiang is meant. But this is only eight and a half miles from Su-chau and has no past or present history of any importance. Any attempt at tracing the localities on etymological grounds seems hopeless. The only place of any importance that is roughly a day's journey from Su-chau (twenty-two miles) is P'ing-wang, which is still a market town of 800 houses. It may once have justified Polo's description as a "great fine city." His "Vughin" is "a great and noble city" with a large trade in silk and other merchandise. This I take to be Hu-chau. His "Ciangan," a rich place with a good trade, I consider can be no other than Ka-shing.

The complete section, therefore, I would give as: Su-chau—P'ing-wang—Ka-shing—Hang-chau.

As Moule (*op. cit.* p. 411) has pointed out, Hu-chau and Hang-chau were very intimately connected, and I imagine Polo to have visited the former on the occasion of one of his numerous stays at Hang-chau.

After leaving Kinsay, the itinerary runs in a general south-westerly direction to Kelinfu (Kien-ning fu), and thence to Zayton on the coast. The difficulty lies in fixing the route between Kinsay and Kelinfu, and again between Kelinfu and Zayton.

In the first section, the places named are: Tanpiju—Vuju—Ghiuju—Chanshan—Cuju. Yule would identify these with: Shao-hsing—Kinhwa—Kiu-chau—Sui-chang—Chu-chau respectively. On looking at the map, we see that Kiu-chau is right away from the itinerary, and no attempt is made to include it in the line of march in Yule's Map VI.

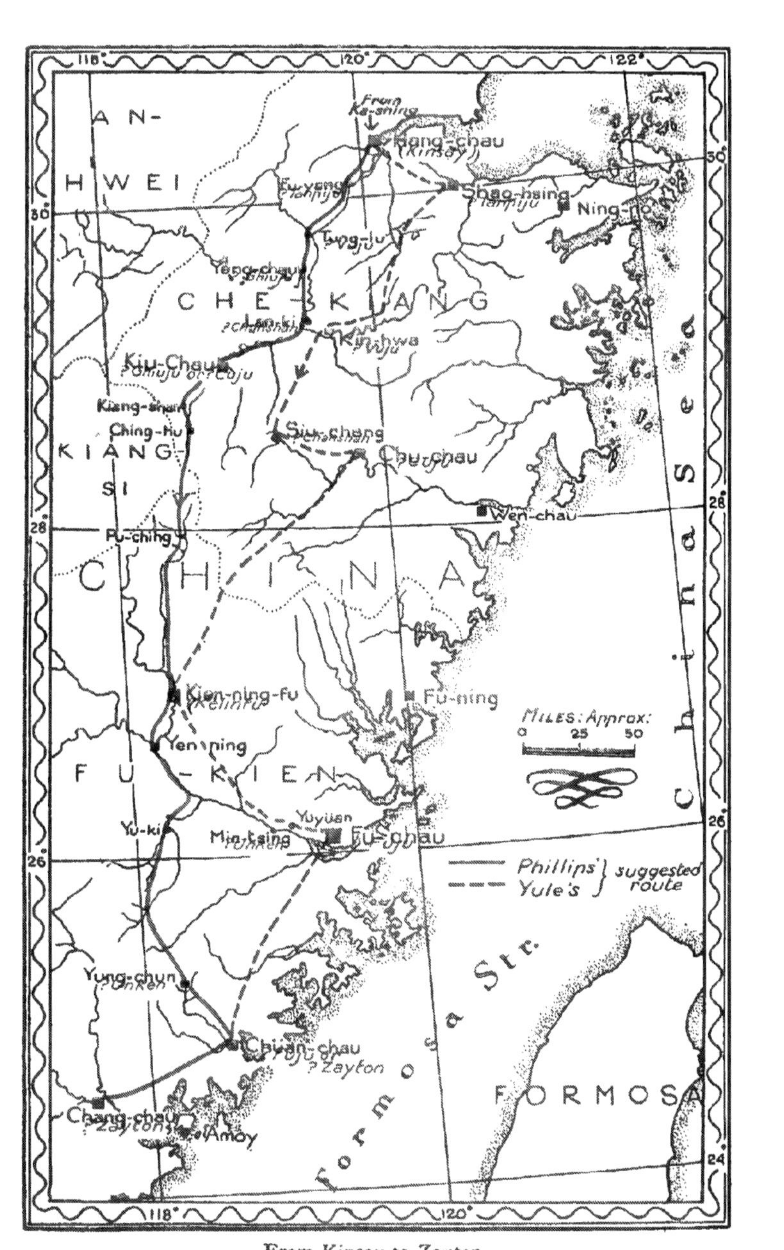

From Kinsay to Zayton

His arguments in support of his choice of Sui-chang and Chu-chau seem to be practically non-existent. In fact, his notes on pp. 221, 222 in no way prepare us for what we find in his map.

Another route has been suggested by Mr Phillips: Fu-yang—Tung-lu—Yeng-chau—Lan-ki—Kiu-chau. Thus Kiu-chau is given as the identification of Cuju and not of Ghiuju.

So far I am inclined to favour Phillips' choice. But to continue—in the second section, i.e. after Cuju, the itinerary runs to Kelinfu whence it goes to Unken, Fuju and Zayton. These three latter places Yule would interpret as Min-tsing, Fu-chau, and Chüan-chau respectively. Phillips, on the other hand, gives them as Yung-chun, Chuan-chau and Chang-chau. The evidence on both sides is given by Yule in his notes (pp. 229–245). The two routes are clearly marked on the map opposite.

An unbiassed survey of the total evidence, aided by the better maps of to-day, convince me that Yule's identification of Cuju with Chu-chau has led him completely off the correct route, and that we should see Cuju in the modern Kiu-chau whence Phillips' itinerary to Kelinfu (Kien-ning fu) *via* Kiang-shan—Ching-hu—Pu-ching is much to be preferred to Yule's most indefinite stretch of country from Chu-chau to Kelinfu.

But *after* Kelinfu, Yule's Min-tsing, Fu-chau, and Chuan-chau seem better than Phillips' Yung-chun, Chüan-chau, and Chang-chau. At the same time, however, I can see no evidence for accepting Min-tsing as the identification of "Unken." As Yule says, the directions here are unusually clear. Polo is shown to be travelling at the rate of thirty miles a day. From Kelinfu to Fuju is three days, and "Unken" is reached after the fifteenth mile on the third day, i.e. seventy-five miles from Kelinfu and fifteen miles from Fuju. This corresponds with Yüyüan better than with Min-tsing. Phillips' suggestion of Yung-chun, despite its similarity to "Unken" and to the fact that it is in the sugar-growing district, would entirely disagree with Polo's clearly recorded details at this point. Moreover, Yung-chun is due south of Kelinfu, while Yüyüan is south-east as the text definitely states.

With regard to the identification of Zayton, Yule's evidence is too strong to give up the idea of Chüan-chau being meant, although the evidence in favour of Chang-chau shows that the harbour of Amoy may be *included* in the term "Zayton." This does not in any way mean that we must accept one theory alone and reject the other entirely. It appears that much shipping to Chüang-chau anchored in Amoy harbour, so that when Polo left "Zayton" with the princess and a fleet of fourteen ships, it is quite possible that either harbour may be meant.

THE MALAY ARCHIPELAGO, INDIA, AND THE HOMEWARD VOYAGE

After telling us of the Khan's expedition against Chipangu (Japan), and of the 7459 islands in the "Sea of Chin" (Frampton has 7448), Polo commences his last itinerary—by sea from the port of Zayton to Venice, *via* the Malay Archipelago, India, Persia and Asia Minor.

Although the course is fairly satisfactorily known, there are many points which hitherto have proved of considerable difficulty to scholars.

In describing Java, it is agreed that Polo is either speaking from hearsay or else had acquired his information on some mission of which he has left us no detailed account. By this time we are used to the introduction of places lying off the main route, and merely regard them as interesting, but quite natural and explicable, interpolations.

It is not easy, however, to determine his exact course through the Straits of Malacca, nor to identify certain place-names on the Sumatran coast.

In order to solve these difficulties as far as possible, it is necessary not only to possess an intimate knowledge of the Straits, but also to be fully acquainted with the various dialects of the Archipelago.

I am, therefore, especially fortunate in obtaining the services of my friend Dr C O. Blagden, the well-known Malay scholar, who has voluntarily offered to pilot the fleet through the Archipelago until it is safely past the Nicobars and Andamans, and well on its way to Ceylon.

The actual place-names mentioned by Polo in connection with his voyage from Zayton to the Sea of Bengal are Chamba, Java, Sondur and Condur, Locac, Pentam, Malaiur, Java the Less, with six of its "kingdoms," viz. Perlec, Basma, Samara, Dagroian, Lambri, and Fansur, the island of Gauenispola, two islands of which one is called Necuveran, and the island of Angamanain. In certain cases he gives the distances from point to point and also the directions, but both are clearly only approximate. Unfortunately, except in the case of Samara, he does not mention where his fleet put in. But it is probable that it did so once or twice before reaching Samara.

Chamba is Champa, roughly the southern half of what is now the coast of Annam. Java is styled by Polo the great island of Java, in contradistinction to his Java the Less, which is Sumatra, though the latter is in fact much the larger. He is, however, not the only authority who uses the name Java for both islands. It is quite certain that he did not visit Java on this journey It would have been ridiculous for the fleet to go so far out of its way, and he himself under-estimates the distance. After coasting

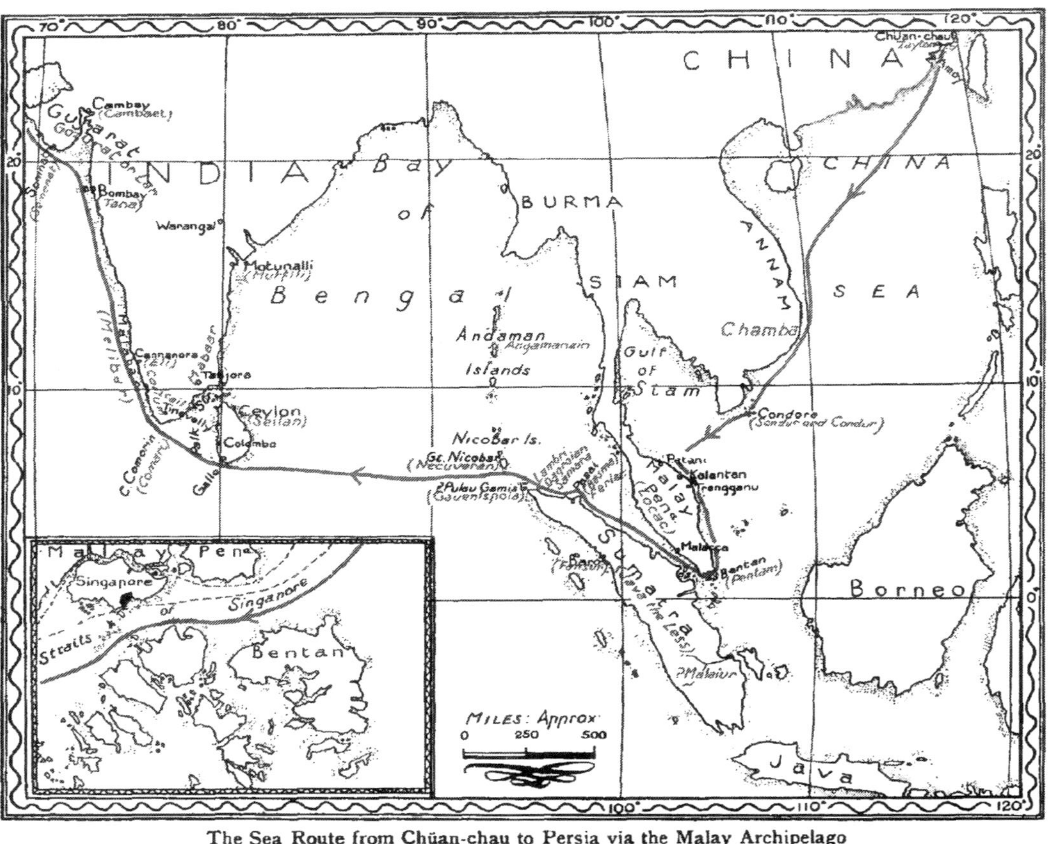

The Sea Route from Chüan-chau to Persia via the Malay Archipelago

along Chamba, the fleet passed Sondur and Condur, a group of small islands off the coast of French Cochinchina, of which the central and largest one is marked Condor or Condore on modern maps. It was a well-known landmark and there are no other islands of note in the neighbourhood. From thence the course lay straight to Locac and the landfall must have been made at some point on the N.E. coast of the Malay Peninsula in the region of Patani, Kĕlantan or Trĕngganu.

The name Locac has been variously and doubtfully explained. Probably the last syllable is the Chinese word *kok*, or *kwok*, "country." The first one may be the same as the first syllable of Lo-yueh, an old Chinese name for the Peninsula, or possibly the end of the term Hsien-lo, which became the Chinese name for Siam after Northern and Southern Siam had been united. But in Polo's time the Northern Siamese of Sukhothai had only recently occupied the isthmus of the Peninsula down to Ligor or Nakhon, about 150 miles N.W. of Patani. The suggestion that Locac is a drastic contraction of Lĕngakasuka, the name of an old state or district in the northern part of the Peninsula, seems improbable in view of the fact that the fuller form is mentioned in the Javanese poem *Nāgarakrĕtāgama* in 1365 and has survived in local popular tradition down to modern times.

At any rate in Polo's terminology Locac is the Malay Peninsula, and the fleet sailed down its eastern coast till it came to the island of Pentam. This can only be Bentan, which lies about fifteen miles to the south of the S.E. promontory of the Peninsula. Here is the eastern entrance of the Straits of Singapore, and Polo says that he proceeded for sixty miles "between these two islands," by which he must have meant Locac and Pentam, for no others have been mentioned in this connection. The mileage is approximately correct (in English miles) as representing the distance between the two extremities (S.E. and S.W.) of the Peninsula or the two ends of the Straits. But in the welter of islands lying about the middle of this space there are numerous channels, three of which are of practical importance.

The first one divides the island of Singapore from the mainland of the Peninsula. The island has much the same shape as the Isle of Wight, but is about a third larger, and the channel dividing it from the mainland is circuitous and subject to strong tides; and though in general about a mile wide, it narrows in some places to little more than three furlongs. It has usually been supposed that this first channel was the ancient traditional course of shipping between the China Sea and the Straits of Malacca, but this erroneous notion was finally exploded by the late W. D. Barnes in the *Journ. Roy. Asiat. Soc. Straits Branch* (1911), No. 60.

The second channel approaches close to the southern point of Singapore Island and there passes between it and some smaller islands, through a passage formerly called New Harbour but in 1900 renamed Keppel Harbour. This is the route now used by large liners calling at Singapore, as they can go alongside their wharves in this harbour, the depth of water near the shore being sufficient for the purpose. This route was in use at least five centuries ago, as attested by Chinese records. It is, of course, the shortest way of approaching Singapore from the west.

The third channel, used by smaller vessels which do not go alongside but lie in the roadstead off the centre of the town, a few miles N.E. of the harbour, runs much further south among the islands to the south of Singapore and is much wider. It is in every way better fitted than either of the others for large sailing ships, particularly if they do not propose to call at Singapore. Polo may have used either the second or the third route, but the probabilities are in favour of the third. Nothing can be certainly inferred from his statement about the shallowness of the course taken by the fleet. His four paces (twenty feet) *throughout* the strait must in any case be an understatement, unless his pilot persistently hugged the shallows or unless the soundings have altered materially in the last six centuries. As for the first channel, if Polo had changed his course to the N.W. and sailed into the strait north of Singapore Island he could not have imagined that he was proceeding between the Peninsula and Bentan, seeing that he had left the latter miles away in his wake and had got behind another large island. On the other hand, if his course lay south of Singapore, it would be perfectly natural for him to take that island for a part of the coast of the Peninsula, while the islands to the south of the third channel, of which Bentan was the first he saw, might well be taken by him to constitute one single island.

The question of the route is somewhat involved with the identification of Polo's Malaiur. He does not say that he went there, but merely that after the sixty miles of Straits one had to go about thirty more in order to reach it. But no place within that space could have been the site of such a fine commercial emporium as he described. If it had been Singapore (which may have existed in his time), its site was at the mouth of the Singapore river on the S.E. edge of the island a little to the N.E. of its southern point and just where the centre of the modern town is, that is to say about half way between the two ends of the Straits, not thirty miles beyond their western end. Malacca, on the other hand, is about a hundred miles away from the western entrance of the Straits of Singapore, and there is no evidence of its existence, let alone of its commercial importance, in Polo's time. Malaiur, a Tamil corruption of the real name Malayu, is well attested as having been for more than six centuries before Polo's day the name of an East Sumatran coast district considerably to the southward, and probably lying along the lower reaches of the Jambi river. It seems likely that this was the place of which he was told; but it is about 150 miles south of the course he followed.

Leaving the Straits of Singapore the fleet proceeded up the Straits of Malacca towards the N.E. corner of Sumatra, and here Polo's information becomes somewhat clearer. His Ferlec (in Malay Pěriak) shows signs of Arabic pronunciation. The fact that it was already Muslim in his time agrees with tradition, which thus, for what that may be worth, helps to support Polo's statement. If the order of his other place names is geographically right, Basma must be Pasai (Achinese Pasè); but the form of the word is difficult to explain. The Portuguese Pacem does not help, for it is merely an example of the common Portuguese tendency to nasalize final vowels (e.g. Tenasserim). It almost looks as if the name of Pasai had been contaminated by Polo's informant mixing it up with such other Sumatran place names as Pasěman, or the still more remote Běsěmah, on the west side of the island, or else the not very distant Pasangan on its N.E. coast. Samara is for Samatra (otherwise Samudra) a place very near to Pasai; it is generally considered that this little port gave its name to the whole island. Polo stayed there for five months under stress of weather. No doubt the S.W. monsoon had set in and the fleet had to wait till it was over. He remained during that time intrenched, for protection against the idolatrous and, as he thought, cannibal natives. But we know for certain that the town, at any rate, was being islamized at this very time, for its first Muslim king died in 1297, as recorded on his tombstone.

The next place, Dagroian, is unidentified but must have been on the same line of coast. Its inhabitants are also accused of cannibalism. Lambri is well known from other sources and cannot have been situated far away from Kota Raja, the capital of Achin, close to the N.W. end of Sumatra. Fansur, on the other hand, is certainly identical with Baros (or Barus), which was celebrated throughout the centuries as a port for camphor. Though Polo associates it also with the production of sago, which he says he himself saw made and ate, it is not likely that he went out of his way down the W. coast of Sumatra to Fansur; he may very well have seen the sago made, and have eaten it, at Samara or elsewhere. Polo locates the last six places together on one side of the island. Had he said "near one end," no fault could have been found with his statement.

Gauenispola is one of the small islands off the N.W. end of Sumatra, and very near to Lambri. It is also mentioned by Arab writers, and its proper name was probably Pulau Gamas (or Gamis). Next come the Nicobars, which he must have skirted (he speaks of two islands and his Necuveran is no doubt Great Nicobar, the main southern island), and lastly the Andamans (which he calls "a very large island"). He cannot have seen them, as they lie to the north of the Nicobars and the fleet must have passed along the south of the Nicobar group. It would then have sailed due west to some southern point of Ceylon, perhaps Galle. Both Dr Blagden and Don M. de Z Wiekremasinghe agree with me that it would be most improbable for a fleet of Polo's time to go *via* the north of Ceylon and through Palk Strait, as he is sometimes represented as doing.

Although Polo mentions many places on the coasts of India (Telingana, Madras, Tanjore, Tinnevelly, Cape Cormorin, Travancore, Cananore, Bombay, Cambay, Somnath and Mekran) and Arabia (Aden, Es-Sheḥr, Dhofar, Kalhāt), as well as Socotra, Madagascar, and Zanzibar, we must, with but few exceptions, regard this as having nothing whatever to do with the homeward journey.

From Ceylon the fleet would round Cape Comorin, and follow the western coast of India and Mekran (called Befmaceian by Frampton and Kesmacoran by Yule) into the Persian Gulf at Hormuz.

We can fairly reasonably assume that Polo derived his information about the Indian coastal regions from one of his earlier missions. The legends concerning the Male and Female Islands, the Roc, etc., are, of course, mere travellers tales picked up in course of conversation.

After the Polos had delivered their charge safely in Persia, they made the final stage of their long wanderings across Armenia to Trebizond and thence by sea *via* Constantinople and Negroponte (Euboea) to Venice.

The moſt noble and famous trauels of *Marcus Paulus*, one of the nobilitie of the ſtate of Venice, into the Eaſt partes of the world, as *Armenia, Perſia, Arabia, Tartary*, with many other kingdoms and Prouinces.

No leſſe pleaſant, than profitable, as appeareth by the Table, or Contents of this Booke.

Moſt neceſſary for all ſortes of Perſons, and eſpecially for Trauellers.

Tranſlated into Engliſh.

AT LONDON,
Printed by Ralph Nevvbery
Anno. 1579.

Title-page of the first edition

THE HOUSE OF MARCO POLO, AT VENICE

In the course of his well-known account of the final return of the Polos to Venice, Ramusio tells us that on their arrival they proceeded to their house in the confine of St. John Chrysostom, and that it became known as the Corte del Millioni The reason given for this name is that Marco Polo, when relating his travels constantly referred to the Great Khan's revenues as amounting to so many millions of gold, while all other references to great wealth or numbers was always made in millions and " So they gave him the nickname of Messer Millioni . the Court of his House, too, at S Giovanni Christomo, has always from that time been popularly known as the Court of the Millioni "

This information has naturally prompted both scholars and travellers in Venice to see if the Court still exists and to try and discover what remains of the house to-day Readers of Yule's *Introduction* (pp. 26–31) are already acquainted with the facts, and no further discovery of any importance whatever has been made since his day

Before referring to the state of what remains of the house at the present time, I may be permitted to restate the facts briefly. Although Ramusio suggests that the house in the San Giovanni Crisòstomo district was the Polo mansion, it has been shown by documentary evidence that the Polo family had always been connected with San Felice, near the famous Cà d'Oro But that a close connection existed between the two parishes is evident from the fact that although the will of Maffeo Polo connects the family with San Crisòstomo, the document itself was drawn up and witnessed by the priests and clerks of S. Felice. Yule concludes that the Palazzo in the former parish was purchased by the Polos after their return from the East. However that may be, it is in the San Crisòstomo district alone that we must seek for what is left of the Polo house. Leaving the Piazza San Marco, and proceeding on foot up the Calle del Fabbri, one soon reaches the Teatro Goldoni, and turning sharp to the right, past S. Bartolomeo and over the R del Pontego, the early Renaissance façade of San Giovanni Crisòstomo looms up into sight. The street is very narrow here, and you are right on top of it almost before you are aware of it Leaving the church on the left hand and passing to the south of it along the Calle Della Chiosa two separate passages appear—that to the left is the Calle del Teatro, and leads to the Teatro Malibran ; while that to the right is labelled " Sottoportico e Corte del Milion." Having passed under the Sottoportico and entered the courtyard, which is very small, we proceed straight on down another passage which opens out into a larger courtyard now called Corte

Seconda del Milion. This is the Corte Sabbionera of Yule's day. It is much larger than the Corte Seconda and is surrounded by tall four or five storey houses. The site of the Polo house lies in the north-west corner, the whole of the north side of the Court backing on to the Teatro Malibran, which is now a cinema. As existing to-day, the remains consist of a double-arched doorway sculptured in the typical Italo-Byzantine reliefs of the thirteenth century. Over the centre of the arch facing the court is a Byzantine cross, which has been engraved by Ruskin, *Stones of Venice*, p. 139, and P. XI, Fig. 4. To the left of the cross is a round sculptured disc representing the Roc of Arabian myth seizing its prey in its talons. It seems probable that some similar disc once existed to the right of the cross, but a window now pierces the house at this spot. According to Ruskin, *op. cit.* iii. 320, other remains of Byzantine sculpture which are doubtless fragments of the decoration of the Polo house, are to be found imbedded in the neighbouring houses. This is due to the great fire that destroyed the Palazzo at the end of the sixteenth century.[1]

As far as I can ascertain, nothing has been done to collect these fragments together, or even to allocate them. Thus the most important relic of the Polo's house is still the double archway.

No inscription, plaque or statue marks the spot, but what is much worse is the fact that the Venetians have thought fit to erect—actually under the very archway itself—an ugly, evil-smelling iron 'convenience,' so that few care to approach the Polo house at all.

What makes it all the more unnecessary even from a utilitarian point of view is that, unbelievable as it may well appear, another similar structure has been built a hundred yards farther on, just outside the Corte Milion, at the very point where the Ponte Marco Polo starts! This is indeed a strange way to honour one's country's mighty ones!

But a memorial tablet *does* exist—if you can find it—and to this point I now turn.

Having dutifully—and most warily—inspected the Byzantine reliefs, so far as your senses of delicacy permit, you continue eastwards across the Court, and passing under the Sottoportico find yourself at the foot of a recently rebuilt brick bridge. This is the Ponte Marco Polo, but on the name-tablet only the 'E' of 'Ponte' and the 'LO' of 'Marco Polo' now remain! The bridge leads in the direction of the parish of Santa Maria Formosa. But if you follow this natural course, you will entirely miss seeing the plaque which has been put above the

[1] See Doglioni, *Hist. Venetiana*, Venezia, 1598, pp. 161, 162.

Doric pediment of what once was the chief entrance to the theatre. This now discarded entrance will be found on turning left at the foot of the Ponte Marco Polo, instead of crossing the bridge. There is merely a small landing-stage for gondolas. In the days before the Teatro Malibran was a cinema, this section of the Rio S. Marina would doubtless have been extensively used by theatre-goers, and the plaque might have been noticed ! Even so, it is high up and almost impossible to read at night, and on seeing it on a theatre façade one would naturally conclude it was either a memorial to Malibran, the famous singer, or some actor or manager of note.

Since those days the lamp which helped to pilot the gondolas to the theatre landing-stage has been moved from its original position at the extreme corner of the building overlooking the canal (see the photograph in Yule, vol. i. p. 28 of the introduction) and has been re-erected immediately between the top of the pediment and Polo's tablet. Thus with the light *below* the inscription and the obstruction of the ironwork itself, it is none too easy to read the inscription at all. But added to this is the fact that the lower ledge of the pediment has proved a handy resting-place for the largest of the cinema advertisements. Thus if a 'special attraction' is showing the unfortunate inscription is entirely hidden. Such was the case when I first tried to photograph it in May 1937, but with the 'change of programme' at the end of the week I was lucky. There was no advertisement large enough to be mounted on the ledge, and so I got my photograph ! The tablet was erected in 1881 by the members of the Venice International Geographical Congress, and reads as follows :

QUI FURONO LE CASE

DI

MARCO POLO

CHE VIAGGÌO LE PIÙ LONTANE REGIONI DELL' ASIA

E LE DESCRISSE

PER DECRETO DEI COMUNE

MDCCCLXXXI

. and that is all !

Quite apart from the appalling condition the Polo area has been allowed to get into, and even overlooking the changes in the theatre entrance, the waterway leading past the inscription was always a secondary one, for all the traffic passes from the Grand Canal to the Canale di S. Marco *via* the Rio del Pantego, the Rio della Fava and the Rio di Palazzo Thus, unless something is done to collect and preserve what relics remain, to repair and clear up the Corte Milion and transfer the inscription to the Polo house where it can be seen and read, there will be nothing left to mark the house of one of Italy's greatest sons.

¶ To the right worſhipfull Mr. Edward Dyar Eſquire,
Iohn Frampton wiſheth proſperous
health and felicitie.

HAVING lying by mee in my chamber (righte Worſhipful) a tranſlation of the great voiage & lõg trauels of *Paulus Venetus* the *Venetian*, manye Merchauntes, Pilots, and Marriners, and others of dyuers degrees, much bent to Diſcoueries, reſorting to me vpon ſeuerall occaſions, toke ſo great delight with the reading of my Booke, finding in the ſame ſuch ſtrange things, & ſuch a world of varietie of matters, that I coulde neuer bee in quiet, for one or for an other, for the committing the ſame to printe in the Engliſhe tongue, perſwading, that it mighte giue greate lighte to our Seamen, if euer this nation chaunced to find a paſſage out of the frozen Zone to the South Seas, and otherwiſe delight many home dwellers, furtherers of trauellers. But finding in my ſelfe ſmall abilitie for the finiſhing of it, in ſuche perfection as the excellencie of the worke, and as this learned time did require, I ſtayed a long time, in hope ſome learned man woulde haue tranſlated the worke, but finding none that would take it in hand, to ſatisfie ſo many requeſts, nowe at laſt I determined to ſette it forth, as I coulde, referring the learned in tongues, delighted in eloquence, to the worke it ſelfe, written in Latine, Spaniſh, and Italian, and the reſte that haue but the Engliſh tong, that ſeeke onelye for ſubſtaunce of matter to my playne tranſlation, beſeeching to take my trauell and good meaning in the beſte parte. And bethinking my ſelfe of ſome ſpeciall Gentleman, a louer of knowledge, to whome I mighte dedicate the ſame, I founde no man, that I know in that reſpecte more worthy of the ſame, than your worſhippe, nor yet any man, to whome ſo many Schollers, ſo many trauellers, and ſo manye men of valor, ſuppreſſed or hindred with pouertie, or diſtreſſed by lacke of friends in Courte, are ſo muche bounde as to you, and therefore to you I dedicate the ſame, not bicauſe you your ſelfe wãt the knowledge of tongues, for I know you to haue the Latine, the Italian, the French, and the Spaniſhe: But bycauſe of youre worthineſſe, and for that I haue ſince my firſte acquaintaunce founde my ſelfe without any greate deſerte on my parte, more bound vnto you than to anye man in *England*, and therefore for your deſert & token of a thankefull minde,

I dedicate the ſame to youre worſhip, moſte humbly praying you to take it in good parte, and to bee patrone of the ſame: and ſo wiſhing you continuaunce of vertue, with muche encreaſe of the ſame, I take my leaue, wiſhing you with many for the cõmon wealths ſake, place with auc̄thoritie, where you maye haue daylye exerciſe of the giftes that the Lorde hathe endowed you withall in plentifull ſorte. From my lodging this .xxvj. daye of Ianuarie .1579.

Your worſhips to commaunde,

IOHN FRAMPTON.

¶ Maifter Rothorigo to the Reader.

¶ An Introduction into Cofmographie.

BIcaufe many be defirous of the knowledge of the partes of the worlde, what names they haue, and in what places they be, and that many and fundry times the holy fcripture doth make mention, and alfo it is profitable for fuche as doe traffique and trade to haue knowledge, I was moued to giue notice to all fuche as are defirous or haue pleafure in reading.

¶ You fhall vnderftande, that a man turning his face to the rifing of the Sunne, that parte that is before hys eies where the Sunne doth rife, is called Orient or Eafte, and his contrarie where the Sunne fetteth, is Occident or Weaft. The courfe or waye of the Sunne is called *Medio die*, or South, whiche is on youre righte hande, his contrarie parte that is on the lefte hande is called *Soptentrion* or North.

¶ Furthermore, you fhall vnderftand, that if a manne ftande in the Ilande of *Cales*, and looke towardes the rifing of the Sunne, he fhall fée thrée principall parts of the worlde, diuided by the Sea called *Mediterraneum*, that cõmeth oute of the greate Occean and Weafte Sea, and runneth towardes the Eafte, and by two very great and principall riuers, the one comming from the South, called *Nilus*, and the other from the North, called *Tanais*.

Affrica

YOu fhall alfo vnderftande, that from the entring of the ftraite called *Iuberaltare*, vppon the right hande to the riuer *Nilus* bordering vppon *Egipt*, is called *Affrica*, the Sea that is towardes vs, is called *Libya*, that whiche is towardes the South, is called *Ethiopia*, whiche is the Occean, the Sea towardes the Weafte, is called *Atlantica*, and is alfo the greate Occean Sea. It hath thefe famous Cities and Prouinces. Ouer againfte *Iuberaltar*, and the coafte of *Mallaga* is *Mauritania*, whiche we call *Barbarie*. It is named *Barbaria*, bycaufe the people be barbarous, not onely in language, but in manners and cuftomes. Following towards the Eaft is *Numidia*, *Getulia*, *Tunes*, a citie in *Affrica*, the name fo giuen by *Afu*, to all *Syria*, and *Aegipt*. On the South parte be the ETHIOPIANS, whiche hereafter fhall be fpoken of.

Europa

EVropa is called al ẏ prouinces againſt *Affrica* towards the North from the greate Occean Sea, that entreth into the ſtreits to the riuer *Tanai*, and the greate lake called *Meotis*, where this riuer entreth into. In this there is comprehẽded *Portugale*, *Britania*, *Spaine*, *France*, *Almaine*, *Italie*, *Grecia*, *Polonia*, *Hungarie*, or *Panonia*, *Valachia*, *Aſia* the leſſer, *Phrygia*, *Turkia*, *Galatia*, *Lydia*, *Pamphilia*, *Lauria*, *Lycia*, *Cilicia*, *Scythia* the lower, *Dacia*, *Gocia* and *Thraſia*.

Aſia

ASia the greater is that that is beyond *Europa* and *Affrica*, that is to ſay, on the other ſide of *Nilus* Southward, and the riuer *Tanais* Northward, following the way Eaſtwarde, and is as bigge as *Europa* and *Affrica*, and compaſſed with thrée Seas, Eaſterly or Orientall, *Indico* to the Southwarde, *Scythia* to the Northwarde, hauing prouinces, *Soria*, *Meſopotamia*, *Parthia*, *Sarmaſia*, *Aſiatica*, *Arabia*, *Perſia*, *Armenia*, *Medea*, *Hircania*, *Carmania*, the *Indias* on thys ſide and beyonde the riuer *Ganges*.

¶ Alſo you ſhall vnderſtande, that the greate Sea called the Occean, doth compaſſe aboute the foreſaid thrée principall partes of the worlde, and ſo doeth compaſſe all the whole worlde, althoughe there be diuers regions and places whereas they be, hauing diuers names.

¶ Moreouer, you ſhall vnderſtande, that in whatſoeuer parts of the Sea that doe anſwere to any parts of ẏ foreſaid Countries, as there be many Ilands inhabited with diuers people, aſwel as the Eaſt parts, whereas is *Taprobane* and *Thyle*, and others infinite number on theyr ſides, aſwel as on the other parts before declared, and thoſe that be betwéene them and al others, are to be vnderſtanded to pertain to one of theſe thrée parts of the world beforeſaid, to whiche it may be moſte properly iudged to be, and lyeth neareſt vnto.

Ethiopia

MOreouer, you muſte note, that *Ethiopia* is a common name to manye Prouinces and Countries, inhabited with blacke people called NEGROS. And to begin with the moſte Weaſte partes, the firſte is *Ginney*, that is to ſaye, from *Cabo Verde* or the gréene Cape, and following the coaſt of the Sea, to the mouth or ſtreits of the Redde Sea. Al thoſe prouinces be called ETHIOPIANS, and of theſe ETHIOPIANS from *Ginney* vnto *Caſa Manſa*, that is to ſaye, the Kings pallace, they be of the ſect of MAHOMET, circumciſed the moſt parte of them. And the chiefeſt and moſt principall of theſe people be the IOLOFOS and MANDINGOS, and be moſte parte vnder the gouernement of a King called MANDIMANSA, for

MANSA is as muche to ſaye as Senior or Lord, and MANDY MANDINGA, ſo by this his title he is Lord MANDINGA. This King is blacke, and his abiding is in the prouince of *Sertano* four hundred leagues within the land, in a Citie compaſſed about with a wall called *Iaga*, which is riche of golde and ſiluer, and of all ſuche merchaundize as is occupied in *Adem* and in *Meca:* and from thence forwarde the ETHIOPIANS be Idolators to the cape called *Buona Eſperanca*, and there turneth againe to the ſect of MAHOMET. Beyonde theſe prouinces following vp into the land of *Sartano* bée greate and highe mountaines or hilles, called mountaines of the Moone, the toppes of them be alwayes couered with Snow, & at the foote of thẽ ſpringeth the riuer *Nilus*, and this Countrie is called *Ethiopia* beſide *Egipt*, and in *Arabia* it is called *Abas*, and the inhabitants *Abaſſinos*, and be Chriſtians, and doe vſe to be marked with an yron in the face: they be not baptized with fire (as ſome doe ſaye) but as we are, but they be HERETIKES, IACOBITES, and HEBEYONITES. They do holde on the olde lawe with the newe, and be circumciſed, and doe kéepe the Sabaoth daye, and doe eate no Porke, and ſome of them doe take manye wiues, and be alſo baptized, and doe ſaye, that their King came and deſcended of King SALOMON, and of the Quéene SABA, and this King hathe continuall warres with the MOORES.

¶ There is another *Ethiopia* called *Aſiatica interior*, which the *Arabians* call *Zenium*, and theſe doe extende from the ſayd hilles of the Moone, and of *Nilus*, to the borders of *Barbarie*. And the ſaying is, that among all Riuers, onelie *Nilus* entereth into two Seas, that is to ſaye, one braunche into the Eaſt Sea, and another braunche into the Weſt Sea. All theſe *Ethiopians* bée *Moores*, and theyr laboure and occupation is digging of golde out of the grounde, where they doe fynde great plentie. There is alſo another *Ethiopia* called *Tragodytica*, and thys dothe reache or extende from the foreſayde *Ethiopia*, to the ſtreyte or mouth of the redde Sea, and theſe bée ſomewhat whyter, and the King and people bée *Moores*, and came out of *Arabia fælix*, for the *Arabians* came ouer the ſtreyte of the redde Sea, and gotte that Countrey of the *Iacobites* by force, and at this daye there is robbing and ſtealing among them ſecretely, for the King of the *Iacobites* is of ſo greate power, that the Souldan of *Babilon* doth giue him tribute.

Nylus

The redde Sea.

Souldan.

Arabia

THat whiche wée doe call *Arabia*, the *Arabians* doe call *Arab*, and is called *Geſyrdelaab*. That whyche is betwéene the redde Sea, and *Sinus Perſicus*, is called the Iland of *Arabia*, and thys is called *Arabia Fælix*, by reaſon of the Incenſe that groweth there.

Arabia Felix.

¶ There bée other two *Arabias* besyde thys, the one of them extendyng from the Mount Sinay, to the dead Sea, where the Children of *Israell* wente fortye yeares, and thys is called *Arabia petrea*, takyng that name of a Citie that is there. The other dothe extende betwéene *Syria* and *Euphrates* towardes the Citie of *Lepo*, and thys they doe call *Arabia desan*, which is as muche to say, as of *Siria*, and our Latines doe call it *Arabia deserta*. And wheras the vulgar people, and men for the most part, do thinke that *Antilla*, or those Ilandes lately found out by commaundemente of the Catholike King DON FERNANDO, and Lady ISABELL Quéene, be in the *Indias*, they be deceyued therein, to call it by the name of the *Indias*. And for bycause that in *Spaniola*, or newe *Spayne*, they do find gold, some doe not let to say it is *Tharsis*, and *Ophin*, and *Sethin*, from whence in the time of SALOMON, they brought gold to *Hierusalem*. And thus augmenting erroures vpon erroures, let not to saye, that the Prophetes when they sayde that the name of oure Lorde God should be pronounced to people that haue not hearde of it, and in places and Countreys very farre off, and aparted, which is sayd to be vnderstanded by those that be called *Indians*, and by these Ilandes, and furthermore doe not let to say to this day, that it is to be vnderstanded by the places mentioned in the holy Scripture, and the Catholike doctors, and that this secret God hath kept hidden all this time, and by finding out these Ilands did reueale it. I séeing how they are deceyued in their vayne inuentions, and greate simplicitie, for zeale and good will of the truth, and to kill this canker, that it créepe no more nor ingender greater erroures, will giue light to this errour, answering to the said muttering talkers, according as to euery of them doth require.

Alepo

Erronious iudgements of the voyages of Salomon

¶ And first you shall vnderstande, that this name *India*, according to all Cosmographers, as well Christians as Infidels, of old time, and of later yeares, the name dothe come of a Riuer named *Hynde*, or *Hyndo*, that going towards the East, is the beginning of the *Indias*, whiche bée thrée in number, that is to say, the first is called the lower or nether *India*, the seconde is called the middle *India*, and the third is called the high or vpper *India*. The first or lower *India* is renamed *Caysar*, and these do extend towards the East, from the Riuer *India*, vnto a Porte or Hauen on the Sea side, of great traffike and trade, called *Cambaya*. And the King of this *India*, and also the most part of the people be *Moores*, and the rest Idolaters. The second or middle *India* is surnamed *Mynbar*, and dothe reache to the borders of *Colchico*, and this hath very faire Hauens, and Portes of greate traffike, where they doe lade Pepper, Ginger, and other Spices and Drugges. The Portes or Hauens be called *Colocud*, *Coulen*, *Hely*, *Fatenor* *Colnugur*, and héere be many Christians Heretikes Nestorians, and

Three Indias, the first is the lower India

The second or middle India

Lading of Spices

many *Indians*, although towards the North they be Idolaters. The thirde *India*, whiche is the hygh *India*, is surnamed *Mahabar*, and dothe extend vnto *Cauch*, whiche is the Riuer *Gange*. Héere groweth plentye of Sinamon, and Pearle. The King and people of thys Countrey worſhip the Oxe. Beſides theſe thrée *Indias*, whiche lye towardes the riſing of the Sunne, there can not be found neyther Author nor Man that hathe trauelled the firme land, neyther the Seas adioyning therevnto, that can ſay, there is anye other Prouince or Ilande named *India*, ſauing that if anye woulde giue to vnderſtand, that going towarde the Weſt, he wente towardes the Eaſt, and that although he came vnto the terrenall Paradiſe, and that theſe Ilands ſhoulde lye in the greate Weaſt Occean Seas, it appeareth playnely, for that thoſe that ſayle thither, ſteame their Shippe towards the Occident, and his direct wind whiche he ſayleth withall, is out of the Orient or the Eaſt. So it appeareth, that they ſayle not vnto the *India*, but that they flye and depart from the *India*. And thus it appeareth that he would ſay, that the firſte name that euer it hadde, or was ſette, naming it *Antillya*, ſéeming, that by the corruption of the vulgar, naming it *Ante India*, as to ſay againſt *India*, euen as Antechriſt is contrary or against Chriſt, or Antenorth againſte the North. And thus it appeareth, that it can not be named *India*, but to vnderſtande it as an antephraſe, cleane contrary, as a *Negro*, or a blacke *Moore* ſhoulde be named white IOHN, or a Negreſſe or blacke woman, to be named a Pearle, or a *Margarita*, that for finding gold in the Iland named *Hiſpaniola*, it ſhould haue the name ſet *Tharſia*, or *Ophin*, or *Sethin*, nor beléeue it ſtandeth in *Aſia* as ſome woulde ſaye, although the thyng is ſo cléere, that it ſéemeth a mockerie to proue it: but reaſon dothe leade, that wée ſhoulde gyue Mylke vnto Children and Infantes. SAINCTE AUSTINE declareth, that the circumſtance of the letter dothe illuminate the ſentence. And it appeareth in the thyrde Booke of Kings, in the tenth Chapter, and the ſecond of Paralipomenon, in the ninth Chapter, do ſaye, that the Seruantes of SALOMON, and of DIRAN, doe ſynde they broughte from *Ophin* and *Sethin*, and *Tharſis*, not onely golde, but alſo Siluer and Timber, called *Thina*, and Elephantes téeth, and Peacocks, and Apes, and Precious ſtones, the whyche thyngs in infinite places of the very true *Indians*, as well in Countreys farre within the lande, as alſo in Countreys vppon the Sea ſyde, and alſo in Ilandes wythout number, that bée in the Oriente or Eaſt Seas, ſhall be founde, as by experience of the Merchantes traffiking into the Eaſt, conforming to the holy Scripture, and to all thoſe that doe write, as well Catholikes, as Prophanes, is manyfeſt. And in the Ilande called *Spaniola*, there can bée found no ſuche Timber, nor all the other thyngs before named, ſauyng

The third India called the higher India.
An Oxe worſhipped.

Golde, the whiche as by this worke wyll appeare, is founde in a greate number of places of the Orientall partes. What is hée that in bringing gold from *Antilla*, will proue it is from *Ophin*, or *Sethin*, or *Tharsis*, from whence it was brought to SALOMON. Firſt hée muſt prooue that it was neuer founde but in one place, and that at thys daye it is not to bée had, but in the ſame place only, from whence it was broughte to SALOMON, the which is a manifeſt vntruth or falſe. And alſo they that vnderſtande that the ſtorie of the holy ſcripture, and the holy prophets, when they do now name countries from whence thoſe things be brought, and farre Ilandes of Idolatours, whereas the name of God was not heard, did not ſpeak but of *Spaniola*, and of the other Weſt parts, he muſt proue there is no other Idolatours in the worlde but thoſe whom he falſely calleth *Indians*, nor other Ilãds but the *Spaniola*, and the other Weſt Ilandes, and thys is of a truth, all falſe, for *Grecia* is Ilandes, *Scicilia*, a noble Ilande, and *Malta*, and *Lipari*, *Yzcla*, *Serdenya*, *Corſica*, *Mallorca*, *Minorca*, *Ybiſa*, *Canarias*, *England*, and others infinite in the foure partes of the world, before now hath bin founde. Of the whiche in the Orient or eaſt, is *Taprobano*, which is the moſt noble Iland in the world, and the Ile which is ſayde to be ſo happie and fortunate, that of neuer trée there falleth a leaſe of in the whole yeare, as alſo by thys Booke of MARCUS PAULUS is to be ſéene in the .106. Chapter, of one Ilande that is in the Orientall ſeas .1500. myles, in the which there is found gold in ſo great abundaunce, that it is ſayde the Kinges Pallace is couered or tyled wyth gold.

¶ And furthermore, it is ſayd, that the ſame is, that in thoſe ſeas be ſeauen thouſand four hundred fortie eight Ilandes, in the whiche there is not founde one trée, but that is ſwéete, pleaſaunt, and fruiteful, and of great profit, wherby we may wel conclud, that in many other Ilands, there is gold to be found: therfore it is not neceſſarie, that the holye Scriptures ſhoulde be ſo vnderſtanded by *Antilla*, when it is ſayde, they went for gold to *Tharſis*, & *Ophyn*, and *Sechyn*, yea and although they wyll not beléeue the other truthes, they can not denye the ſaying of the holye Scripture in the Seconde Chapter of GENESIS, where it is ſayde that the firſt riuer that goeth out of Paradiſe is *Bhyſon*, which doth compaſſe the whole countrey of *Euilath*, where golde doth growe, and that the golde of that countrey is very good and pure, nor it was not néedefull to haue thrée yeres from Ieruſalem to *Antilla*, as it is for the Ilands of the *Indians*, whiche is more further off, by a great deale, and with much more difficultneſſe to prouyde the precious ſtones, and all other things they brought frõ thence, and alſo the wayes be more difficulte and ſtrange, by reaſon of contrarye windes, and manye other incumbraunces. And that this was not vnderſtanded that the people a farre of are theſe Ilandes now founde, it appeareth by Saint Paule in the fiftéenth Chapter to the Romaynes, where is expounded the ſaying of Eſay in the .52. Chapter, wher it is ſayde, That thoſe to whom it was not pronounced vnto, ſhoulde ſée, and thoſe that did not heare of him, ſhould vnderſtande. And this, as a lyttle aboue is ſayde, is vnderſtanded, that from Ieruſalem to the Iles of *Grecia*, to the ſea *Illyrico* which is the end of *Grecia*, and the beginning of *Italy*, by *Slauonia*, or *Dalmatia*, and *Venice*, where before they had not hearde the name of Chriſt declared. And bycauſe the holye ghoſt hath interpreted thys ſentence by Saint Paul, applying that prophecie with other like of his workes, there remayneth no licence for other to apply it to *Antilla*.

¶ But now let vs come to the ſumme of this reckning, and ſay, that if for the golde that is founde in *Antilla*, wée ſhould beléeue that it is *Tharſis*, and *Ophyn*, and *Sethyn*, by ẏ other things that be founde in *Ophyn*, *&c*, and not in *Antilla*, we muſt beléeue that it is not thoſe, nor thoſe it. And moreouer, it appeareth that *Aſia* and *Tharſis*, *Ophyn*, and *Sethyn*, be in the Eaſt, and *Antilla* the *Spanyola* in the weſt, in place and condition much different.

FINIS.

¶ Here foloweth a Table of the Chapters conteyned in this Booke.

FINIS TABVLÆ.

The Prologue.

TO all Princes, Lordes, Knightes, and all other persons that this my Booke shall sée, heare, or reade, health, prosperitie, and pleasure. In thys Booke I do mind to giue knowledge of strange and maruellous things of the world, and specially of the partes of *Armenia*, *Persia*, *India*, *Tartaria*, and of many other prouinces and Countreys, whiche shall be declared in this worke, as they were séene by me MARCUS PAULUS, of the noble Citie of *Venice:* and that which I saw not, I declare by report of those that were wise, discrete, and of good credite, but that which I saw, I declare as I saw it, and that which I knew by others, I declare as I heard it. And for that this whole worke shall be faithfull and true, my intente is not to write any thing, but that which is very certaine. I do giue you all to vnderstande, that sithence the birth of our Sauioure and Lorde Iesus Christ, there hathe bin no man, Christian, nor Heathen, that hathe come to the knowledge and sight of so manye diuers, maruellous, and strange things, as I haue séene and hearde, whiche I will take in hande the laboure to write, as I did sée and heare it. For me thinke I shoulde do a great iniurie to the world, in not manifesting or declaring the truth. And for better information to them that shall reade or heare this worke, I do giue you to vnderstand. that I trauelled in the foresayd Prouinces and Countreys, and did sée those things that I will declare, ẏ space of sixe and twentie yeares, & caused thẽ to be written to Mayster VSTACHEO of *Pisa*, the yeare of our Lorde God .1298. He and I then being prisoners in *Ianua*. 1298

RAigning in *Constantinople* the Emperoure BALDOUINO, and in his time in the yeare of oure Lord .1250. NICHOLAS my father, and MAPHEO my vncle his brother, Citizens of *Venice*, went to *Constantinople* with their Merchandises. And béeyng there a certayne tyme, wyth councell of theyr friendes, passed wyth such wares and iewels as they had boughte in the Countrey of the Souldan, where they were a long time, 1250 determining to goe forwarde, and trauelling a long iourney, came to a Citie of the Lorde of the *Tartarians*, which is called BARCACAN, who was Lord of a greate parte of *Tartaria*, *Burgaria*, and *Aſia*. And this Lord BARCACAN, tooke greate pleaſure to ſée my father NICHOLAS and my Uncle MAPHEO, and ſhewed them greate friendſhip, and they preſented to hym ſuch iewels as they broughte with them from *Conſtantinople*, who receyued them thankefully, and gaue them giftes double the valew, whiche they ſent into dyuers partes to ſell, and they remayned in his Courte the ſpace of one yeare, in which tyme warres beganne betwéene the ſayde BARCACAN and ALAN, Lord of the *Tartares* of the Eaſt, and there was betwéene them many great battayles, and muche ſhedding of bloud, but in the end, the victorie fell to ALAN. And bycauſe of theſe warres, my father and vncle coulde not returne the way they went, but determined to go forwarde to the Eaſtward, and ſo to haue returned to *Conſtantinople*, and following their way, came to a Citie in the Eaſt partes, called *Buccata*, whiche is within the precinct of the Eaſt Kingdome. And departing from this Citie, paſſed the Riuer which is called *Tygris*, whiche is one of the foure that commeth out of Paradiſe terrenall, and goyng ſeauentéene dayes iourneys through a Deſerte, not finding anye Citie or Towne, yet méeting with manye companyes of *Tartares*, that went in the fields with their Cattel: béeing paſt thys Deſert, they came to a great & noble Citie called *Bocora*, and the ſame name hadde that Prouince, which the Kyng of that Countrey had, and the Citie was called *Barache*, and this is the greateſt Citie in *Perſia* In thys Countrey, were theſe two bréethren thrée yeares. And in this time came an Embaſſadoure from HANUL Lorde of the Eaſte, whiche wente to the greate ALAN Lorde of the *Tartares*, that before was ſpoken of. This ALAN is otherwiſe called the greate CANE. Thys Embaſſadoure maruelled muche to ſée theſe twoo Bréethren béeyng Chriſtians, and tooke greate pleaſure at them, bycauſe they hadde neuer before that tyme ſéene any Chriſtians, and ſayde to them, Friends, if you wyll followe or take my councell, I will ſhewe you wayes or meanes whereby you ſhall gette greate riches and renowme. Oure Lorde the King of the *Tartares*, didde neuer ſée anye Chriſtians, and hathe great deſire to ſée of them, if you will goe with me, I will bring you to his preſence, where you ſhall haue greate profite and friendſhippe of hym.

¶ They hearing thys, determined to goe with hym, and trauelling the ſpace of one yeare towardes the Eaſt Southeaſt, and after turning to the lefte hande towards the Northeaſt, and after towardes the North, in fine, they came to the Citie of the great CANE, in the whyche trauell they ſawe manye ſtraunge and maruellous things, whyche ſhall be declared in thys

Booke. And theſe two bréethren, béeyng preſented to the great CANE, were receyued by him very fauourably, ſhewing to them greate friendſhippe, demaundyng of them of the Emperoure of the Chriſtians, of hys ſtate, and howe hée ruled and gouerned hys Countreys, and kepte them in peace and iuſtice. And when hée made anye warres, howe and after what manner hée broughte hys people into the fielde, and he demaunded of them the ſtate and order of other Kyngdomes and Dukedomes in Chryſtendome, of theyr conditions, and afterwarde wyth greate diligence, hée enquyred of them of the POPE and the Cardinalles, and of theyr fayth, and of the Catholike Church, and of all other conditions of the Chriſtians, to the which demaundes the two bréethren aunſwered in order very diſcretely and wiſely, who hadde vnderſtanding, and could ſpeake the *Tartarie* language. The great CANE vnderſtandyng theyr anſweres, had grat pleaſure therein, and ſpeaking to his Lords, ſaying, that hée woulde ſende an Embaſſadour to the Pope, the head Biſhop of the Chriſtians, and requeſted the ſaid two bréethren, that it woulde pleaſe them to be his Embaſſadors to the Pope, with one of his Lordes: they aunſwered, they were readie to doe all that he woulde commaund them. Streight way the great CANE cauſed to bée written Letters of beliefe in the *Tartarian* tong to the Pope, and alſo commaunded by worde of mouth to hys ſayd Embaſſadors, that they ſhoulde ſaye, and deſire hys holyneſſe, that it would pleaſe him to ſend him a hundred men, diſcrete, wiſe, and learned Chriſtians in the Catholike faith, to inſtruct him and his Subiects, whereas then they did all worſhip Idols, and would gladly receyue the true faith. And alſo, the great CANE requeſted them to bring him ſome of the Oyle that did burne before the Sepulchre of Ieſus Chriſte in *Ieruſalem*. This done, the great CANE commaunded to be broughte to him a Table of gold, and wrote in it, commaunding expreſly to all hys ſubiects that ſhoulde ſée that his Table, that they ſhoulde receyue thoſe Embaſſadors with all frendſhippe, and to ſhew them honour and obedience, and to do al things that ſhoulde be neceſſarie, and to deliuer them money, and to prouide them what they woulde demaunde, as well for ſhipping, as alſo Horſes, or any other thing, in as ample maner, as if it were for his owne perſon. When the ſayd NICHOLAS and MAPHEO, and COCOBALL, Embaſſador to the great CANE, were at a poynt to depart, taking their leaue of ẏ great CANE, they rode with their cõpany thirtie days iourney, and at the ende of them, the ſaide COCOBALL fell ſicke and dyed, and the two bréethren followed on theyr iourney, and in euery Towne where they came ſhewing the foreſayd Table of gold, were very honourably receyued and enterteyned, as the perſon of the King. And continuing their iourney, they came to a towne called *Giaza*, and from thence departed, and came to *Acre* in the moneth of Aprill, in the yeare of our Lord 1272 God .1272. whereas they vnderstoode that the Pope CLEMENT was dead, and finding there a Legate of the Popes, which was called MISER THEBALDO, that was there for the defence of the holy Church, at the vttermost partes of the Seas, to him they did theyr Embassage of the greate CANE, and when MISER THEBALDO vnderstoode their Embassage, he prayed them to tarrie the creation of a newe Pope, and hearing this aunswere, the two bréethren departed incontinente, and went to *Nigro Ponte*, and from thence to *Venice*, to sée their houses, and founde the wife of NICHOLAS dead, and had left behinde hir a sonne, whose name was MARCUS, of the age of fiftéene yeares, which neuer saw his father before, for he left hir with child of him at his departing, and this is the same MARCUS that made thys Booke, as héereafter followeth. These two bréethren remayned in *Venice* the space of two yeares, tarying the creation of a newe Pope, and séeyng howe long they had taryed, departed from *Venice* to *Ierusalem*, for to gette some of the Oyle that burned in the Lampe before the holy Sepulchre of oure Lorde God, for to carrie with them to the greate CANE, according as he commaunded, and caryed with them MARCUS, sonne to the saide NICHOLAS, and after they had taken of the sayd oyle, returned to *Acre*, whereas the Popes Legate THEOBALDO was, and taking leaue or licence of him to returne to the great CANE, for whome the sayde Legate gaue them Letters, séeing they woulde not tarrie to do their Embassage to the Pope, and sayde, as soone as there was a new Pope created, he would doe their Embassage to the Pope, and that he should prouide that which should be conuenient, and so departed the two bréethren, and MARCUS, and trauelled till they came to a Towne called *Giaza*. And in this time the Legate receyued Letters from *Rome*, that there was a new Pope created, called GREGORIE of *Placentia*. The sayd Legate incontinent sent his messenger after these two bréethren, that they should returne to *Acre*, certifying thẽ, ẏ there was a new Pope created: and they vnderstanding this, requested the King of *Armenia* to commaunde to arme forthe a Galley, wherein they sayled incontinente to the Pope, of whome they were well receyued, who hauyng hearde their Embassage, streighte way gaue them two Friers, of the order of SAINCT DOMINIKE, being greate Clearkes, to go with them to the greate CANE, the one of them was called Frier NICHOLAS of *Venice*, and the other Frier WILLIAM of *Tripolle*, the whiche were well séene and exercised in disputations in the defense of the holy Catholike faith. And these two religious men with NICHOLAS and MAPHEO, and MARCUS, trauelled, till they came to a Towne called *Giaza.* And in this time the Souldan of *Babylon* came into *Armenia*, and did there greate hurte, and for that caufe, fearing to paffe anye further, the two Friers taryed there, and wrote to the greate CANE, that they were come thyther, and the caufe wherefore they wente not forwarde. The fayd NICHOLAS and MAPHEO, and MARCUS hys fonne wente on theyr iourney, and came to a Citie called *Bemeniphe*, where the great CANE was, but in the way they paffed in greate daunger of their bodyes, and faw many things, as fhall héereafter be declared, and taryed in going betwéene *Giaza* and *Bemeniphe*, a yeare and a halfe, by reafon of great Riuers, rayne, and cold in thofe countryes: and when the greate CANE hadde knowledge that NICHOLAS and MAPHEO were returned, he fent to receyue them, more than fortie dayes iourney, and at their comming receyued them with gret pleafure, and they knéelyng down, making great reuerence, he commaũded them to arife vp, demaũding of them how they fpedde in their voyage, and what they had done with the Pope, and after they had made their anfweare to al things, deliuered to him the Friers letters that remayned in *Giaza*, and the oyle they had taken out of the Lampe that burned before the holy Sepulchre of Iefus Chrifte, whiche he receyued with great pleafure, and put it vp, and kept it in a fecrete place, with alfo the letters, and demaunding of them, who MARCUS was, they aunfwered, he was NICHOLAS fonne, of the which the great CANE was glad, and toke him into his feruice, and gaue order to place him in his Court among his Lordes and Gentlemen.

Here foloweth the diſcourſe of many notable
and ſtrange things, that the noble and worthy
MARCUS PAULUS of the Citie of *Venice* did
ſee in the Eaſt partes of the world

¶ Howe MISER MARCO POLO vſed himſelfe in the Court of the Great CANE

CHAPTER I

[*Marsden*: Bk. I. Ch. I. Sect. IV (from line 19). *Pauthier*: Chs. XV, XVI (Prologue). *Yule*: Chs. XV, XVI (Prologue). *Benedetto*: Chs. XVI, XVII]

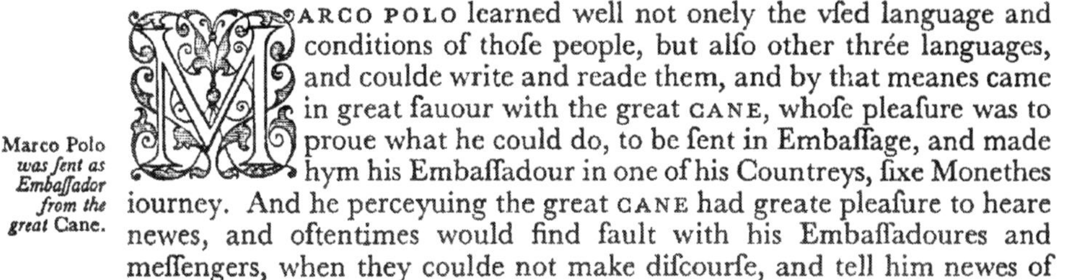

Marco Polo *was ſent as Embaſſador from the great* Cane.

MARCO POLO learned well not onely the vſed language and conditions of thoſe people, but alſo other thrée languages, and coulde write and reade them, and by that meanes came in great fauour with the great CANE, whoſe pleaſure was to proue what he could do, to be ſent in Embaſſage, and made hym his Embaſſadour in one of his Countreys, ſixe Monethes iourney. And he perceyuing the great CANE had greate pleaſure to heare newes, and oftentimes would find fault with his Embaſſadoures and meſſengers, when they coulde not make diſcourſe, and tell him newes of the Countreys and places they trauelled into, he determined with himſelfe to note and vnderſtand in that iourney all that could be ſpoken, as well of the Townes, Cities, and places, as alſo the conditions and qualities of the people, noting it in writing, to be the more readie to make his aunſwere, if any thing ſhould be demaunded of him: and at his returne declared to the great CANE the aunſwere of the people of that Countrey to his Embaſſage: And withall declared vnto hym the nature of Countreys, and the conditions of the people where he had bin, and alſo what he had heard of other Countreys, which pleaſed well the great CANE, and was in great fauoure with him, and ſet great ſtore by him, for which cauſe, all the noble men of his Courte had him in great eſtimation, calling him SENIOR or Lorde. He was in the greate CANES Court .xvij. yeares, and when anye greate Embaſſage or buſineſſe ſhoulde be done in any of hys Countreys or Prouinces, he was alwayes ſente, wherefore, diuers great men of the Court did enuie him, but he alwayes kepte thys order, that whatſoeuer he ſawe or heard, were it good or euill, hée alwayes wrote it, and had it in minde to declare to the great CANE in order.

Marco Polo *was in the great* Canes *Court ſeauẽteene yeares.*

The manner and wayes that the two breethren, and MARCUS PAULUS had for their returne to *Venice*

CHAPTER 2

[*Marsden*: Bk. I. Ch. I. Sect. V (in part). *Pauthier*: Ch. XVII (Prol.). *Yule*: Ch. XVII (Prol.). *Benedetto*: Ch. XVIII]

THe ſayd NICHOLAS and MAPHEO, and MARCUS PAULUS, hauyng bin in ẏ greate CANES Court of a long time, demaunded licence for to returne to *Venice*, but he louing and fauouring them ſo well, would not giue them leaue. And it fortuned in that time, that a Quéene in *India* dyed, whoſe name was BALGONIA, and hyr Huſbande was called Kyng ARGON. This Quéene ordeyned in hir Teſtamente, that hyr Huſbande ſhoulde not marrie, but with one of hyr bloud and kynred, and for that cauſe the ſayde Kyng ARGON ſente hys Embaſſadors with great honor and companye to the Greate CANE, deſiring hym to ſende hym for to bée hys Wife, a Mayde of the lignage of BALGONIA his firſte Wife. The names of theſe Embaſſadors were called ONLORA, APUSCA, and EDILLA. When theſe Embaſſadors arriued at the Courte, they were very well receyued by the Great CANE. And after they hadde done theyr meſſage, the Greate CANE cauſed to bée called before him a Mayden, whiche was called COZOTINE, of the kindred of BALGONIA, the whyche was verye fayre, and of the age of ſeauentéene yeares. And as ſhe was come before the Great CANE, and the Embaſſadors, the great CANE ſayde to the Embaſſadors, thys is the Mayden that you demaunde, take hyr, and carrie hir in a good houre: and wyth thys the Embaſſadors were very ioyfull and merrie. And theſe Embaſſadors vnderſtandyng of NICHOLAS and MAPHEO, and MARCUS PAULUS, *Italians*, which before that tyme had gone for Embaſſadors vnto the *Indians*, and were deſirous to depart from the greate CANE, deſired hym to gyue them licence to goe, and accompanye that Lady: and the Greate CANE, although not wyth good will, but for manners ſake, and alſo for honour of the Ladye, and for hyr more ſafegarde, in paſſing the Seas, bycauſe they were wiſe and ſkilfull menne, was content they ſhould goe.

How they ſayled to *Iaua*

CHAPTER 3

[*Marsden*: Bk. I. Ch. I. Sects. V (last 23 lines), VI. *Pauthier*: Ch. XVIII (Prol.). *Yule*: Ch. XVIII (Prol.). *Benedetto*: Ch. XIX]

Marco Polo *and his Father & Uncle had leaue to depart, and went without Embaſſadors.*

HAuing licence of the Greate CANE, the ſayde NICHOLAS & MAPHEO, and MARCUS PAULUS, as aforeſayde, as his cuſtome was, gaue them two Tables of golde, by the whiche he did ſignifie that they ſhould paſſe fréelie through all his prouinces and dominions, and that theyr charges ſhould be borne, and to be honourably accompanyed. And beſides this, the great CANE ſent diuers Embaſſadors to the POPE, and to the Frenche King, and to the King of *Spayne*; and to many other Prouinces in Chriſtendome, and cauſed to be armed and ſette forth fouretéene great Shippes, that euery one of them had four Maſtes. To declare the reaſon wherefore he did this, it were too long, therefore I let it paſſe. In euery Shippe he put ſixe hundreth men, and prouiſion for two yeares. In theſe Shippes wente the ſayd Embaſſadors, with the Lady and NICHOLAS, and MAPHEO bréethrẽ, and MARCUS PAULUS aforeſayd, and ſayled thrée Monethes continually, and then arriued at an Ilande called *Iaua*, being in the South partes, in the which they found maruellous and ſtrange things, as héereafter ſhall be declared. And departing from this Iland, ſayling on the *Indian* Seas .xviij. Moneths before they came to the place they would come to, founde (by the way) many maruellous and ſtrange things, as héereafter ſhall be declared.

Fouretеene great Ships with foure Maſtes in a Shippe, and ſixe hundred men in euery Shippe, and vittayled for two yeares. Within three Monethes ſayling, they arriued at Iaua.

How NICHOLAS and MAPHEO, and MARCO POLO returned to *Venice*, after they had ſeene and heard many maruellous thinges

CHAPTER 4

[*Marsden*: Bk. I. Ch. I. Sects. V (last 23 lines), VI. *Pauthier*: Ch. XVIII (Prol.). *Yule*: Ch. XVIII (Prol.). *Benedetto*: Ch. XIX. (All continued)]

AFter their arriuall with this foreſayde Lady to the Kingdome they went vnto, they found that the King ARGON was dead, and for that cauſe, married that mayde to his ſonne: and there did gouerne in the roome of the Kyng, a Lorde, whoſe name was ARCHATOR, for bycauſe the King was very yong. And to this Gouernoure or Uiceroy, was the Embaſſage declared, and of him the two Bréethren and MARCO POLO demaunded licence to goe into their Countrey, whiche he graunted, and withall gaue them foure Tables of gold, two of them were to haue Ierfawcons, and other Hawkes with them. The thirde was, to haue Lyons. And the fourth was, that they ſhoulde goe frée, withoute paying any charges, and to be accompanyed and enterteyned as to the Kings owne perſon. And by this commaundement, they had company and gard of two hundreth Knightes from Towne to Towne, for feare of manye Théeues vppon the wayes: and ſo much they trauelled, that they came to *Trapeſonſia*, and from thence to *Conſtantinople*, and ſo to *Nigro Ponto*, and finallie, to *Venice*, in the yeare of oure Lord God .1295.

¶ This we doe declare, for that all men ſhall knowe, that NICHOLAS and MAPHEO bréethren, and MARCO POLO, haue ſéene, hearde, and did knowe the maruellous things written in this Booke, the which declaring in the name of the Father, and the Sonne, and the holy Ghoſt, ſhall be declared as héereafter followeth.

The returne of the two breethrẽ and Marco Polo *to* Venice *in Anno* .1295.

Of *Armenia* the leſſer, and of many things that there is made

CHAPTER 5

[*Marsden*: Bk. I. Ch. II. *Pauthier*: Bk. I. Ch. XIX. *Yule*: Bk. I. Ch. I. *Benedetto*: Ch. XX]

Irſt and formoſt, I will beginne to declare of the Prouince of *Armenia*, noting ſuche commodities as there is. You ſhall vnderſtand, there be two *Armenias*, the greater, and the leſſer. In the leſſer, there is a King ſubiect to the *Tartar*, and he dothe maynteyne the Countrey in peace and iuſtice. In this Countrey be many Cities and Townes, and greate abundance of all things. In thys Countrey they take great pleaſure and paſtime in Hawking and Hunting, as well of wilde beaſtes, as of Fowles of all ſortes. In that Countrey be many infirmities, by reaſon the ayre is yll there, and for that cauſe, the men of that Countrey, that were wonte to be valiant and ſtrong in armes, bée turned nowe to be vile, and giuen to ydleneſſe and drunckenneſſe. In this Prouince vpon the Sea ſide, there is a Citie called *Gloza*, wherevnto is greate trade of Merchandiſe, and all Merchantes that doe traffique thither, haue their Cellers and Warehouſes in that Citie, as well *Venetians*, and *Ianoueys*, and all other that do occupye into *Leuant*.

They take great pleſure in Hawking and hunting.

A Citie vpon the Sea ſide, called Gloza.

Of the *Torchomanos* in *Armenia* the leſſer

CHAPTER 6

[*Marsden*: Bk. I. Ch. III. *Pauthier*: Bk. I. Ch. XX. *Yule*: Bk. I. Ch. II. *Benedetto*: Ch. XXI]

Good Horſes called according to the Countrey Torchomanos *and good Moyles. Goodly rich and faire carpets made heere. Cloth of ſilke of Crimſon, and other couloures made heere. Heere was* Saint Blaſe *martyred.*

I Haue declared vnto you of *Armenia* the leſſer, and now I will ſhewe you of *Torchomania*, whiche is a part of *Armenia*, in the which ther be thrée maner of people, the one called *Torchomanos*, and thoſe bée *Mahomets*, and ſpeake the *Perſian* language, and they liue in the Mountaynes and fieldes, whereas they may finde paſture for their Cattell, for thoſe people liue by ẏ gaines of their Cattell. There be very good Horſes called *Torchomani*, and good Moyles of great value. The other, or ſecond maner of people be *Armenians* and *Greekes*, and thoſe dwell and liue togither, and liue by occupations and trade of Merchandiſes. There they doe make very goodly and rich Carpettes, large and fayre, as you ſhall finde in any place. Alſo, they worke there, cloth of Crymſon Silke, and other goodly couloures. The chiefeſt Cities in that Countrey be *Chemo*, *Iſiree*, and *Sebaſto*, whereat SAINT BLASE was martired. There be alſo many Townes, of which I make no mention, and they bée ſubiecte to the TARTAR of the Eaſt, and he ſetteth gouernoures there.

Of *Armenia* the greater, and of the Arke of NOE

CHAPTER 7

[*Marsden*: Bk. I. Ch. IV. *Pauthier*: Bk. I. Ch. XXI. *Yule*: Bk. I. Ch. III. *Benedetto*: Ch. XXII]

A*Rmenia* the greater is a greate Prouince or Countrey. In the beginning thereof is a greate Citie called *Armenia*, where they doe make excellente BOCHACHIMS or Buckrams. In this Citie be very good Bathes naturallye. And this Countrey is ſubiect to the TARTAR, & there is in it many Cities & Townes, and the moſt noble Citie is called *Archinia*, which hath ioyning to it two prouinces, the one called *Archeten*, the other *Arzire*. In this Citie is a Biſhop. The people of this Countrey in ẏ ſommer time bée in the paſtures & meddowes, but in ẏ winter they can not, by reaſon of ẏ great cold, ſnow, & waters, for then it is ſo colde, ẏ ſcant the cattell and beaſtes can liue there, and for this cauſe they do driue their cattel into warmer places, wher they haue graſſe plẽty. In this gret *Armenia* is ẏ Arke of NOE on a high Mountain towards ẏ South, which doth ioyne to a Prouince towardes the Eaſt called *Mauſill.* And in that Prouince dwell Chriſtians, which be called *Iacobites*, and *Neſtorians* Heretikes, of the which hereafter ſhall be ſpoken. This Countrey towards the North doth ioyne vpon the *Georgians*, of the whych ſhall be ſpoken in the next Chapter. In this part towards the *Georgians* there is a well, the water wherof is like oyle, and is of great abundance & quantitie, that ſometimes they lade .100. Ships with it. And this oyle is not good to eate, but for Lamps and Candles, and to annoint Camels, Horſes, and other beaſtes that be galled, ſcabbie, and haue other infirmities, and for this cauſe it is fetched into diuers places.

Heere on a high Mountayne reſted the Arke of Noe *after the floud.*

Heere be Chriſtians of the ſect of the Neſtorians *and* Iacobites

Here is a wel that the water is like to Oyle, and is occupyed for diuers purpoſes.

Of the *Georgians*, and of the Tower and gate of yron

CHAPTER 8

[*Marsden*: Bk. I. Ch. v. *Pauthier*: Bk. I. Ch. XXII. *Yule*: Bk. I. Ch. IV. *Benedetto*: Ch. XXIII]

IN *Georgiania* is a king called NAND MALICHE, which is as much to ſay as DAWNID, and is ſubiect to the TARTAR. The ſaying is that in the olde time, the Kings of that Prouince were borne with a token or ſigne vnder their right ſhoulder. In this Countrey ẏ men be faire of body, venterous & valiant in armes, and good archers, and are Chriſtians & Gréekes mingled togither, & they go all with their heare like Prieſtes. This is the Prouince ẏ King ALEXANDER could not paſſe, whẽ he woulde haue come towards the Weſt parts, bycauſe ẏ wayes were dangerous & narrow, & compaſſed on ẏ one ſide with ẏ Sea, & on the other ſide with high Mountaines, that no Horſe can paſſe, or go for ẏ ſpace of four leagues, for ẏ way is ſo narrow & ſtrõg, ẏ a few mẽ be able to kéepe it againſt al the hoſtes of ẏ world. And K. ALEXANDER perceiuing ẏ by no meanes he coulde paſſe, would likewiſe make prouiſion, that the people of that Countrey might not paſſe to him. And made there a greate & ſtrong Tower, which is called the Tower and gate of yron. In this Prouince of the *Georgians* be many Cities and townes, & there they do make great plẽty of cloth of gold, & of ſilke in great abundance, for they haue greate plentie of ſilke. And there doe bréede the goodlyeſt and beſt Hawkes in the world. And the Countrey is plentifull of all things néede-

Heere was King Alexander *put backe and could not be ſuffered to paſſe.*

In this countrey be many fayre Cities and Townes where is made great plenty of cloth of gold, and of ſilke.

Excellente good hawkes. Great trade of Merchandise. *A Monastery of Monckes of the order of* S. Bernard. *A water or lake of syxe hundred miles compasse, wherein is no fish, but only in the Lent.* Euphrates.

full. They liue there by the trade of Merchandise, and by labour of the Countrey. Through all this Countrey is greate Mountaynes, and the way narrow and strong, and many welles, and for this cause the TARTARS can neuer haue the vpper hand of them. There is a Monasterie of Monckes of the order of SAINT BERNARDE, and hard by the Monasterie there is a water that descendeth from the Mountayne, in the which they find no fishe, but in Lent, and then they do take it in greate plentie from the firste day of Lent, till Easter euen. The place is called *Geluchelan*, and hath sixe hundred Miles compasse, and it is from the Sea twelue dayes iourney, and this water entreth into *Euphrates*, whyche is one of the foure principall Riuers whiche come from Paradice terrenall, and commeth out of *India*, and is deuided into many branches, and doth compasse those hilles. From thence they bring a silke called *Gella*. Now I haue declared vnto you the partes of *Armenia* which be towards the North, and now I wil declare vnto you of others their neyboures which be towards the South and West.

Of the parties of *Armenia* towards the South, and of the Kingdome of *Mosull*

CHAPTER 9

[*Marsden*: Bk. I. Ch. VI. *Pauthier*: Bk. I. Ch. XXIII. *Yule*: Bk. I. Ch. V. *Benedetto*: Ch. XXIV]

M*Osull* is a great Kingdome, in the which dwell many generations of people called *Arabies*, and all be of the secte of MAHOMET, although there be some Christians, called *Iacobites*, and *Nestorians*, and these haue by themselues a Patriarke, called IACOBIA, and he dothe institute Bishops, Archbishops, Abbots, Priestes, and other Religious men. There is made cloth of gold, and of silke, which be called by the name of the Kingdome *Mosulinus*, and there is greate plentie and abundance of it, and also greate plentie of spices and good cheape, and of other Merchandise. In the Mountaynes of this prouince dwell people called *Cordos*, and others called *Iacobinos*. The rest be *Moores* of the sect of MAHOMET, and be good men of warre, and be all rouers and robbers of Merchants.

Here is made cloth of golde and silke, called Mosulinus.

Of *Baldach*, and of many goodly things that be there

CHAPTER 10

[*Marsden*: Bk. I. Chs. VII, VIII (in part). *Pauthier*: Bk. I. Ch. XXIV. *Yule*: Bk. I. Ch. VI. *Benedetto*: Ch. XXV]

B*Aldache* is a very great Citie, in the whych is resident one that is called CALIPHO, whiche is among ẏ *Moores*, as it were chiefe gouernour & head. Through the middest of the Citie runneth a great Riuer, and goeth into the *Indian* Sea. And there is from this Citie to the place where it entreth into the Sea .xviij. dayes iourney. From this Citie to the Sea, and from the Sea to this Citie, there dothe passe dayly by this Riuer, in many and diuers vessels, diuers kinds of Merchandise, and they haue to their neyboure the *India*. And in this Countrey is a Citie called *Chisi*. By this Riuer they goe to the *Indian* Sea. Betwéene *Baldach* and *Chisi* vppon the Riuer is a Citie called *Barsera*, compassed with greate Mountaynes of Palmes and Date trées perfect good. In *Baldach* they doe make cloth of golde of diuers sortes, and cloth of silke, called cloth of Nasich, of Chrimson, and of diuers other coloures and fashions. There is great plentie of foure footed Beastes, and of Fowles. This Citie is one of the best and the noblest in the worlde. There was in this Citie a CALIPHO of the *Moores*, wonderfull and maruellous rich of gold and pretious stones. And in the yeare of our Lorde God .1230. the King of the *Tartars* called ALAN, ioyned a greate company, and went and sette vpon this Citie, and toke it by force, being in the Citie one hundred thousande Horsemen, besides infinite number of footemen. And there he founde a great Tower full of golde, siluer, and pretious stones. And King ALAN séeing this great treasure, maruelled much, and sent for the CALIPHO, and sayd vnto him: I do much maruell of thy auarice, that hauing so great treasure, didst not giue parte of it to mainteyne valiant men, that might defend me from thée, knowing that I was thy mortall enimie. And perceyuing the CALIPHO knewe not how to make him an answere, said vnto him, bycause thou louest this treasure so well, I will thou shalte haue thy fill of it, and caused him to be shut fast in the same Tower, where he liued foure dayes, and died miserably for hunger, and from that time forwards the *Moores* woulde haue no more CALIPHOS in that Citie.

Thorough this Citie Baldach *goeth a Riuer, and entreth into* Sinos Persicus.

Great trade vp and down this Riuer, to and from the Indians.

Here is made cloth of golde and of silke, called cloth of Nasich.

Calipho *is among the* Moores, *as the* Pope *is in Christẽdome. This Citie was wonne in Anno* .1230. *by* Alan *King of the* Tartars, *and he put the* Calipho *into a Tower among his treasure, and so was famished.*

Of a Citie called *Totis,* and of other notable things

CHAPTER 11

[*Marsden*: Bk. 1. Ch. IX. *Pauthier*: Bk. 1. Ch. XXIX. *Yule*: Bk. 1. Ch. XI. *Benedetto*: Ch. XXX]

This Citie Totis *is a noble Citie, and of great trade of merchandiſe. There is made cloth of gold and of ſilke very rich. To this City there commeth Merchants from diuers countreys.*

TOtis is a greate Citie of the Prouince or Countrey of *Baldach,* in the whiche Prouince there be manye Cities and Townes, but the moſt nobleſt is *Totis.* The people of thys Citie bée Merchantes, and handycraftes men. There they do make cloth of golde, and of ſilke, very riche, and of greate value. And this Citie is ſette in ſo good a place, that they doe bryng thyther all Merchandiſes of *India,* and of *Baldach,* and of *Oſmaſeilli,* and of *Cremes,* and of many other Cities and Countreys, and alſo of the Latines. There is greate plenty of pretious ſtones, and for that cauſe the Merchants gette muche. Thyther trade the *Armenians, Iacobites, Neſtorians, Perſians,* and theſe in a manner bée all Mahomets. Rounde aboute this Citie be many fayre Gardens full of ſingular good frutes, although the *Moores* that there doe dwell be very ill people, robbers and killers.

Of a great miracle that hapned in *Moſull*

CHAPTER 12

[*Marsden*: Bk. 1. Ch. VIII (in part). *Pauthier*: Bk. 1. Chs. XXV, XXVI, XXVII, XXVIII. *Yule*: Bk. 1. Chs. VII, VIII, IX, X. *Benedetto*: Chs. XXVI, XXVII, XXVIII, XXIX]

IN *Moſull,* a Citie in the Prouince of *Baldach* was a CALIPHO, a great enimie of the Chriſtians, whoſe ſtudie daye and night was how he might deſtroy them, and to make them forſake their faith in Ieſus Chriſt, and vpon this, ioyned in councell diuers times with hys wiſe men, and in the ende one of them ſaid, I will tell you a way how you ſhal haue good cauſe to kill, or force them to renounce their Faith. Ieſus Chriſt ſayth in hys Goſpell, *If you haue ſo much Faith as the grayne of a Muſtard ſeede, and ſaye to thys Mountayne paſſe from this place to another place, it woulde do,* therefore cauſe to be called togither all the Chriſtians, and commaund them by their beléefe, that ſuch a hill doe paſſe from that place to ſuche a place: truly it is not poſſible for them to doe it, and not doing it, you may iuſtly ſaye to them, that eyther theyr Goſpell dothe not ſaye truth, and by that meanes they follow lyes, or elſe they haue not ſo much Fayth as a grayne of Muſtarde ſéede. And thus as well for the one, as for the other, you maye iuſtly putte them to death, or elſe force them to forſake theyr Fayth they holde. This councell pleaſed well the CALIPHO, and thoſe of hys ſect, beléeuing, that nowe they hadde good occaſion to performe their euill purpoſe, and incontinent he commaunded all the Chryſtians that were in hys Countrey, to come togither, whiche was a great number, and they being come before hym, he cauſed thẽ to reade thoſe Scriptures of Ieſus Chriſt. And after that euery one of them had hearde it, he aſked them if they beléeued that theſe ſayings were true, and they anſwered yea. Incõtinent ſaid the CALIPHO to them, I wil giue you fiftéene days reſpite, to make either yõder hil to paſſe to ſuch a place, or elſe to renounce youre fayth in Ieſus Chriſte as falſe, and to turne *Moores*, and if you will not doe this, you ſhall all die. And the Chriſtians hearing this cruell ſentence, were ſore troubled, yet on the other part they comforted themſelues, with hope in the faith they had in the truth they beléeued. And incontinent the Biſhops, and Prelates, and Miniſters that were among the Chriſtians, commaunded all the Chriſtians, men, women, and children, to fall to continuall Prayer to oure Lorde Ieſus Chriſt, that he would helpe and councell them howe to rule and gouerne themſelues in that greate trouble and néede.

¶ And after eyght dayes were paſt, appeared an Angell to a holy Biſhop, and commaunded him that he ſhould ſay vnto a Shomaker that was a Chriſtian, that had but one onely eye, that he ſhould make Prayers to God, the which for his fayth and Prayers, ſhoulde make that hill remoue from his place, into the place the CALIPHO had appoynted. And incontinente the Biſhop ſente for that Shomaker, and with great deſire prayed him to make Prayers to oure Lord God, that for hys mercie and pitie he woulde remoue that hill as the CALIPHO and *Moores* had appoynted. The poore Shomaker excuſed himſelfe, ſaying, he was a greate Sinner, and vnworthy to demaund that grace of God: and this excuſe he made with great humilitie, like a iuſt and chaſt man, full of vertue and holyneſſe, and a kéeper of Gods commaundements, deuoute, and a great almes man, according to his abilitie.

¶ You ſhall vnderſtande, that thys Shomaker dyd pull out his eye by this meanes: He hadde hearde manye times this ſaying in the Goſpell, *If thy eye offende thee pull it out, and caſt it from thee.* He being a ſimple man, thought, that ſo corporally and materially the Scriptures ſhoulde be vnderſtanded. For it chanced on a time, there came a Mayde into his Shoppe to beſpeake a payre of Shoes, and to take the meaſure of hir foote, put off hir hoſe, and he withall was tempted to lye with hir, remembring himſelfe, and thinking vpon his ſinne and yll intent, ſent hir away, without diſcouering any thing of his yll thoughte and intente, and remembring the ſaying of the holy Goſpell, being ouercome with zeale, and yet not hauing the true knowledge, plucked out his eye. And ſo this Shomaker being ſo deſired by the Biſhop, and other Chriſtians, did graunt, and promiſed to praye vnto our Lord God for the ſayd cauſe. And the time of the .xv. dayes being come, that the CALIPHO had appoynted, he cauſed to come togither all the Chriſtians, whiche came in Proceſſion with their Croſſe, into a faire playne, hard by the hill and Mountayne. And to that place came the CALIPHO, with muche people armed, with intention, that ſtreight way, if the Mountayne did not remoue, to kill them all. Incontinente the Shomaker knéeled downe vppon the earth vpon his bare knées, and very deuoutely prayed to oure Lorde, lifting vp his hearte and handes to Heauen, praying to Ieſus Chriſte to ſuccour and helpe them his Chriſtians, that they ſhoulde not periſhe: and for that his faith was cléere, makyng an end of his Prayer, the power of the Almightie God Ieſus did cauſe the Mountayne to remoue and goe from the place it ſtoode, into the place the CALIPHO and his Councell hadde commaunded. ¶ And the *Moores* ſéeyng thys greate and manyfeſt miracle, ſtoode wonderfully amazed, ſaying, Great is the God of the Chriſtians, and the CALIPHO, with a greate number of the ſame *Moores* became Chriſtned. And after this CALIPHO dyed, the *Moores* that were not Chriſtned, would not conſente that this CALIPHO ſhould be buried, wheras the other *Caliphoes* were buried, for bycauſe that after that myracle, he lyued and dyed like a true and faythfull Chriſtian.

A great miracle. A Mountain remoued frō one place to another.

The Calipho *became chriſtned, and a great nūber of his* Moores.

Of *Perſia*, and of the Countreys of the *Magos*, and of other good things that be in them

CHAPTER 13

[*Pauthier*: Bk. I. Chs. XXX, XXXI. *Yule*: Bk. I. Chs. XIII, XIV. *Benedetto*: Chs. XXXI, XXXII]

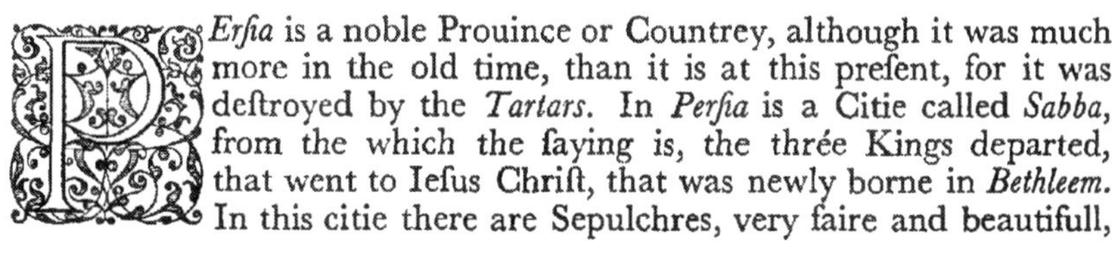

PErſia is a noble Prouince or Countrey, although it was much more in the old time, than it is at this preſent, for it was deſtroyed by the *Tartars*. In *Perſia* is a Citie called *Sabba*, from the which the ſaying is, the thrée Kings departed, that went to Ieſus Chriſt, that was newly borne in *Bethleem*. In this citie there are Sepulchres, very faire and beautifull, and I MARCUS PAULUS was in that Citie, and asked of the people of that Countrey what they could say or knewe of the thrée Kings, to the which they could say nothing, but that they were buried in those thrée Sepulchres. But ẏ other people out of the Citie thrée dayes iourney, talked of this matter in thys maner following, for the which you shal vnderstãd, that thrée days iourney frõ the Citie *Sabba* is a Towne, which is called *Calassa Tapezisten*, which in our language is as much as to say the Towne of them that worship the fire for their God. And these people say, that whẽ the thrée Kings departed frõ ẏ prouince, for to go to the land of the Iewes, which was *Bethleem*, to worship the great Prophet there newly borne, they carried with thẽ Golde, Incense, and Myrre, and when they came to *Bethleem* in *Iudea*, found a child lately borne, and did worshippe him for God, and presented to him the foresaide thrée things: and that the said child did giue thẽ a little Boxe, closed, or shut fast, commanding thẽ they should not open it. But they, after they had trauelled a long iourney, it came in their mindes to sée what they carried in the said Boxe, and opened it, and foũd nothing in it but only a stone: and they taking it in ill parte, that they sawe nothing else, did cast it into a well, and by and by descended fire from Heauen, and burnt all the Well wyth the stone. And the Kings séeing this, each of them toke of the same fire, and carried it into their Countreys: and for thys cause they do worship the fire as God. And when it chanceth in any place in that Countrey that they lacke fire, they goe to séeke it in another place where they cã get of it, and so do light their Lampes. And sometimes they goe and séeke it eyght or tenne dayes iourney, and not finding of it, they goe ofttymes to the Well aforesayd, to haue of the same fire Of all this before written, you shall take ẏ which doth agrée with the holy Gospell, in saying the thrée Kings went to worship our Lord Iesu, and did offer those giftes aforesaide. All that is declared besides that, be erroures, and reacheth not to the truth, but augmẽted with lyes vpon lyes, as the vulgar people without knowledge are accustomed to do.

In this Citie Sabba *the three Kings met that wẽt to worship Christ, and heere they were buried.*

The three Kings offered Gold, Incense, and Myrre.

A miracle if it be true.

Of eyght Kingdomes in *Perſia*, and the commodities of them

CHAPTER 14

[*Marsden*: Bk. 1. Ch. xi. *Pauthier*: Bk. 1. Ch. xxxii. *Yule*: Bk. 1. Ch. xv. *Benedetto*: Ch. xxxiii]

IN the Prouince of *Perſia* be eyght Kingdomes, the firſt is called *Caſun*, the ſecond which is towardes the South is *Curdiſtan*, the third *Lore*, the fourth *Cieſtan*, the fifth *Iuſtanth*, the ſixth *Iciagi*, the ſeauenth *Corchara*, the eyght *Tunchay*. All theſe Kingdomes be in *Perſia*, in the partes towards the South, ſauing *Tunchay*. In theſe Kingdomes be very faire Horſes and Moyles, & courſers of great value, and Aſſes the greateſt in the worlde, & of great price, that wil go and runne very ſwiftely, and theſe the Merchants of *India* do commonly buy in the Cities of *Atriſo*, & of *Arcones*, which do ioyne by Sea vpon the *India*, and do ſel thẽ as Merchandiſe. In this Kingdome *Tunchay* be very cruell mẽ, ẏ wil kill one another. If it were not for feare of ẏ TARTAR of the Eaſt, which is their Lord and King, neyther Merchant nor other could paſſe, but ſhould be eyther robbed or taken priſoner. They be ſtrong people, and be of the ſect of MAHOMET. There they do worke, and make greate plentie of cloth of gold and ſilke in great abundance and rich. In that Countrey groweth greate plentie of Cotten wooll. Alſo, there is greate abundance of Wheate, Barly, Dates, and other grayne, and Wine, and Oyles, and frutes.

Heere is great plenty of fayre Horſes, Moyles, and Aſſes.

Heere is made great plentie of rich cloth of gold & ſilke.

Of *Iaſoy*, and of many maruellous things there

CHAPTER 15

[*Marsden*: Bk. 1. Chs. xii, xiii, xiv, xv (first 12 lines). *Pauthier*: Bk. 1. Chs. xxxiii, xxxiv, xxxv, xxxvi (first 10 lines). *Yule*: Bk. 1. Chs. xvi, xvii, xviii, xix (first 12 lines). *Benedetto*: Chs. xxxiv, xxxv, xxxvi, xxxvii (first 7 lines)]

Heere they do make gret plentie of cloth of golde and ſilke.

I*Aſoy* is a goodly Citie and bigge, full of Merchants. There they do make great abundance of cloth of gold, and ſilke. They be called accordyng to the Citie *Iaſoy*. The people of this Countrey be of the ſect of MARTIN PINOL, that is, MAHOMET, and do ſpeake another language than the *Perſians*. And going forward eyght dayes iourney from this Citie, through a playne Countrey, but not peopled, or anye Towne, ſauing Mountaynes, where is great plentie of Partriches, and wild Aſſes, at the ende of this, is the Kingdome of the *Crerina*, that is, a Kingdome of the *Perſians*, of a great and long inheritance.

¶ In this Countrey they doe finde greate plentie of pretious ſtones, and of Turkies great ſtore in the Mountaynes, in the whiche Mountaynes, is greate plentie of Uayne, or Ore of Stéele, and of CALAMITA. In this Citie, they do make greate plentie of coſtly ſaddles, bridles, and harneſſes for Horſes, and for noble men Swords, bowes, and other riche furniture for Horſe and Man. The Women of this Countrey doe nothing, but commaunde their Seruauntes They make alſo there very riche cloth of gold and ſilke. And in thoſe Mountaynes be excéeding good Hawkes, valiaunte, and ſwifte of wings, that no fowle can ſcape them. And departing from *Crerina*, you ſhall goe eyght dayes iourney in playne way, full of Cities and Townes, very faire, and there is pleaſaunte Hawking by the way, & great plentie of Partriches. And being paſt the ſayd eyght dayes iourney, there is a going downe the hil of two dayes iourney, whereas there is great plenty of frutes. In the olde time there was manye Townes and houſes, and now there be none but heardmen, that kéepe the Cattell in the field. From the Citie of *Crerina*, to this going down, al the winter is ſo great cold, that although they go very wel clothed, they haue ynough to do to liue. And being paſt this going downe two dayes iourney forwarde, you ſhall come into a faire playne way, the beginning whereof is a great faire Citie, called *Camath*, the whiche was in the old time noble and greate, and nowe is not ſo, for that the *Tartars* haue deſtroyed it That playne is very hote, and that Prouince is called *Reobarle* There be apples of Paradiſe, and Feſtucas, and Medlars, and diuers other goodly frutes in great abundance. There be Oxen maruellous great, the heare ſhort and ſoft, and the hornes ſhort, bigge, and ſharp, and haue a greate rounde bunche betwéene the ſhoulders, of two ſpannes long. And when they will lade theſe Oxen, they do knéele downe on theyr knées like Camels, and being ladẽ, do riſe, and they carrie great weight. There the Shéepe be as greate as Aſſes, hauing a greate tayle, and thicke, that will weigh .32. pound, and be maruellous good to eate. In that playne be many Cities & townes with walles, and Towers of a great heigth for the defence of the enimies, called *Caraones*, which be certaine Uillages. The people of that Countrey their Mothers be *Indians*, and their fathers *Tartars*. When that people will go a robbing, they worke by enchantment by the Deuill, to darken the aire, as it were midnight, bycauſe they woulde not bée ſéene a farre off, and this darkeneſſe endureth ſeauen dayes. And the Théeues that know well all the wayes, goe togither, withoute making anye noyſe, and as many as they can take, they robbe. The olde men they kill, and the yong men they ſell

Pretious ſtones, as Turkiſes and others

Sadles and bridles, and other coſtly furniture for Horſes

Cloth of gold and ſilke

Excellent good hawkes.

Sheepe as great as Aſſes

Enchantmẽt

for ſlaues. Their King is called HEGODAR, and of a truth I MARCUS PAULUS do tell you, that I eſcaped very hardly from taking of theſe robbers, and that I was not ſlaine in that darkeneſſe, but it pleaſed God, I eſcaped to a towne called *Ganaſſalim*, yet of my companie they toke and ſlewe many. This playne is towardes the South, and is of ſeauen dayes iourney, and at the end of them is a mou̅tayne, called *Detuſtlyno*, that is eightéene miles long & more, and is alſo very daungerous with théeues, that do rob Merchauntes and all trauellers. At the ende of this mountaine is a faire playne, called the goodly playne, which is ſeauen dayes iourney, in the which there be many wels, and date trées, very good, and this playne bordereth vpon the Ocean Sea, and on the riuer of the ſea, is a Citie called *Carmoe*.

Of the Citie *Carmoe*, and of many maruellous and ſtraunge things that be there

CHAPTER 16

[*Marsden*: Bk. I. Chs. XV (cont.), XVI (in part). *Pauthier*: Bk. I. Ch. XXXVI (cont.). *Yule*: Bk. I. Ch. XIX (cont.). *Benedetto*: Ch. XXXVII (cont.)]

Great trade of Merchãts.

CArmoe is a greate Citie, and is a good porte of the Ocean ſea. Thither do occupie Merchãts of the *Indeas* with ſpices, cloth of gold & ſilke, and with precious ſtones, and Elephantes téeth, and is a Citie of great trade, with merchaundize, and is heade of that kingdome, and the king is called MINEDANOCOMOYTH. It is very hote there, and ẏ ayre infectious. When there doth dye any Merchaunt, they doe make hauocke of all his goods. In this Citie they do drinke wine made of Dates, putting good ſpices to it, yet at the beginning of dinner it is daungerous, for thoſe that be not vſed to it, for it will make them very ſoluble, ſtreight waye, but it is good to purge the body. The people of that Countrey do not vſe of our victuals, for when they eate bread of wheate and fleſhe, by and by they fall ſicke. Their victuals is Dates & ſalte, Tonny, Garlike & Onyons. The people of that Countrey be blacke, and be of the ſect of MAHOMET. And for the great heate in the Sommer, they dwell not in the town, but in the fields, and in gardens, and Orchyards. There be many riuers and Wels, that euery one hath faire water for his garden: and there be manye that dwell in a deſart, wheras is al ſande, that ioyneth to that playne. And thoſe people aſſoone as they féele the great heate, they goe into the waters, and there tarrie till the heate of the daye be paſt. In that countrey, they do ſowe their wheate and corne in Nouember, and gather it in Marche. And in thys time the fruites be greater than in any place. And after March is paſte, the graſſe, hearbes, and leaues of trées doe drie, ſauing of Date trées, which continue till Maye. And in that countrey they haue this cuſtome, that when the huſband doth dye, the wife and hir friendes doe wéepe once a day, for the ſpace of foure yeares.

When the huſband dyeth the wife & the friends do weep once a day for the ſpace of four yeares.

Of the Citie of *Crerima*, and the death of the Olde man of the Mountaine

CHAPTER 17

[*Marsden*: Bk. I. Chs. XVII (in part), XXI (last 10 lines). *Pauthier*: Bk. I. Chs. XXXVI (cont.), XLII. *Yule*: Bk. I. Chs. XIX (cont.), XXV. *Benedetto*: Chs. XXXVII (cont.), XLIII]

LEauing here this Citie, and not declaring any more, of the INDIANS, I retourne to the Northwardes, declaring of thoſe prouinces turning another way, to the Citie *Crerima*, aforeſayde, for bycauſe that way, that I would tell of, could not be trauelled to *Crerima* for the crueltie of the king of that countrie, whiche is called *Reu me cla vacomare*, from whome fewe coulde ſcape, but eyther were robbed or ſlayne. And for this cauſe manye kings did paye him tribute, and hys name is as muche to ſaye, as the olde man of the mountayne. But I wyll nowe declare vnto you, howe this cruell King was taken priſoner in the yeare of our Lord .1272. ALAN King of the *Tartars* of the Eaſt, hearing of the greate crueltie of this olde man of the Mountayne, that he did, ſent a great hoſt of men, and beſette his Caſtell rounde about, and thus continued thrée yeares, and coulde neuer take it, till that victuals did fayle them: for it was very ſtrong, and vnpoſſible to be gotten. At the length ALAN toke the Caſtell, and the old man of the Mountayne: and of al his Souldioures and men he cauſed the heads to be ſtricken off, and from that time forwarde that way was very good for all trauellers.

What is found in that Countrey

CHAPTER 18

[*Marsden*: Bk. I. Ch. XXII (first half). *Pauthier*: Bk. I. Ch. XLIII. *Yule*: Bk. I. Ch. XXVI. *Benedetto*: Ch. XLIV]

DEparting from the foresayd Castell, you shall come into a very faire playne, full of grasse, with all things in it fitte for mans sustenance. And this playne dothe last sixe dayes iourney, in the whiche there is many fayre Cities and Townes. The people of that Countrey speake the *Persian* language, and haue greate lacke of water, and sometimes they shall fortune to go .40. miles, and not finde water. Therfore it shall be néedefull for those that do trauell that way, to carrie water with them from place to place. And being past these sixe dayes iourney, there is a Citie called *Sempergayme*, faire and pleasaunte, with abundance of victuals. There be excellente good Mellones, and the best Hunters for wilde beastes, and taking of wilde Fowle, that be in the world.

Of the Citie of *Baldach*, and of many other things

CHAPTER 19

[*Marsden*: Bk. I. Chs. XXII (second half), XXIII (first half). *Pauthier*: Bk. I. Chs. XLIV, XLV (in part). *Yule*: Bk. I. Chs. XXVII, XXVIII (in part). *Benedetto*: Chs. XLV, XLVI (in part)]

TRauelling forward in this Countrey, you shall come to a Citie called *Baldach*, in the whiche King ALEXANDER married with the daughter of DARIUS king of ỳ *Persians*. This Citie is of the Kingdome of *Persia*, & they do there speake the *Persian* tong, and be all of the sect of MAHOMET. And this Countrey dothe ioyne with the TARTAR of the East, betwéene the Northeast, and the East. And departing from this Citie towardes the Countreys of the said TARTAR, you shall goe two dayes iourney, withoute finding any Towne, bycause the people of that Countrey do couet to the strong Mountaynes, bycause of the ill people that be there. In that Countrey be many waters, by reason whereof is greate plenty of wild Fowle, and of wylde Beasts, and there be many Lions. It is néedefull for the trauellers that way, to carrie prouision with them that shall be néedefull for themselues, and for their Horses those two dayes iourney.

And being paſt that, you ſhall come to a Towne called *Thaychan*, a pleaſaunt place, and well prouided of all vittayles néedefull, and the hilles be towardes the South faire and large. That prouince is .xxx. dayes iourney. And there is great plẽtie of ſalt, that all the Cities and Townes thereaboutes haue their ſalt from thence.

Great plenty of ſalte.

Of that Countrey

CHAPTER 20

[*Marsden*: Bk. I. Ch. XXIII (second half). *Pauthier*: Bk. I. Ch. XLV (cont.). *Yule*: Bk. I. Ch. XXVIII (cont.). *Benedetto*: Ch. XLVI (cont.)]

DEparting from that towne, and trauelling Northeaſt, and to the Eaſt for the ſpace of thrée dayes iourney, you ſhall come to faire Cities and Townes well prouided of victuals and frutes in great abundance, and theſe people do ſpeake the *Perſian* language, and be MAHOMETS. There be ſingular good wines, and great drinkers, and yll people. They go bareheaded, hauing a Towell knit aboute their browes. They weare nothing but ſkinnes that they do dreſſe.

Good wines and great drinkers.

Of the Citie *Echafen*

CHAPTER 12 [21]

[*Marsden*: Bk. I. Ch. XXIV (lines 1–11). *Pauthier*: Bk. I. Ch. XLV (cont.). *Yule*: Bk. I. Ch. XXVIII (cont.). *Benedetto*: Ch. XLVI (cont.)]

AFter that you haue trauelled forwarde foure dayes iourney, you ſhall come to a Citie called *Echaſen*, on a playne, and there is not farre from it manie Cities and townes, and great plentie of woods about it. There goeth through the middeſt of this Citie a gret riuer. There is in that countrie, many wilde beaſtes, and when they be diſpoſed to take anye of them, they will caſt dartes, and ſhoote them into the flancks and into the ſides. The people of that countrey doe ſpeake the *Perſian* tong, and the huſbandmen, with their cattayle do liue in the fieldes and in the woods.

Of the manner of the Countrie

CHAPTER 22

[*Marsden*: Bk. I. Ch. XXIV (lines 11–end). *Pauthier*: Bk. I. Ch. XLV (cont.). *Yule*: Bk. I. Ch. XXVIII (cont.). *Benedetto*: Ch. XLVI (cont.)]

DEparting from this Citie, you ſhall trauayle thrée dayes iourney, without comming to any towne, or finding any victuals eyther to eate or drinke, and for thys cauſe the trauellers do prouide themſelues for ÿ time, & at the end of theſe thrée days iourney, you ſhal come to a prouince called *Ballaſia*.

Of the prouince called *Ballaſia*, and of the commodities there

CHAPTER 23

[*Marsden*: Bk. I. Ch. XXV. *Pauthier*: Bk. I. Ch. XLVI. *Yule*: Bk. I. Ch. XXIX. *Benedetto*: Ch. XLVII]

B*Allaſia* is a great prouince, & they do ſpeake the *Perſian* tong, & be MAHOMETS, and it is a great kingdome, and auncient. There did raygne the ſucceſſours of king ALEXANDER, and of DARIUS king of *Perſia*. And their king is called CULTURI, which is as much to ſay, as ALEXANDER, and is for remembraunce of the great king ALEXÃDER. In this countrey grow the precious ſtones, called *Ballaſſes* of greate value. And theſe ſtones you can not carrie out of the countrey without ſpeciall licence of the king, on pain of léeſing life and goods. And thoſe that he doth let paſſe by, eyther he doth forgiue tribute of ſome king, or elſe that he doth ſell: and if they were not ſo ſtraightlye kept, they would be little worth, there is ſuch great plentie of them. This countrie is very colde, and there is found greate plenty of ſiluer: there be very good courſers, or horſes, that be neuer ſhod, bycauſe they bréede in the mountaines and woods. There is great plentie of wilde foule, and greate plentie of corne, and *Mylo*, and *Lolio*. In this kingdome be great woods & narrow ways, ſtrong men, and good Archers, and for this cauſe they feare no bodie. There is no cloth, they apparell themſelues with ſkinnes of beaſtes that they kil. The women do weare wrapped aboute their bodies like ÿ neather part of garments, ſome an hundreth fathom, & ſome foureſcore, of linnen very fine and thinne, made of flaxe and Cotton wool, for to ſéeme great and fayre, and they doe weare bréeches very fine of ſilke, with Muſke put in them.

For lacke of cloth the people weare ſkinnes of ſuch beaſtes as they kil.

Of the Prouince of *Abaſsia* where the people be blacke

CHAPTER 24

[*Marsden*: Bk. I. Ch. XXVI. *Pauthier*: Bk. I. Ch. XLVII. *Yule*: Bk. I. Ch. XXX. *Benedetto*: Ch. XLVIII]

AFter you be departed frõ *Ballaſia* eyght dayes iourney towards the South, you haue a prouince called *Abaſſia*, whoſe people be blacke, and do ſpeake the *Perſian* tong, and doe worſhip Idolles. There they do vſe Negromancie. The men do weare at their heares iewels of golde, ſiluer, and pretious ſtones. They be malicious people, and leacherous, by reaſon of the great heate of that Countrey, and they eate nothing but fleſh and Rice.

Of the Prouince called *Thaſsimur*, and of many things there

CHAPTER 25

[*Marsden*: Bk. I. Ch. XXVII. *Pauthier*: Bk. I. Ch. XLVIII. *Yule*: Bk. I. Ch. XXXI. *Benedetto*: Ch. XLIX]

WIthin the iuriſdiction of this Countrey, betwéene the Eaſt and the South, there is a Prouince called *Thaſſymur*, and the people do ſpeake the *Perſian* tong. They be Idolaters, and great Negromancers, and do call to the Spirits, and make them to ſpeake in the Idols, and do make their Temples ſéeme to moue. They doe trouble the ayre, and doe many other diueliſh things. From hence they may go to the *Indian* Sea. The people of that Countrey be blacke and leane, and do eate nothing but fleſh and Rice. The Countrey is temperate. In this Countrey be many Cities and Townes, and rounde about many hilles and ſtrong wayes to paſſe. And for this cauſe they feare no body, and their King dothe mainteyne them in peace and iuſtice. There be alſo Hermites, that do kéepe great abſtinence in eating & drinking. And there be Monaſteries, and many Abbeys, with Monkes, very deuout in their Idolatrie and naughtineſſe.

Negromancers.

Of the ſaide prouince of *Thaſſymur*

CHAPTER 26

[*Marsden*: Bk. I. Ch. XXVII (cont.). *Pauthier*: Bk. I. Ch. XLVIII (cont.). *Yule*: Bk. I. Ch. XXXI (cont.). *Benedetto*: Ch. XLIX (cont.)]

I Minde not now to paſſe further in this prouince, for in paſſing of it I ſhould enter into the *Indeas*, wherof for this time I wil not declare any thing, but at the returne, I wil declare of it largely, as wel of the commodities there, as alſo of their manner, and vſages.

Of a prouince called *Vochaym*

CHAPTER 27

[*Marsden*: Bk. I. Ch. XXVIII. *Pauthier*: Bk. I. Ch. XLIX. *Yule*: Bk. I. Ch. XXXII. *Benedetto*: Ch. L]

A Citie of .3. days iourney long.

DEparting from *Balaſſia*, you ſhall goe thrée dayes iourney betwéene the Northeaſt, and by a riuer that is neare to *Balaſſia*. In thys prouince be many Cities and townes. The men of this prouince be valiaunt in armes, and ſpeake the *Perſian* language, and be MAHOMETS. At the ende of this thrée dayes iourney is a Citie called *Vochayn* very long, of thrée dayes iourney on eyther ſide. The people of this prouince, be ſubiecte to the king of *Balaſſia*, and there be greate hunters of wilde beaſtes, and taking of wilde foules in great number.

Of the nouelties of this Countrey

CHAPTER 28

[*Marsden*: Bk. I. Ch. XXVIII (cont.). *Pauthier*: Bk. I. Ch. XLIX (cont.). *Yule*: Bk. I. Ch. XXXII (cont.). *Benedetto*: Ch. L (cont.)]

Sheep that haue hornes of foure or fiue and ten ſpans long.

THrée dayes iourney going forewarde, you ſhall goe vp an hill, vpon the whiche is a riuer, and goodly fruitefull paſtures, that if you put in your cattell there, very leane, within tenne dayes they wil be fat. There be greate plentie of wylde beaſtes, and among them wilde ſhéepe, that ſome of them haue their hornes of foure and ſome of ſeuen, and ſome of tenne ſpannes long. And of theſe hornes the heardemen there doe make diſhes, and ſpones. In the valey of this mountaine called *Plauor*, you ſhall trauell tenne dayes iourney, without comming to anye towne, or anye grasse, therefore it shall be néedefull, for the traueylours that waye, to carrie prouision with them, as wel for themselues, as for their horses. There is greate colde in that Countrey, that the fire hath not the strength to séethe their victuals, as in other Countries.

Of the Desert *Bosor*, and of manye maruellous things there

CHAPTER 39 [29]

[*Marsden*: Bk. I. Ch. XXVIII (cont.). *Pauthier*: Bk. I. Ch. XLIX (cont.). *Yule*: Bk. I. Ch. XXXII (cont.). *Benedetto*: Ch. L (cont.)]

AFter that you be departed from thence, within thrée daies iourney you shal be faine to trauell fortie dayes iourney continually vpon Mountaines, Heathes, and Ualleys, betwéene the Northeast and East, and passing ouer diuers riuers and deserts. And in all this waye, you shall come to no towne nor habitation, nor grasse, and therefore it is néedefull for those that do trauell that waye, to carrie with them prouision and victuals for themselues and their horses. And this Countrey is called *Bosor*. The people there liue on the high hils, & be called people of the Mountaines. They be Idolaters, and liue by their cattel, and be cruell people.

Fortie dayes iourney and haue no habitation.

Of the prouince *Caschar* and of other Nouelties

CHAPTER 40 [30]

[*Marsden*: Bk. I. Ch. XXIX. *Pauthier*: Bk. I. Ch. L. *Yule*: Bk. I. Ch. XXXIII. *Benedetto*: Ch. LI]

LEaue this prouince, and let vs goe to another called *Caschar*, that in olde time was a kingdome, although nowe it be subiect to the greate CANE. In this prouince are manye faire Cities and townes, the best is *Caschar:* they be all Mahometes. This proüince is betwéene the Northeast & the East. In it be many great Merchants, faire possessions and Uines, they haue much Cottenwooll there, and very good. The Merchaunts of that countrey bée neare, and couetous. In this prouince which endureth fiue dayes iourney, be Christians called Nestorians, and haue Churches, and speake the *Persian* tong.

Of *Sumarthan*, and of a miracle

CHAPTER 31

[*Marsden*: Bk. I. Ch. xxx. *Pauthier*: Bk. I. Ch. LI. *Yule*: Bk. I. Ch. xxxiv. *Benedetto*: Ch. LII]

S*Vmarthan* is a Citie great and faire, in the which dwell Chriſtians, and Moores, that be ſubiect to ẏ great CANE: but this king beareth them no good will. In this Citie chaunced a maruellous thing. A brother of the greate CANE, that was Lorde of that Countrey, became a Chriſtian, by meanes whereof, the Chriſtians there, receyued great comfort, and buylded them a Churche, in the name of Saint Iohn Baptiſt. And it was builded in ſuch ſorte, that one Piller of Marble ſtanding in the middeſt, did beare vp all the rouſe of the Church, and the Chriſtians did put vnder the ſayde piller a goodly Marble ſtone, whiche was the MOORES, and for bycauſe the king was a Chriſtian, they durſt ſay nothing of it. This king died, and one of his ſons ſuccéeded him in the kingdome, which was no Chriſtian, and on a time the Moores demaunded their ſtone of ẏ Chriſtians, thinking that in taking away that ſtone, the whole rouſe of the Church would fal downe: and the Chriſtians did offer to pay the Moores for the ſtone, what they woulde demaunde: but they woulde not by anye meanes, but haue their ſtone, and in the ende, the new king commaunded the Chriſtians to reſtore the ſtone to the Moores, and the time appointed being come, that the Moores would haue it, the ſayde Piller lifted it ſelfe vp, thrée ſpannes aboue the ſtone, and ſo hãged in the ayre, that the Moores might take away their ſtone, and yet the Church fell not, and ſo doth the Piller remayne til this day.

Of the prouince of *Carcham*

CHAPTER 32

[*Marsden*: Bk. I. Ch. xxxI. *Pauthier*: Bk. I. Ch. LII. *Yule*: Bk. I. Ch. xxxv. *Benedetto*: Ch. LIII]

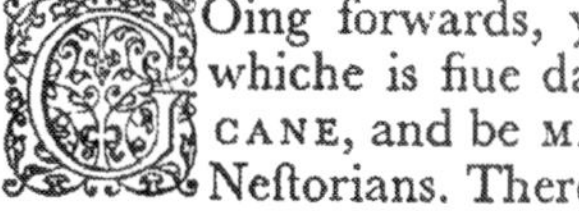

Oing forwards, you ſhall come to a prouince called *Carcham*, whiche is fiue dayes iourney long, and is ſubiect to the greate CANE, and be MAHOMETS, but there is among them Chriſtians Neſtorians. There is in this prouince aboundaunce of all things.

Of the prouince *Chota* and of their manners

CHAPTER 33

[*Marsden*: Bk. I. Ch. XXXII. *Pauthier*: Bk. I. Ch. LIII. *Yule*: Bk. I. Ch. XXXVI. *Benedetto*: Ch. LIV]

C*Hota* is a prouince betwéene the Northeaſt, and the Eaſt, and is of fiue dayes iourney, ſubiect to the gret CANE, and be MAHOMETS. In this prouince there be diuerſe cities and towns, but the chiefeſt is *Chota*. In this prouince be goodly poſſeſſions, and faire Gardens and Uines, plentie of Wine and fruites, and Oyles, Wheate, Barley and all other victuals, great plentie of Cotton-wooll. In this Countrey be rich Merchaunts, good and valiaunt men of armes.

Of the prouince of *Poym* and of their vſages

CHAPTER 34

[*Marsden*: Bk. I. Ch. XXXIII. *Pauthier*: Bk. I. Ch. LIV. *Yule*: Bk. I. Ch. XXXVII. *Benedetto*: Ch. LV]

P*Oym* is a ſmall prouince of fiue dayes iourney, it is betwéene the Northeaſte and the Eaſt, and be ſubiect to the great CANE, and be MAHOMETS, and the principall Citie is called *Poym*. In this prouince there is a riuer, in the whiche there is founde precious ſtones, called IASPES and CALCEDONIES, there is great plentie of all kinde of victuals, and great trade of Merchandizes. In this prouince there is this cuſtome, that when the huſband departeth from his houſe for fiftéene or thirtie dayes, or more or leſſe, if the wife can get another huſbande for the time, ſhe taketh him, and the huſbande taketh another wife til he returne home to his houſe.

Iaſpes and Calcedonies.

Of the prouince of *Ciarchan* being in great Turkie

CHAPTER 35

[*Marsden*: Bk. I. Ch. XXXIV (in part). *Pauthier*: Bk. I. Ch. LV (in part). *Yule*: Bk. I. Ch. XXXVIII (in part). *Benedetto*: Ch. LVI (in part)]

AL the prouinces beforeſayde, from *Caſchar*, to this, be ſubiectes to the greate CANE, and were of greate Turkie, in ẏ which there is a great Citie called *Ciarchan* in a prouince alſo called *Ciarcham*, ſet betwéene the Northeaſt & the Eaſt, and the people of that Countrey ſpeake the *Perſian* tong, and be Mahomets. In this prouince be many Cities, townes, and riuers, wherein be found many pretious ſtones, called *Calcedonies*, whiche Merchauntes carry all the worlde ouer to ſell, and get muche money by them. In this Countrey is aboundaunce of all things néedefull: And this prouince for the moſt part is ſandie, and the waters there, for the moſt part, pleaſaunt and ſwéete, yet in ſome places brackiſh. And the people of that Countrey, fering the ill people, do flie with their houſeholde ſtuffe, and cattell, two or thrée dayes iourney, till they maye come to ſome good place, whereas is water and graſſe for their cattel, and by reaſon the way is ſandie, their tracte is ſoone filled, by reaſon whereof, the théeues knowe not howe to follow in that Countrey.

Of a great deſerte, and of the Citie called *Iob*

CHAPTER 36

[*Marsden*: Bk. I. Chs. XXXIV (end), XXXV. *Pauthier*: Bk. I. Chs. LV (end), LVI. *Yule*: Bk. I. Chs. XXXVIII (end), XXXIX. *Benedetto*: Chs. LVI (end), LVII]

The Citie Iob.

DEparting from *Ciarchan*, you ſhal trauayle fiue dayes iourney in ſande, and in the waye, freſh and ſwéete waters, and ſome ſaltiſh. Being paſte theſe fiue dayes iorney, you ſhal finde a great deſert, and at the beginning of it a gret Citie called *Iob*, betwéene the Northeaſt and the Eaſte. They be vnder the obedience of the great CANE, & be Mahomets. And they that wil paſſe this deſert, had néede to be in thys Citie a wéeke, for to prouide them victuals and other neceſſaries for them and theyr horſes for a moneth, for in thys deſert, you ſhall finde nothing to eate or drinke: and there be many ſandie hils, and greate. After you be entred into it one dayes iourney, you ſhall finde good water, but after that neyther good nor badde, nor beaſtes, nor foules, nor any thing to eate: and trauelling that waye by nighte, you ſhall heare in the ayre, the ſound of Tabers and other inſtruments, to putte the trauellers in feare, and to make them loſe their way, and to depart from their company, and looſe themſelues: and by that meanes many doe die, being deceiued ſo, by euill ſpirites, that make theſe ſoundes, and alſo do call diuerſe of the trauellers by their names, and make them to leaue their companye, ſo that you ſhall paſſe this deſert with great daunger.

Of the prouince of *Tanguith*, and of the Citie *Sangechian* and of many ſtraunge things there

CHAPTER 37

[*Marsden*: Bk. I. Ch. XXXVI. *Pauthier*: Bk. I. Ch. LVII. *Yule*: Bk. I. Ch. XL. *Benedetto*: Ch. LVIII]

AFter you be paſte the ſayde thirtie dayes iourney by the deſerte, you ſhall come to a Citie called *Sangechian*, ſubiect to the greate CANE. And this prouince is called *Tanguith*, in the whiche al be Idolaters, ſauing ſome be Chriſtians, Neſtorians, and ſome Mahomets. The Idolaters ſpeake the *Perſian* tong, and doe liue by the fruites of that Countrey. There be among them manye Monaſteries of the Idolaters, wher with great deuotion they bring their children, and with euerie of them a ſhéepe, and do preſent to their Idols: and euerie yeare they come with theyr children and make great reuerence to their Idols, & bryng with them their ſhéepe, and kill them, and ſéeth them, and preſent them there, before their Idols, ſaying to them, they muſt eate their meate, the which they can not doe, for they haue neyther mouth nor ſenſe, and ſéeing their Idols doe not eate it, they carrie it home to their houſes with greate reuerence, and call theyr kyndered togyther, and do eate of it, as meate ſacrificed to their Gods, and put the bones in a baſket. When anye man or woman dieth, they burne the body: and this they accuſtome to doe with al the Idolators. And in the way that the deade bodies ſhall paſſe to be burnte, ſtande all their friendes and kinſfolkes to accompany the body to the ſepulchre, all clothed in cloth of golde and ſilke: and after the burnte bodye is put into the grounde, they cauſe to be brought thither meate & drinke, and there they do eate and drinke with greate myrth, ſaying: Theſe bodies ſhall be receiued in the other worlde with like honour. When they burne the bodies, they do alſo burne with them diuers papers paynted, of men, women, and beaſtes, ſaying, that as many pictures of men, women, and beaſtes, as they do burne with them, ſo many ſeruaunts they ſhall haue in the other world to doe them ſeruice: and when they cary them to bury, there goeth before them diuers kinds of inſtruments playing. And whẽ one of theſe Idolators dieth, his friendes incontinentlye declare to the Aſtrologers, the day and the houre hée was borne in, and wil not bury him before the day & houre the Aſtrologers doe commaunde: by that meanes ſome they bury ſtraightways, and ſomtimes, they tarry ten, twenty, and thirtie dayes, and ſometime ſixe

A rich mourning & good cheare.

moneths, according as the Aſtrologers doe commaunde: and in the meane time, they do fire the body with ſpices, and put it in a coffin, and nayle it faſte, and lay a cloth ouer it, and euerye day they ſet their table ouer the Coffin, and there do eate and drinke, and pray the dead body to eate with them. And when the day appointed is come for to bury him, the Aſtrologers do ſay, that if he hath layne there one month, it is not good to take him oute of that place, by the iudgement of the Conſtellations, and for that cauſe muſte firſt remoue him to ſome other ſide of the houſe, & from thence carry him to bury.

Of the prouince *Chamul*, and of the euill cuſtomes there

CHAPTER 38

[*Marsden*: Bk. I. Ch. XXXVII. *Pauthier*: Bk. I. Ch. LVIII. *Yule*: Bk. I. Ch. XLI. *Benedetto*: Ch. LIX]

C*Hamul* is a prouince in the whiche be manye Citties and Townes, whereof the chiefeſt is called *Chamul*, and this prouince is towards the winde called *Maiſtral*, which is Northeaſt, and hath two Deſerts: on the one ſide, the Deſerte is of thrée dayes iorney, and on the other ſide as muche. The people of this Countrey worſhip Idols, and doe ſpeake the Perſian tongue. They liue by their labor in the Countrey, and haue plentie of al things néedefull. They be people giuen much to their owne pleaſure, as playing on inſtrumentes, dauncing, and ſinging. And if any ſtraunger doe goe to ſée their paſtime, they receiue him, and make very much of him, with feaſting and cheare, and the goodman commaundeth his wife to make hym the beſte cheare ſhe can, and to obey him in al things he will commaunde or deſire, and ſo the goodman goeth to his laboure into the fieldes, and leaueth the ſtraunger with hys wife, willing hir to obey hym as to his owne perſon: and this cuſtome the menne and the women vſe there, & be not aſhamed therof. The women be very faire there. In the time of the greate CANE that is paſte, for the greate diſhoneſtie hée heard of the people of that countrie, and the greate hurte they ſuſteined in their houſes, commaunded them that they ſhoulde receiue no ſtraungers into their houſes, wherewithall the people were ſore offended, and thinking themſelues not well vſed, ſent Embaſſadors to the greate CANE, requeſting him, that he woulde not reſtraine them from their auntient liberties and cuſtomes, that their anticeſſors hadde euer vſed, and they for their partes woulde continue the ſame, otherwiſe they ſhoulde be vnthankefull to their Idolls. After the greate CANE hadde hearde their Embaſſage, aunſwered them, ſéeyng they had pleaſure in ſuche ſhamefull vſages, and woulde not leaue it, he alſo was contented with it.

Of the prouince *Hingnitala*, and of the Salamandra that is founde there

CHAPTER 39

[*Marsden*: Bk. I. Chs. XXXVIII, XXXIX. *Pauthier*: Bk. I. Chs. LIX, LX. *Yule*: Bk. I. Chs. XLII, XLIII. *Benedetto*: Chs. LX, LXI]

H*Ingnitala* is a prouince ſet betwéen the North and the Eaſte, and is a long prouince of ſixetéene dayes iourney, and is ſubiect to the great CANE, and there is manye Cities and Townes. There is alſo in that prouince, thrée linages of people, to ſaye, Idolators that be Chriſtians, Neſtorians and Iacobites, and the other Mahomets. At the ende of this prouince towardes the North is a greate hill, on the whiche there is neither beaſtes nor Serpent, and from thence they doe gather that whiche is called Salamandra, which is a thréede they doe make cloth of. They gather it after this manner, they digge a certaine vayne that they doe there finde, and afterwardes they beate it in a morter of a ſofer, and afterwarde waſhe it, and there remaineth ſmall fine thréedes faire and cleane, and after they haue caſte out that which they doe waſhe it withall, they ſpinne it, and weaue it, and make table clothes and napkins of it, then they caſte them into the fire for a certaine time, whereas it waxeth as white as ſnowe: and the great CANE once in thrée yeres doth ſend for ſome of them that be made of *Salamandra.* And they wer wont for to ſẽd of theſe napkins, for to hang before the vernacle of oure Lorde Ieſus Chriſt, whome the people of *Leuant* do take for a great prophet. Departing from this prouince, and going betwéen the Northeaſt and Eaſt, you ſhal trauaile tenne dayes iourney and come to little habitation, and at the end of the tenne dayes iourny, you ſhall find a prouince called *Sachur,* in it be Chriſtians and Idolators, ſubiects to the great CANE. The two prouinces beforeſaide, to ſay, *Chamul,* and *Hingnitala* be called *Tanguth,* with the prouince of *Sachar.* In all the hilles of this prouince is found greate plentie of Rewbarbe, and there the Merchauntes do buy it, and carry it to all places to ſel. There they doe not vſe any occupation, but the moſte parte doe liue by the laboure of the Countrey.

Of the Citie called *Campion*, and of many euill vſages there

CHAPTER 44 [40]

[*Marsden*: Bk. I. Ch. XL. *Pauthier*: Bk. I. Ch. LXI. *Yule*: Bk. I. Ch. XLIV. *Benedetto*: Ch. LXII]

C*Ampion* is a greate Citie and fayre, & is the heade of the prouince of *Tanguth*. In this Citie be thrée ſortes of people, that is to ſay, Chriſtians, Idolators, and Mahomets. The Chriſtiās haue thrée great Churches and faire, and the Idolators haue alſo Monaſteries, Abbeys, and religious houſes, more chaſte and comly than the other, and they do kil no beaſt nor fowle there till the fifth day of the Moone, and in thoſe fiue days they liue more honeſt, deuout, and chaſt, than in any other time of the yeare. Theſe Idolators may haue thirtie wiues apéece, or more, if they be able to maintaine them, but the firſte wife is chiefe, and if anye of them doe not contente him, he may put hir away. They do mary in kinreds, and liue like beaſtes. In this Citie was MAPHEO NICHOLAS and MARCUS PAULUS ſeauen yeres, vſing the trade of merchaundize.

Of a Citie called *Euſina*, and of many notable things in *Tartaria*

CHAPTER .xlj. [41]

[*Marsden*: Bk. I. Chs. XLI, XLII, XLIII, XLIV (in part). *Pauthier*: Bk. I. Chs. LXII, LXIII, LXIV (in part). *Yule*: Bk. I. Chs. XLV, XLVI, XLVII (in part). *Benedetto*: Chs. LXIII, LXIV, LXV (in part)]

DEparting from the foreſayde Cittie *Campion*, and trauailing twelue dayes iorney, you ſhall come to a Citie called *Euſina*, the whyche is in a fielde of the Deſert called *Sabon*, toward the North, and is of the prouince *Tanguth*. In this Citie they bée al Idolators, and haue great abundaunce of Camels and other cattell withall: they gette their liuing by labouring the ground. In this Citie thoſe that do trauaile, do prouide them of victualles, and other neceſſaries, for fortie dayes iourney, whyche they muſt paſſe through a great Deſert, wheras be no towns nor houſes, nor graſſe, but in the mountaines about dwel people, and alſo in the valleys beneath the Deſert. There be many Aſſes and other wild beaſts of the mountaines, and greate Pine apple trées. At the ende of this Deſerte there is a Citie called *Catlogoria*, whiche is towarde the North, and of this Citie was the firſt Prince or Lorde among the TARTARS, and his name was *Catlogoria*. The TARTARS dwel towards the North, wheras is but few cities & Townes, but true it is, there be fayre playnes, paſtures, riuers, and very good waters. There dwell TARTARS that haue no King nor Lorde, they doe gouerne themſelues in common, and do pay tribute to PRESTER IOHN. It fortuned, that theſe TARTARS multiplyed to ſo greate a number, that PRESTER IOHN did feare, that they woulde riſe againſt him, therefore he determined with himſelfe to ſende certaine Lordes of his that ſhoulde be among them to kéepe them aſunder, and alſo to kéepe the countrey in good order, and to baniſhe or diminiſhe parte of them, bycauſe they ſhould not be of ſo greate a power. And the TARTARS perceyuyng thys, ioyned themſelues togither, and tooke councell, determined to leaue that countrey, and to goe and dwell vpon the mountaines and in the deſerts, by meanes whereof from that time forwarde they ſtoode in no feare of PRESTER IOHN, nor woulde pay him tribute. And at the end of certaine yeares, that they were not vnder the obedience of PRESTER IOHN, they did elect and chooſe among themſelues a King whiche they called CHENCHIS, a valiaunt and wiſe man: and this was in the yeare of oure Lorde God .1187. and crowned him for King of the TARTARS aforeſaide. And all the TARTARS that were in *Perſia*, and other Countreys thereaboutes, came to him, and put themſelues vnder his gouernement, and obeyed him as their King, and he receiued them very friendly, gouerning them iuſtely and diſcréetely. And after that CHENCHIS was confirmed, and had the whole gouernment, within a ſhort time he made war, and in ſhorte time conquered eighte Kingdomes or Prouinces, and when he hadde gotten anye Prouince or Citie, he did iniurie to no man, but lette them remaine wyth their goods, ſauing to thoſe that were able and fitte menne for him, them he tooke with him into the warres, and by this meanes he was welbeloued, and all men were content to goe with him.

Of the beginning of the raigne of the Tartars, and of many maruellous and ſtraunge thinges

CHAPTER 42

[*Marsden*: Bk. I. Chs. XLIV (in part), XLV, XLVI. *Pauthier*: Bk. I. Chs. LXIV (in part), LXV, LXVII (in part), LXVIII. *Yule*: Bk. I. Chs. XLVII (in part), XLVIII, L (in part), LI, LII. *Benedetto*: Chs. LXV (in part), LXVI, LXVIII (in part), LXIX]

CHENCHIS perceyuyng himſelfe to be of ſuche power, minding to ioyne himſelfe in kindred or ſtocke with PRESTER IOHN, ſente to him his Embaſſadoures, requiring his daughter in marriage: and this was in the yeare of oure Lord God .1190. PRESTER IOHN diſdained that Embaſſage and aunſwered, that he maruailed muche that CHENCHIS being his Subiecte ſhoulde preſume to demaunde his Lordes daughter to be his wife, ſaying he woulde rather kil hir: ſo the matter remayned thus. CHENCHIS hearing this aunſwere of PRESTER IOHN, was ſore troubled and vexed in minde againſte hym, and incontinent ſent him defiaunce, ſaying, he woulde warre vppon him, and of this PRESTER IOHN made ſmall reckning ſaying, that the TARTARS were but ſlaues, and not menne of warre, notwithſtanding he made himſelfe in a readineſſe, and came vpon CHENCHIS, who had alſo made himſelfe in a readineſſe, and came oute againſte him and encountred togither in a great plaine called *Tanguth*, where it was appointed the battaile ſhoulde be of both parties, & thus ioyned togither in a fierce & lõg battel, for both parts was ſtrong, but in the end, PRESTER IOHN being ſlaine, and many of both parts, the field remayned to CHENCHIS, who conquered all the prouince, Cities, and townes of PRESTER IOHN, and raigned after his death ſixe yeares, and at the end of ſixe yeares, laying ſiege to a Caſtell, was hurte in the knée with an arrowe, and of that wounde dyed. After the death of this CHENCHIS, was made Lord of the TARTARS one called CANE, and this was the firſte that was called Emperoure and Greate CANE. And after hym raigned BATHE CANE, and the fourth was called CHENCHIS CANE, & the fifth was CUBLAY CANE, which raigneth nowe. This CUBLAY CANE is the greateſt and of moſt power of anye of al his predeſſors, for among the Chriſtians and Heathen, there is not a greater Prince than he is, nor of ſo great a power, and that ſhall you cléerely perceyue hereafter, by that which followeth. All the CANES, ſucceſſors of the firſt CHENCHIS, were buryed in a mountaine called *Alchay*, and there dwelled the greate CANE. And when the greate CANE dyeth, they cary hym to be buryed

Preſter Iohn *ſlaine in battel by* Chenchis *King of the* Tartars.

The firſt Emperour of the Tartars *called Great* Cane.

In this mountaine Alchay, *be al the gret* Canes *buryed.*

there. Thofe that do cary him, or go with him, kill as manye as they méete withall in the waye or ftréete, and when they kill them, they faye Go, and ferue our Lorde in the other worlde, & they beléeue certainely, that they go, and doe him feruice And likewife by this reafon, when the greate CANE dieth, they kill all his Camels, Horfes, and Moyles, beléeuing that they fẽd them to ferue their Lord in the other worlde. When MONGUY CANE Lorde of the TARTARS dyed, there was flaine .300000. men that they encountred in the way, by thofe that wente wyth hym to hys buriall to the faide mountaine.

¶ The habitation of the TARTARS in the Winter, is in the plaine fieldes, where it is warme, and good graffe and pafture for their Cattell, and in the Sommer in the mountaines and wooddes, where it is frefhe and pleafaunt aire: and they make rounde houfes of tymber, and couer them with feltes, and thefe houfes they carry with them at all times when they do remoue. and alwayes they fette their doore in the Sommer time towards the South, and in the Winter towardes the North. Thefe Tartars haue theyr cartes or Wagons couered with blacke feltes, that neuer any water can paffe through, and in thefe Cartes or Wagons go their wiues, children, and family, and their Cammels do drawe thefe Wagons The Tartars wiues doe buy and fell al manner of things belonging to houfeholde, or any thing néedefull: their hufbands take no care for it, but onely in hawking, hunting, and going on warrefare They do eate all manner of flefhe, and drinke milke of all kinde of beaftes and mares The Tartars maye take as manye wiues as they will, and maye marry with anye of their kinred, excepting no degrée: but their firfte wife is the chiefeft, and is mofte made of. the women doe gyue their downes to their hufbandes. There is none of them will haue conuerfation with an other mannes wife. And when the father dyeth, his eldeft fonne doeth marry wyth his mother in lawe, and when the fonne dyeth, his brother marryeth with hys fifter in lawe, and for the time do kéep great folemnitie and feaftes at the wedding.

Of the cuſtome, orders, faith and honoring the great CANE, and howe he goeth to the warres

CHAPTER 43

[*Marsden*: Bk. I. Chs. XLVII, XLVIII (in part), XLIX (in part). *Pauthier*: Bk. I. Ch. LXIX (in part). *Yule*: Bk. I. Chs. LIII, LIV, LV (in part). *Benedetto*: Ch. LXX (in part)]

THe Greate CANE Emperour of the Tartars, doth worſhippe for his God, an Idoll called NOCHYGAY, and they ſaye and beléeue, that he is the eternall God, that taketh care to preſerue hym, hys wiues, children, familie, cattell, and corne, and hathe him in great reuerence, and euery one hath the figure of that Idoll in his houſe. And this Idoll is made of feltes, or of other cloth, and of the ſame felte or cloth they doe make wiues and children for their Idols, and the women be ſette on the lefte ſide of the Idols, and the children before them. When they thinke it dinner tyme, then they doe annoynte the mouthes and lippes of theyr Idols, and wiues and children, with the fatte of the ſodden fleſh, and do poure out the broath vpon the floore, ſaying, that theyr Idols, their wiues, and children doe fill themſelues with it, and they do eate the ſodden fleſh, and their drinke is the milke of Mares trimmed with ſpices, that it is like white wine, and it is very good, and is called with them *Cheminis*. The Lordes and men of power and riches, goe apparelled in cloth of golde, and cloth of ſilke, furred with riche furres. Their harneſſe is the Hydes of Buffe, or other thicke and ſtrong Skynnes. The TARTARES be valiant men of armes, and ſtrong to abyde any trauell or laboure, and can well ſuffer hunger and thirſt, for in the warres they be many times one moneth, and eate nothing, but of wylde beaſtes they doe kill in the field, and drinke Mares Milke. When they be in the field day and night they be on Horſebacke, and the bridle in their hands they giue the Horſes meate. When their King ſetteth forward with his hoſt, before and on euery ſide of him they do ſet foure battels of the beſt and moſt valiant men, for bycauſe their King ſhoulde not bée put in feare. And when he goeth a warrefare a farre off, he caryeth nothing with hym but hys armoure, and a thing to couer him when it doth rayne, and two flaggons with Milke for to drinke, and a Potte to ſéeth his meate in when néede is. In a tyme of néede hée will ride tenne days iourney, without eating any ſodden meate. For his drinke, they will carrie Milke made like dry paſte, and when hée is diſpoſed to drinke, he will take a little of that paſte, and diſſolue it in

The Tartares doe make them Idols of feltes, and other baggage.

The Nobilitie & Gentlemen go in cloth of gold and ſilke, furred with rich furres.

fayre water, and ſo drinke it: and when thys ſhall fayle hym, and that he can gette no other drinke, hée letteth hys Horſe bloud, and drinketh of it. When the TARTARES wyll ſkyrmiſhe wyth theyr enimies, they hyde their Sallets ſecretely, and as they doe beginne to ſkyrmiſhe, ſtreightway they ſhewe as though they woulde runne away, and that they were ouercome of theyr enimies, and thus fleeing, putte on theyr Sallets, and ſtreyght way they returne valiantly vpon their enimies, and by this meanes commonly they doe breake the array of theyr enimies. The TARTARES haue thys cuſtome, that if one of theyr ſonnes dye being yong, and alſo of another man his daughter, after they be dead, they marrie them, ſaying, they ſhall be maried in the other worlde. And of thys Matrimonie they doe make a publike writing, and this writing they burne, ſaying to the dead, that as the ſmoke thereof aſcendeth on high, ſo doe they ſende them that writing, declaring theyr mariage. And at ſuche mariages they make great feaſting and ſolemnitie, and do ſéeth muche victuals, and poure out the broath vppon the floore, ſaying, that thoſe which be dead in this world, and maried in the other, do eate of the victuals prepared for the wedding. And beſides all this, they cauſe to be painted the figure of the ſonne and daughter, vppon the backſide of the foreſayde writing, and withall the pictures of manye Camels, and other diuers beaſts, and apparell and money, and many other things, ſaying, that as that writing dothe burne, all thoſe things therein goe ſtraight way to their chyldren, after the ſmoke as aforeſayde, and the fathers and mothers of theſe children that dyed, doe take hands togither, and be alwayes after friendes, and Grandfathers and Grandmothers, and Couſens, euen as though they had bin maried aliue.

The Tartares going a warfare, carrie with them a thing made in paſte of Mares milke and other compounds, and do ſerue for his drinke.

When any of the Tartares ſonnes dye, and alſo a daughter of another, then they do marrie theſe two togither, ſaying, they ſhall be ſo in the other worlde.

Of a plaine called *Barga*, and of the cuſtoms of the people of that Countrey

CHAPTER 44

[*Marsden*: Bk. I. Ch. L. *Pauthier*: Bk. I. Ch. LXX. *Yule*: Bk. I. Ch. LVI. *Benedetto*: Ch. LXXI]

DEparting from the Citie called *Cuthogora*, aforeſade, and the mountaine called *Acay*, where they bury theyr Kings of the Tartars, whiche is the greate CANE, you ſhall trauell through a great plaine called the plaine of *Barga*, fortie dayes iorney towards the North. The people of that country be called MECRITH. They be ſauage people, and doe lyue the mofte parte by killyng of redde Deare called Stagges, and other wilde beaftes, and doe ride and trauaile vppon harts or ftagges, as they doe in other places vppon horfes. They haue neyther breade nor wine, and be fubiectes to the greate CANE.

Of the greate Sea called the Occean

CHAPTER 45

[*Marsden*: Bk. I. Ch. L (cont.). *Pauthier*: Bk. I. Ch. LXX (cont.). *Yule*: Bk. I. Ch. LVI (cont.). *Benedetto*: Ch. LXXI (cont.)]

AFter you haue trauailed fortie daies iorney, you fhal come to a greate Sea called the Occean Sea, and alfo greate mountaynes, in the which you fhal haue goodly Hawkes greate plentie, and fpeciall good, called PEREGRINOS. And in the Ilandes of the Sea bréedeth great plentie of Gerfalcons. In this Sea be two great Ilandes, whiche fhall be fpokẽ of hereafter, and lye towardes the North, and haue the Sea out of the South.

Of the Kingdome *Erguyl*, and of many other Kingdomes, and of Mufke, and other fweete and pleafaunte thinges that be there founde, and many other things

CHAPTER 46

[*Marsden*: Bk. I. Ch. LI. *Pauthier*: Bk. I. Ch. LXXI. *Yule*: Bk. I. Ch. LVII. *Benedetto*: Ch. LXXII]

I Haue declared vnto you of the prouinces of the North, till you come to the mountaines, and the Occean Sea: and now I will compte to you of the other prouinces belonging to ỳ great CANE, til you come to his country, returning to the country called *Campion*, where you fhal paffe .5. days iorny in length, in the which many times you fhal hear the voices of euil fpirits. At ỳ end of thefe fiue days iorny towards ỳ Eaft, there is a kingdom called *Erguil*, of ỳ prouince of *Tanguth*, fubiect to the greate CANE, and in this prouince there liue thrée forts of people, that is to fay, Chriftians that be Neftorians, and Idolators and Mahomets: and there be many Cities and Townes, but the principall Citie is called *Erguyl*. From this Citie trauelling Eaft Southeaft, you fhall come to a Countrie whiche is a greate prouince, in the whiche there is a great Citie called *Syrygay*,

The voice of euil fpirites heard.

that hath neare vnto it many Cities and Townes, all ſubiect to the greate CANE, and there be in it Chriſtians, Idolators, and Mahomets. There be wild Oxen as bigge as Elephants, very faire beaſts to ſée, white and blacke, al couered with haire, ſauing a ſpanne long vpon the necke, whyche is called *Del Eſpinazo*, whiche is bare, and hath no haire, and many of theſe Oxen they do make tame, and doe laboure and till the grounde with them. They will carrye greate waighte, by reaſon they be ſo great bodyed. There is the beſt Muſke in the worlde. The Beaſt that they haue it off, is bodyed like a Catte, with foure téeth, two aboue, and two beneath, of thrée fingers long, they be ſlender of body, and haue heare like a redde Déere, and féete lyke a Catte, and they haue a thing like a poſhe, or bagge of bloud, gathered togither néere to their nauell, betwéene the ſkinne and the fleſhe, whiche they cutte and take away, and that is the Muſke: and there be many of thoſe Beaſtes there. The people of that Countrey do liue by their occupations and trade of Merchandiſe, and haue good plenty of corne. This Countrey is long, of .25. days iourney. There be plenty of Feyſants, and very greate, for one of them is as bigge as two of oures, with tayles of eyght, nine, and tenne ſpannes long. The people of that Countrey be fatte, and of lowe browes, and blacke heared, and haue no beardes, but a fewe heares about the mouth. The women be faire and white, and well bodyed. The people of that Countrey bée gyuen muche to the pleaſure of the body, for a riche man to obteyne the fauoure of a woman, wyll gyue hir a ioynter. They bée all Idolaters.

Monſtrous greate Oxen as bigge as Elephants.

Heere is the beſt Muſke in the world.

I think theſe be Peacocks.

Of the Citie called *Calacia*, and of many things they do make there

CHAPTER 47

[*Marsden*: Bk. 1. Ch. LII. *Pauthier*: Bk. I. Ch. LXXII. *Yule*: Bk. I. Ch. LVIII. *Benedetto*: Ch. LXXIII]

DEparting from *Erguill*, and trauelling towardes the Eaſt eyght dayes iourney, you ſhall come to a Prouince called *Egregia*, that hathe vnder it many Cities, and is of the Prouince of *Tanguthe*, and the principall Citie of it is called *Chalacia*, and is ſubiecte to the greate CANE, in the which be thrée Churches of Chriſtians Neſtorians, and all the reſt be Idolaters. There they make excellent good Chamlets of Camels heare of white wooll, and from thence Merchantes carrie them to ſell into other Countreys.

Heere be Chamlets made.

Of the Prouince called *Tanguthe* which is ſubiect to PRESTER IOHN, and of a ſtone called *Lapis laguli*, that is there found, and of *Gog* and *Magog*

CHAPTER 48

[*Marsden*: Bk. I. Ch. LIII. *Pauthier*: Bk. I. Ch. LXXIII. *Yule*: Bk. I. Ch. LIX. *Benedetto*: Ch. LXXIV]

Eparting from *Arguill*, and entring into the Kingdomes of PRESTER IOHN, you ſhall come to a Prouince called *Tanguthe*, which is vnder a King of the lignage of PRESTER IOHN, whiche is called GEORGE by his proper name, and he holdeth that Countrey of the great CANE, eſpecially thoſe that were taken of PRESTER IOHN. And the greate CANE dothe alwayes take the chiefeſt daughters of this Kyng commonly, ſince that CHENCHIS the firſt King of the TARTARES ſlewe PRESTER IOHN in battell, as before is declared. In this Countrey is found *Lapis laguli*, whiche is a ſtone, that maketh a fine blew. The moſt part of this prouince be Chriſtians, and they be gouernoures, and chiefe of the Countrey. There be alſo Mahomets, whiche doe liue by Cattell, and labouring of ẏ ground. In this Prouince be another kind of people called *Argarones*, or *Galmulos*, this they do ſay, for bicauſe they do deſcẽd of two ſeueral nations, ẏ is to ſay, of ẏ chriſtiãs of *Tãguthe*, & the Mahomets. They be faire mẽ, wiſe and diſcret more than the others of ẏ countrey. In this prouince was ẏ imperiall chayre or ſeate of PRESTER IOHN, when he raygned ouer the TARTARS: and yet there doe raygne in that prouince, of the ſtocke of PRESTER IOHN, of whome came this GEORGE King of thys prouince. Here is that place that the holye Scripture ſpeaketh of, called *Gog* and *Magog*.

Heere is founde the ſtone called Lapis laguli, *wherewith they do make a fyne blewe.*

Heere was the imperiall ſeate of Preſter Iohn.

Of the Citie *Sindathoy* in *Cataya*, where ſiluer is founde

CHAPTER 49

[*Marsden*: Bk. I. Ch. LIV. *Pauthier*: Bk. I. Ch. LXXIII (cont.). *Yule*: Bk. I. Ch. LIX (cont.). *Benedetto*: Ch. LXXIV (cont.)]

Auing paſſed ſeauen dayes iorney in thys prouince towards the Eaſt, you ſhal come to *Cataya*, a broade Countrey, in the which there be many Chriſtians, and many Idolators, and many of the ſect of MAHOMET, and they be al handi-crafts men and Merchauntes. There they make great plentie of cloth of gold, and alſo of cloth of ſilke verye fine. In this prouince is a Citie of the greate CANES called *Sindathoy*, where they doe worke and make all manner and kinde of armour for the wars, and in the mountaines of this prouince be vaines of fine ſiluer, and plentie, called there *Idica*.

Of a Citie called *Giannorum*, and of many nouelties

CHAPTER 50

[*Marsden*: Bk. I. Ch. LV. *Pauthier*: Bk. I. Ch. LXXIII (cont.). *Yule*: Bk. I. Ch. LX. *Benedetto*: Ch. LXXIV (cont.)]

PArting from this Citie, and trauelling .iij. dayes iorny, you ſhall come to a Citie called *Gianorum*, in the which there is a meruellous goodly Pallace of the great CANES to lodge him and his Court when he commeth to that Citie, and in this Citie he is deſirous to be with good will, for bycauſe that neare vnto it is a good countrey, in the which be great plentie of wyld Géeſe, and Duckes, and of Cranes, of fiue ſortes or manners: the firſt be great and all blacke like Crowes: the ſecond all whyte, ſauing the heades that be all red: the thirde al black, ſauing the heade is white and ſhyning: ẏ fourth gréene, with blacke heads: they be farre bigger than ours: the fifth be little with all their feathers redde. Neare vnto this Citie is a great valley, where the great CANE hath many wilde beaſtes, great and ſmal, and among thẽ great plentie of Partridges, to ſerue for his prouiſion, when hée goeth into that Countrey.

Here be Cranes of fiue ſorts or colours.

Of a maruellous Citie called *Liander*, and of many maruellous and faire things they haue there

CHAPTER 51

[*Marsden*: Bk. I. Ch. LVI. *Pauthier*: Bk. I. Ch. LXXIV. *Yule*: Bk. I. Ch. LXI. *Benedetto*: Ch. LXXV]

DEparting thrée dayes iourney from this Citie, betwéen the Northeaſt and ẏ North you ſhall come to a Citie called *Liander*, which CUBLAY CANE buylded. In this Citie is a maruellous goodlye Pallace made of Marble and flint ſtones, called *pedras viuas*, al gilded wyth gold, and neare to this Pallace, is a wall which is in compaſſe fiftéene miles, and within this wall be faire riuers, Wels, and gréene Meadowes, where the

The wall of this houſe is gilded.

Here ŷ Emperor hath great ſtore of Haukes of all ſortes.

great CANE hath plentie of all kinde of wilde ſoule and beaſtes, for to finde his Hawkes, called Faulcons, and Gerſaulcons, that bée there in mew, which be at ſometimes more than .40000. ŷ which many times he goeth thyther to ſée. Whẽ he doth ride in theſe Meadowes, he carrieth behinde him on the buttockes of his horſe, a ruſſet or graye Lyon tame, and ſetteth him to the ſtagges, or redde Déere, and to other wylde beaſtes, and vppon theſe beaſtes do the Gerſaulcons and Faulcons ſeaſon. In the middeſt of theſe Meddowes is a great houſe, where the great CANE doth reſort to dinner, and to banquet, and to take his reſte and pleaſure in, when he goeth that waye. And this houſe is compaſſed about with greate Canes, that be gilded and couered with Canes that be varniſhed, and cloſed all in one, in ſuch ſort, that no water can paſſe through it. Euerye Cane is at the leaſt thrée ſpannes compaſſe, and from tenne to fiftéen paces long. And this houſe is ſo made, that at al times they maye take it downe and ſet it vp againe, vpon a ſodayne. It is tyed with aboue .200. cordes of ſilke, after the manner of tentes, or pauilions. And the greate CANE repayreth thither for his pleaſure, in Iune, Iuly, and Auguſt, and there by commaundement of his Prophets, Idolaters maketh ſacrifice with milke to his Idols, for to preſerue and kéepe his wiues, and ſonnes, and daughters, and his ſubiectes, and ſeruauntes, and cattell, and ſoules, corne, vines, fruite, and all other things in his countries. All the Mares that the great CANE rideth on, be as white as milke. Among the which, he hath alwayes ten Mares that no body doth drinke of their milke, but onlye he and ſome greate men of his Courte, and ſome others that bée called honourable and noble, bycauſe of a victorie had againſt the enemies of CHENCHIS the firſt king of the TARTARS.

Here ŷ Cane doth make ſacrifice with milke to his Idols.

Al ŷ Mares the great Cane do ride on, be white.

Of the ſacrifice and other maners, of the life of the greate CANE

CHAPTER 52

[*Marsden*: Bk. I. Ch. LVI (cont.). *Pauthier*: Bk. I. Ch. LXXIV (cont.). *Yule*: Bk. I. Ch. LXI (cont.). *Benedetto*: Ch. LXXV (cont.)]

A ſuperſtitious beliefe ŷ great Cane hath.

WHen the great CANE will make ſacrifice, he poureth out the Mares milke vpon the ground, and in the ayre, and the Prophets of his gods ſay, that milke poured out, is the holye Ghoſte, of the which all the Idols be ful, and do beléeue, that this ſacrifice is the cauſe of his confirmation, and of his ſubiects, & of al his other things. And this ſacrifice he doth euery yere ẏ .29. day of Auguſt. And to thoſe white horſes and Mares whereſoeuer they do go, they do great reuerence. This greate CANE hath in his Court certaine Negromanciers, whiche by arte of the Diuel, when it is foule & troubleſome weather, it ſhal be fayre and cleare weather in his Pallace. And do gyue to vnderſtande to the people, that the clearneſſe is ouer the Pallace where the great CANE is, only for his deſerts and holy life, and by vertue of his Idols. When anye one is iudged to dye, as ſoone as he is deade they ſéeth him, and eate him, but thoſe that dye by natural death, be meat for their Idols.

¶ And beſides thys, when the great CANE is at hys table, theſe inchaunters doe worke by arte of the Diuel, that Cuppes doe riſe from the table tenne Cubits into the ayre, and do ſet themſelues down again, and whẽ they wyll doe this, they demaunde of the greate CANE a blacke ſhéepe, and the wood of Alloe and Incenſe, & other ſwéete ſpyces, wherof there is great plenty, bicauſe their ſacrifice ſhould ſéeme the more ſwéeter, and he commaundeth to be deliuered to them, what they will haue, for bycauſe they beléeue that their Idols doe preſerue and kéepe him and all his companie. Theſe Prophets and Prieſtes, do cauſe the fleſh to be ſodden with ſpices in preſence of their Idols, & do put incenſe therin, and poure the broth into the ayre, & they ſay the Idol taketh of it what pleaſeth him: and thys they do with gret ſinging. Euery Idol hath his name, and to euery one they do this worſhip on their dayes, as we do on our ſaints dayes. They haue many Monaſteries deputed to the names of their Idols. There is in that countrey one Monaſterie as big as a good Citie, in the which there be .400. Monkes that goe honeſtly apparelled, and their beardes and heads ſhauen. Vpon their feaſte dayes they kepe great ſolemnity, with ſinging, and prayſing, and lights, and ſome of theſe religious men haue many wiues, and ſome of them liue chaſte: the chaſt do eate the branne and the meale kneaded togither, with a little hote water, and do faſt oftentimes in reuerence of their Idols, and do weare garments made of Canuas died blacke or blewe, & ſome white, and do lye in Almadraques, ſharpe and harde beds, and the other religious that be maried, they go well apparelled, and do eate and drinke wel, and doe ſaye that thoſe which liue the ſtreight life be Heretickes and fooles, bycauſe they do puniſh their bodies, by meanes whereof they cannot honor their Idols as they ought to do, and as reaſon is. All the Idols of theſe married religious men, they do name by the name of women, bycauſe they be ſuch leacherous people.

Here his enchaunters do worke by the Diuel.

A great Monaſterie of Monkes.

Of a victorie the great CANE had

CHAPTER 53

[*Marsden*: Bk. II. Ch. I. Sects. I, II (abridged). *Pauthier*: Bk. II. Chs. LXXV–LXXIX (abridged). *Yule*: Bk. II. Chs. I–V (abridged). *Benedetto*: Chs. LXXVI–LXXX (abridged)]

HEre, for your better information, I wyll declare vnto you of a victory the gret CANE had, wherby you ſhal the better vnderſtand and know of his ſtrength and power. It was he that now raigneth, which was called CUBLAY CANE, whiche is as muche to ſaye, as Lorde of Lordes. You ſhall vnderſtande that this CUBLAY CANE deſcended lineally of the imperiall ſtocke, from CHENCHIS CANE, from whence he muſt deſcende, that ſhall be Lorde of the TARTARES: and this CUBLAY CANE, beganne his raigne in the yere of our Lord God .1256. And as CHENCHIS CANE by his prouidence and wiſedome, made himſelfe the firſte Lord of the TARTARES, as is before declared, ſo likewiſe this for his wiſdome and prouidence, contrarie to the good will of his kinred, that would haue put him out of it, did ſo cõſerue and gouerne his Dominions and Countries, til the yeare of our Lord God .1298. ſo that he raigned two and fortie yeares, and was fiue and forty yeares old when he was made Emperor, and euerye yeare hadde warres, for he was valiant and expert in the warres, but he himſelfe after he was made Emperour, neuer went to the warres but one time, but alwayes ſent his ſonnes, or ſome noble men, whom he thought beſt. And the cauſe wherefore hée went at that time in perſon, was this. In the yeare of our Lord God .1286. a nephew of his, of the age of thirtie yeares Lord of many prouinces, Cities, and townes, perceyuing himſelfe to be ſubiecte to the greate CANE, as his predeceſſors had ben, determined in himſelf not to be ſubiect to anye, and concorded with another kinſeman of the great CANES, whyche was called CARDIN, whyche mighte well make .100000. Horſemen, and was mortall enimie to the greate CANE hys vncle, and did moue warre both of them with theyr hoſtes agaynſte the great CANE, and hée hauyng knowledge thereof, dyd not feare, for hée was a Prince of maruellous greate power: but incontinent he called hys people togither for to go againſt hys enimies, and toke an oth, that the crowne ſhoulde neuer come on his head, till that he had cruelly reuenged hymſelfe on them as Traytors and Rebels, ſo that within two and twenty days, he had ioyned particularly a great hoſt of thrée hundred thouſand fighting men, of horſemen and footemen, and woulde ioyne no greater an hoſt, nor haue it publiſhed abrode, that

Thrée hundred thouſand fighting men.

his enimies ſhoulde haue knowledge of it, and alſo for that he had many of his men of warre abroade in other places on warfare, and coulde not bring them togither in ſo ſhort a time. But you ſhall vnderſtande that when the greate CANE will make his power, and take time to doe it, he may ioyne ſo greate a number, that it were a greate trouble to number them. Theſe thrée hundred thouſande of fighting men, be not all menne of experience, for there were aboue foure thouſande Falconers, and Seruants, and Courtiers that attended vppon the Kings perſon, and ſerued in his Courts. But thus hauing his hoſts ioyned, he commanded to be called before him his Aſtrologers, and would know of them in what ſort and time he ſhoulde ſet forward on this enterpriſe, and they anſwered him that the time was good, and that he ſhoulde haue victorie ouer his enimies, and ſo incontinent ſet forwarde on his way with his people, and came to a playne, where as was NAUIA with .200000. men tarrying there the comming of CAYDU with another hundred thouſand of horſemen, for to ſet on the Countreys of the great CANE. The Lordes of the great CANE had beſet all the wayes, and taken all the ſtreytes, that neither ſuccoure ſhoulde come, nor his enimies flée, bycauſe he would take them all priſoners. NAUIA knowing nothing of this, or that the great CANE had prepared himſelfe for any warre, for the greate CANE had before beſet all the wayes and paſſages, that no mã could paſſe to carrie any newes to NAUIA, and by this meanes, not thinking nor ſtãding in any doubt, thought he might well take his reſt that nighte, and all his people but the greate CANE was ſtirring in the morning betimes with all his hoſtes, and did ſette his Campe hard by the place where as NAUIA had his, and founde them all vnarmed, and vnprouided, not thinking any thing of it, and perceyuing it, he was in greate feare. And the great CANE had made a great frame vpon an Elephant, wherin his ſtanderdes were caried, and before and behinde, and by the ſides went his battels of Horſemen and footemen, that is to ſay .25000. in a battell. And with theſe battels be ſette all the hoſt of NAUIA round, and when NAUIA ſawe thys, he lept on horſebacke, and cauſed his trumpets to blowe, and ſet his armie in as good order as he could, and ſo ioyned battell, whereas was a great and ſtrong fighte, and continued from morning till nighte, and greate number ſlayne on both parties, but at the end NAUIA and his company were not able any longer to withſtande the furie of the greate CANES armed men, and beganne to flée, in ſuch ſort, that NAUIA was taken priſoner, and his people not being able to doe anye good, ſubmitted themſelues to the great CANE: and NAUIA being preſented aliue to the great CANE, he cauſed him to be bounde vp in a Carpet, and ſo long hée vſed him to bée caried,

The pollicie of the great Cane

A ſtrange kind of death to his couſin.

that hée dyed, and thys deathe hée gaue hym, for that hée woulde not haue the bloud of NAUIA béeing of his kindred, fall to the grounde, nor that the ayre ſhoulde ſée hym dye an euill deathe. After that NAUIA was deade, all his Lordes and other priſoners became ſworne to the great CANE, to be obediẽt to him. Theſe foure prouinces were vnder the obedience of NAUIA, that is to ſay, *Furciorcia*, *Guli*, *Baſton*, *Scincinguy*.

¶ Now that I haue ſhewed you of the great CANE, howe he paſte with NAUIA, I will alſo declare vnto you, of hys manner, condition, and perſon, and of his wiues and children, and of other things.

Of the perſonage of the great CANE, and of his wiues and children

CHAPTER 54

[*Marsden*: Bk. II. Chs. IV, V. *Pauthier*: Bk. II. Chs. LXXXI, LXXXII. *Yule*: Bk. II. Chs. VIII, IX. *Benedetto*: Chs. LXXXII, LXXXIII]

THe great CANE that was called CUBLA CANE, was a manne of a middle ſtature, well fleſhte, and of good complexion, and wel proportioned in al his mẽbers, well coloured of face, his eyes black, his noſe well made: he hath four that be his Legitimate wiues, and his eldeſt ſonne, that he hath by his firſt wife, doth kepe Court by himſelfe, and euerye one of theſe foure Quéenes, haue in their Courtes .300. wayting women, and many maydens, with alſo many mẽ and women, that do ſeruice in the Courtes: for euery one of theſe foure Quéenes haue in their Courtes more than .4000. perſons, of men, women, maydens, and ſeruaunts. Alſo the greate CANE hath many Concubines of TARTARS, which be called *Origiathe* and be of a good and honeſt behauiour, and of theſe the greate CANE hath a hundreth maydens choſen out for himſelfe, which be in a pallace by thẽſelues, and haue auntient women to kéepe them. And of theſe hundreth, euery thrée dayes ſixe of them doe ſerue and attend vpon the great CANE in his Chamber, and the thrée dayes being paſt, they doe returne to their Pallace agayne, and other ſixe come for to kéepe the great CANES Chamber. And thus they do remoue from thrée dayes to thrée dayes. The ſayd great CANE had by his ſayd wiues two and twentie Sonnes, the eldeſt of them is called CHINCHIS, in remembrance of the firſt King of TARTARES, and alſo to renue that name, this firſte ſonne is called CHINCHIS CANE, and ſhoulde haue ſuccéeded his father in the

The great Cane hath foure wiues and they kepe great Courts.

The great Cane hath many Concubines.

The greate Cane had by his foure wiues two and twenty ſonnes, after his eldeſt ſon dyed, who ſhould haue bin King. His ſonne was heyre, and kept a great Court.

Kingdome, but bycaufe he dyed beforé his father, his eldeſt ſonne called THEMUR CANE, and this his ſonnes ſonne, bycauſe he ſhould raigne after him, kepte a greate Court by himſelfe.

Of a greate Citie called *Cambalu*, and of all the goodly and maruellous things that be done there

CHAPTER 55

[*Marsden*: Bk. II. Chs. VI–XIII (abridged). *Pauthier*: Bk. II. Chs. LXXXIII–LXXXIX (abridged). *Yule*: Bk. II. Chs. X–XVII (abridged). *Benedetto*: Chs. LXXXIV–XCI (abridged)]

NOw I will declare vnto you of the worthy and noble Citie called *Cambalu*, the whiche is in the prouince of *Cathaya*. This Citie is foure and twenty myles compaſſe, and is foure-ſquare, that is, to euery quarter ſixe miles compaſſe. The wall is very ſtrong, of twenty paces high, and battlements of thrée paces high. The wall is fiue paces thicke. This Citie hathe twelue gates, and at euery gate is a very faire pallace. And vpon the toppe of euery corner of the ſaid wal is alſo a faire pallace, and in all theſe pallaces ioyning to the wall be many people appoynted for to watch and kéepe the Citie. And in thoſe pallaces be all maner of armour and weapons for the defence and ſtrength of the Citie. The ſtréetes of this Citie be ſo faire and ſtreight, that you may ſée a Candle or fire from the one ende to the other. In this Citie be manye fayre Pallaces and houſes. And in the middeſt of it is a notable greate and faire Pallace, in the whiche there is a great Toure, wherein there is a greate Bell, and after that Bell is tolled thrée times, no body may goe abroade in the Citie, but the watchmen that be appoynted for to kéepe the Citie, and the nurſes that doe kéepe children newly borne, and Phiſitions that goe to viſit the ſicke, and theſe may not go without light. At euery gate nightlye there is a thouſand men to watch, not for feare of any enimies, but to auoyde théeues and robbers in the Citie, which many times do chance in the Citie. And this great watche the greate CANE doth cauſe, to conſerue and kéepe his people and ſubiects, that no man ſhould do them hurt. Without this Citie be twelue ſuburbes very greate, and euery one of thẽ anſwereth to his gate of ẏ Citie. And in theſe be many Merchantes and men of occupations: and thyther do reſort all people that come out of the Countreys, and ſuch Lordes as haue to do with the King or his Courtes. And in theſe ſuburbes be moe than twentye thouſande ſingle or common

Cambalu. *This is a goodly Citie, and well ordered.*

At euery gate is a thouſand men that do watch.

No common woman may dwell within the Citie

Aboue a thousande Cartes with silke goeth euery daye out of this Citie

The greate Cane is garded nightly, with twentye thousande Horsemen.

The manner of the greate Cane at hys dinner with his wiues and children

women, and neuer a one of them maye dwell within the Citie on payne of burning. Out of this Citie goeth euery daye aboue a thousande Cartes with silke. The great CANE is garded euery night with twentie thousande Gentlemen on Horsebacke, not for any feare, but for dignitie. They be called *Chisitanos*, which is as much to say, as Knightes for the body, or trustie Knights. The manner of the great CANE for his dinner, is this: They make ready all the Tables rounde about the Hall, and in the middest of the Hall, is made ready the Table for the greate CANE, setting his backe towardes the North, and his face towardes the South. His firste wife sitteth next vnto him on hys lefte hande, and his other wiues following orderly. On his other side do sitte his sonnes, and his sonnes children, one after another, according to his age. Those that be of the imperiall lignage, do sitte downe afterward at another table more lower. And the other Lords and their wiues do sitte at other Tables more lower, according to their degrées, dignities, offices, estates, and age. At the saide Tables commonly do sitte foure thousand persons, or very néere, and euery one may sée the great CANE as he sitteth at his dinner In the middest of the Hall is a very greate vessell or cesterne of fine gold, that will holde tenne Hoggesheads, which is alwayes kept full of perfect good drinke. And néere vnto that vessell be other foure vessels of siluer bigger than that, full of good wine, with many other vessels and pottes by them, of gold, and of siluer, which may be of pottels a péece, or as muche as will serue foure men for a dinner. At dinner, out of the vessell of golde, wyth pottes of golde, they drawe wine for to serue the greate CANE his Table, for him, his wiues, children, and kindred: and out of the vesselles of siluer, with Iars and Pottes of siluer, they drawe wine to serue the Lordes and the Ladies, and all others sitting at the Tables, as well women as men. And euery one that sitteth at the tables hathe a cuppe of golde before hym to drinke in. And euery one that bringeth anye seruice to the greate CANES Table, hathe a towell of golde and silke before his mouth, bycause his breath shall not come vppon the meate and drinke they bring. When the great CANE will drinke, all the Musitians that bée in the Hall doe play, and euery one that serueth, knéeleth downe tyll hée haue drunke. In the Hall be alwayes Iesters, Iuglers, and fooles, attending vpon the Tables, to make pastime all dynner tyme, and after Dinner is done, and the Tables taken vppe, euerie man goeth aboute his businesse. All the TARTARES kéepe greate feasting and chéere euery yeare on the daye that CUBLAY CANE was borne, which was on the eight and twentith day of September, and that is the greatest feast they make in all the yeare saue one, that héereafter shall be spoken of The greate CANE doth apparell

Commonly foure thousand persons do sitte in that Hall at a dinner.

A vessell of fine gold that will holde tenne Hogsheads of Wine, and four of siluer bigger than that

Euery one that sitteth at the tables, hath a cuppe of gold before him

Euery one that bringeth meate or drinke to the Table, hath a towell of golde and silke before his mouth

Great feast is made euerie yeare, the day when the great Cane was borne.

himſelfe that day he was borne on in cloth of golde maruellous rich, and .12000. Barõs be apparelled with him after the ſame ſorte touching the cloth of gold, but not ſo rich and precioufe, and euery one of thẽ hath a great girdle of gold, and that apparell and girdles the great CANE giueth them. And there is neuer a one of thoſe garments with the girdle, but it is worth .10000. Biſancios of golde, whiche may be a thouſand Markes. By this you may perceyue, that he is of great power and riches. And on the ſayde day, all the TARTARES, and Merchantes, and ſubiects, and thoſe that dwell in his Countreys, be bounde to preſente vnto hym euery one ſomethyng, according to his degrée and abilitie, in knowledging him to be their Lorde. And whatſoeuer he be that doth begge any office or gift of him, muſt giue him a preſent, according to the gift he doth aſke. And all his Subiects and Merchantes, and trauellers, or anye other that be founde in his Countreys or Prouinces, be vſually bounde to pray for the greate CANE to hys Idols, to preſerue hym and hys Countreys, whether they be Tartares or Chriſtians, or Iewes, or Moores. The TARTARES begin their yeare the firſte day of February, and do kéepe a great feaſt that day. And the greate CANE and hys Barons, with all the reſt of the Citie, doe apparell themſelues in white that daye, making greate paſtymes, ſaying, the greate CANE is bleſſed and fortunate, and ſo doe deſire a ioyfull yeare. And on that daye there is preſented to the great CANE more than .10000. Horſes and Mares al white, and more than fiue thouſand Elephãts, with two greate baſkettes vpon them full of prouiſion neceſſarie for hys Courtes. And beſides thys, there is preſented to hym a great number of Camels, couered all with white cloth of ſilke, for ſeruice of their K. And when they giue theſe preſents, they doe all paſſe by, where the great CANE doeth ſtande and ſée them. On the ſame daye that this feaſt is, in the morning betimes, before the Tables be couered. all the kings, Dukes, Marqueſſes, Lords, Captaynes, Gouernours, and Iuſtices of his countryes, & other officers, come into the Hal before ẏ preſence of the great CANE, and thoſe that can not come in, be in another place, where as the great CANE may ſée them all: and thus being altogither as though they woulde make ſome requeſt, there goeth one vppe vpon a buylding or ſcaffolde that is made for the ſame purpoſe, in the middeſt of the hall, & with a loude or high voyce, biddeth them al knéele downe vpon their knées, and giue laudes and thankes to their Lord, and ſtreight wayes euery one doth honor him as if he were an Idoll: and this they doe foure times, and thys being done, euery one goeth and ſitteth downe in his place, and afterwardes do riſe one after an other, and goe to an aulter, whiche is ſet in the middeſt of the hall, and vpon it is a table ſet, written on with letters of

He giueth a rich Liuerie.

Euery Liuerie is worth a thouſande Markes.

The Tartares begin their yeare the firſt day of February.

Tenne thouſand white Horſes and Mares preſented to the great Cane.

Al his nobilitie do knele and worſhip the Cane *as if he were an Idol*

gold, and garniſhed with pretious ſtones of greate value, and the writing is the proper name of the greate CANE, and wyth Senſors of fine golde full of incenſe and fire, they incenſe that table in honour of the great CANE. And after that, euery one in preſence of the great CANE, doth offer great and precious giftes according to his ſtate, condition, and abilitie, and this being done, they go all and ſitte downe at the tables to dinner. And the great CANE thirtéene times in the yeare doeth giue apparell to his Barrons, in thirtéen great feaſtes he doeth make, and at euery time he doeth chaunge this apparel, and this apparel that he doth giue, is of greater and leſſer value, according to the degrée of him that he giueth it vnto. And to euerye one he giueth a girdle, or a payre of hoſen, or a hatte, garniſhed wyth golde, and ſet with pearles and pretious ſtones, according to the degrée of the parties: and of this apparell is euerye yeare .156000. and this he doth for to honour and magnifie his feaſtes. And at euery ſuch feaſt the gret CANE hath lying at his féete a tame Lyon, vpon a rich Carpet. And the great CANE is reſident, during the ſayde thrée moneths, in *Camballo*, that is to ſay, December, Ianuarie, and Februarie. And during the ſayd thrée months, the whole country thereabout, to ſay thirtie dayes iourney, is kept for hawking, hunting, and fouling, only for to ſerue the Courtes, and what they do take and kil, is preſented and broughte to the greate CANES Courte, and ſuch as dwell further of in other prouinces that kill wilde beaſtes, not able to bée brought to the Court, they do trimme and dreſſe the ſkins thereof, and bring them to the Courte for to dreſſe, make, and trimme armour and munitions, for the wars, which he hath infinite number.

A great and rich offering.

The great Cane *doth giue liueries .13. times in a yeare, and euery time he changes his colours.*

Four [Thrée] moneths he doth continue in Camballo.

No man may hunt no hauk nor foule within thirtie days iourney of his Citie.

Of the manner the great CANE doth vſe in his hunting

CHAPTER 56

[*Marsden*: Bk. II. Ch. XV. *Pauthier*: Bk. II. Ch. XCI. *Yule*: Bk. II. Ch. XIX. *Benedetto*: Ch. XCIII]

THis CUBLAY CANE, or great CANE, hath wyth him two noble men, that be his brethren, the one called BAIAN, and the other MYTIGAN, and they be called *Cinitil*, whych is as muche to ſay, as maiſters or gouernours of the dogs or Maſties of theyr Lordes, eyther of theſe two noble menne, hath tenne thouſande menne all apparelled in one liuerye of whyte and redde, and euerye one of theſe twentie thouſande menne

Two noble men be maiſters of his dogs, and they haue ten thouſand mē apeece.

hath charge and gouernemente of two Maſtyes, or at the leaſt one, and when the great CANE wyll go on hunting, theſe two noble men go wyth him with theyr twentye thouſande men, or with the moſte parte of them, and ſo beginne their hunting with thoſe men and dogges, who be well vſed to it, and the great CANE goeth into the middeſt of the fields, hauing his two Lordes with their men and dogges on eche ſide of him, and diuideth them into companies, in ſuch ſorte, that there ſhal no game riſe, that ſhall ſcape them, what kynde of beaſte ſo euer it bée.

Of the manner of his hauking for wildefoule

CHAPTER 57

[*Marsden*: Bk. II. Chs. XVI, XVII (in part). *Pauthier*: Bk. II. Ch. XCII and a few lines from Chs. XCIII and XCIV. *Yule*: Bk. II. Ch. XX and a few lines from Chs. XXI and XXII. *Benedetto*: Ch. XCIV and a few lines from Chs. XCV and XCVI]

THe firſt day of March, the great CANE departeth from *Cambalu* and goeth with his Court and Barons, towards the South ſeas, named the *Occean*, that lyeth two dayes iourney from *Cambalu*, and he carrieth with him ten thouſande Faulcons, fiue thouſand Gerſaulcons, and other kinde of Haukes a great number, which are very ſingular and good, aboue all other, and are bred in his Seniories, and al thoſe that they take in his countries are preſented to the great CANE, for his own vſe, Court, and Barrons, that alwayes kepe his companie, which are neuer leſſe than .15000. and they bée called *Tuſtores*, which is as much to ſay, as the Lords gard, & all theſe do practiſe hauking, and euery one of them doth carry his reclayme or lewer, and haukes hood, that when he hath néede he may take vp his Hauke. They doe neuer léeſe one of theſe Faulcons, for euery one of them hath faſtned vnto hys Belles a Scutchion of gold, wherin is written the name of hys Mayſter, and when ſoeuer one of thẽ is loſte, he that findeth him ſtreyghte wayes doeth preſent him vnto the great CANE, or to one of thoſe barrõs his brethren, and he cauſeth hym to be deliuered agayne, to him that before had charge of him, for he is knowen by the Scutchion that the Hauke hath vpon his belles.

The great Cane *hath with him ten thouſand Faulcons & fiue thouſand Gerſaulcons.*

They do neuer léeſe Faulcon nor Gerſaulcon.

Of the manner that the great CANE hath in trauelling in his countrey, and how he abydeth in the fields in his tents and pauilions

CHAPTER 58

[*Marsden*: Bk. II. Chs. XVI, XVII (in part) (cont.). *Pauthier*: Bk. II. Ch. XCII and a few lines from Chs. XCIII and XCIV (cont.). *Yule*: Bk. II. Ch. XX and a few lines from Chs. XXI and XXII (cont.). *Benedetto*: Ch. XCIV and a few lines from Chs. XCV and XCVI (cont.)]

A ſtraunge going a hauking.

WHen the greate CANE maketh any iourney in his countrey, he goeth in a fayre lodge or edification, hauing a verye faire chamber made vpon foure Elephants, which is couered with the ſkinnes of Lions, and in this chamber he hath twelue Gerfaulcons, and certain of the Barrons in his company to giue him pleaſure and paſtime: and round about theſe Elephants there be on horſebacke very many barrons, and as ſoone as they ſée anye foule, or Crane fly, they declare it vnto their Lord, and he immediately, letteth theſe Gerfaulcons flye: and after this ſort he goeth through his countrey: and when the greate CANE commeth to any broade and faire fields, which they do call *Caziamon*, which he doth finde ready ſet with tents and pauilions for him and his wiues, and for his children and barrons, and theſe tentes and pauilions, are at the leaſt .10000. and the tentes of the great CANE are ſo large, that when they are ſet vp, there may be vnder and walke at theyr eaſe .2000. knights, and the entring into them openeth towardes the South, and one of the tentes is for the Barrons and Knightes that are of the Lordes garde, and in a ſmaller tente that ſtandeth by it, opening towardes the Septentrion, edified wyth faire chambers, wrought all with golde, ordayned for ẏ great CANE where he kéepeth Courtes, and audience to all them that come: and in this tent there be two chambers with faire Halles, and the ſéelings is ſuſteined vppon thrée pillers of a maruellous worke, and are couered with Lions ſkinnes, and of other beaſts, wroughte and painted of diuers coloures, ſo that neyther wind nor raine can enter or paſſe through, for they are made onely for that purpoſe: and theſe chambers and halles, are furred with Ermines and Iebelines or Sables, whiche Sabels is ſo pretious, that one furre for a Knighte are or is worth .2000. Byſancios of gold. All the cordes of theſe tents are of ſilke, and theſe twoo tentes are of ſuche value, that a meane King thoughe he do ſell all his lande, is not able to buy them. And rounde aboute theſe two tentes ſtande manye other tentes being verye faire, for the Barons, and for the other people, ſo wel ſet and ordayned,

There be at the leaſt ten thouſand tẽts and pauilions ſet vp in the fielde.

Theſe two [te]nts bee of a good value.

that it séemeth to be a greate Citie: & from euery place there commeth people to sée the mightinesse & pleasure of the greate CANE. There goeth with the greate CANE all his Courte that he kéepeth in *Cambalu*, and in the place he remayneth hunting and hawking vntil al the moneth of Aprill, for there they finde greate plentie of wildefoule, for that there be great lakes and riuers. When the greate CANE goeth on hawking for wilde foule, there may no man hawk néere him, not within twentie dayes iorny, vppon a great penaltie. And from the beginning of March vntill October, there is no Baron nor subiecte vnto the great CANE, that dare take any wild beast or foule, though there be very greate plentie in that countrie, vppon great penaltie, and when the time of his hawking is ended, hée returneth vnto the Citie of *Cambalu*, hawking by the way, and néere vnto the Citie he doth kéepe solemne cheare .iij. dayes. Within the saide Citie they lodge no straungers, nor bury any dead corps. There commeth vnto this Citie merchandize from all parts of the world, cloth of gold and of silke, pretious stones and pearles, and great plentie of other notable thinges to maintaine the magnificence of the greate CANES Courte that he hathe, and for the greate resorte of people that come thither: and this Citie is scituated in the middest of his prouinces and countries.

Three dayes he doth make great cheare after his hunting is ended.

Of the money that is vsed in all that countrey

CHAPTER 59

[*Marsden*: Bk. II. Ch. XVIII. *Pauthier*: Bk. II. Ch. XCV (abridged). *Yule*: Bk. II. Ch. XXIV (abridged). *Benedetto*: Ch. XCVII (abridged)]

THe greate CANE causeth his money to be made in this manner, causing the rine of a Mulberry trée to be cut very thinne, whiche is betwéen the vtter rine and the trée, and of this he maketh mony both small and great, whiche some of them is worth halfe an ounce, some an ounce, some ten groats, some twentie, some thirtie, and some worth a Bisanco of golde, and some of twoo Bisancoes, and so they rise vntil tenne Bisancios of gold. This money is stãped with the signe of the Lord, & it is currant in al his Country, and in al the prouinces which are subiect vnto him, & no man may refuse this mony, for if he do he must léese his head, & he that doth counterfet hys coine shall be destroyed vnto the third generation. There commeth sometimes vnto the Courte of *Cambalu*, Merchants that bring golde and pretious stones for to buy the cloth of golde and silke, and other Merchaundizes in quantitie of thrée thousand

The money that is vsed in those countries.

He that doth counterfaite hys coyne shall be destroyed to the thirde generation.

Bifancios of golde, and many times the greate CANE commaundeth, that all the golde, filuer, and pretious ftones, that may be founde in the Merchauntes handes, and fubiectes of his dominions, fhoulde be deliuered to his treafurers, and fo they doe, and they be paid for it in this faide money, which is made of the rine of a Mulbery trée, that they may fée how al the gold, filuer, pearle & pretious ftones is clofed vp in his treafury being boughte for this vile money of no value, fo that little golde, filuer, pearles and pretious ftones commeth out of his country: and after this forte he maketh himfelfe the richeft Prince of the worlde.

Of the order and rules that he hath in his dominions

CHAPTER 60

[*Marsden*: Bk. II. Ch. XIX. *Pauthier*: Bk. II. Ch. XCVI. *Yule*: Bk. II. Ch. XXV. *Benedetto*: Ch. XCVIII]

The noble men that doe fet order for all the greate Canes *affaires.*

THe great CANE hathe fette tenne Barons or noble men of greate eftimation to gouerne .64. prouinces and countries fubiects vnto him, and they euer remaine in hys Citie imperial of *Cambalu*, and thefe tenne Barons doe appoynt Iudges, and Notaries ouer the Countries that are vnder their guiding, of the which euery one of them doth exercife his office in the country that he hath charge of, and thefe Iudges remaine alfo in the Citie of *Cambalu*, vnder the obedience of thofe Barons. Thefe tenne Barons do conftitute gouernours and officers throughe all the Countries, and doe chaunge them when they lifte, and when they haue putte them in the roome, they doe prefent them before the greate CANE, and hée doeth accepte them, and giueth them Tables of Golde, and by writing the order howe to vfe themfelues: and thefe gouernoures and officers doe gyue them knowledge by letters and meffengers vnto the Iudges which are deputies ouer them, and thofe Iudges doe notifie all things vnto thofe ten Barons, and they do make declaration of it vnto the great CANE, fo that after this manner, he knoweth what is done in hys Countries, and prouideth for all things neceffarie.

Of the ſaide order

CHAPTER 61

[*Marsden*: Bk. II. Ch. XIX (cont.). *Pauthier*: Bk. II. Ch. XCVI (cont.). *Yule*: Bk. II. Ch. XXV (cont.). *Benedetto*: Ch. XCVIII (cont.)]

THeſe ten Barons are called SENICH, which is to ſay, the principalles of the Court: and theſe doe prouide for the preſeruation of the great CANES eſtate, and they do ordain his warres and hoſtes, and Knightes, and they doe treate and make peace betwéene the Lordes, and they doe make prouiſion in euery manner of thing that toucheth their Lordes eſtate, and to all his dominions, but they lette nothing paſſe, vntill ſuche time as their Lorde do vnderſtande it.

Of the Citie *Cambalu*

CHAPTER 62

[*Marsden*: Bk. II. Chs. XX (first 4 lines), XXIII (in part). *Pauthier*: Bk. II. Chs. XCVII (first 4 lines), CI. *Yule*: Bk. II. Chs. XXVI (first 5 lines), XXX. *Benedetto*: Chs. XCIX (first 3 lines), CIII]

THe Citie of *Cambalu* hathe manye outlettes and gates, that thoroughe them they maye goe vnto diuers prouinces and countries, & when they goe from thence, for to goe vnto *Cataya*, they finde a great mountaine, where there is blacke ſtones, & they burne like wood, when they be well kindled they will kéep a fire from one day to an other, which I ſuppoſe be of the nature of oure Sea-coles, and they do burne of them in that Country, thoughe they haue woodde, but the woodde is more dearer than are the ſtones or ſeacoales.

Of the meruailous things that be founde in that countrey

CHAPTER 63

[*Marsden*: Bk. II. Ch. XXVII. *Pauthier*: Bk. II. Ch. CIV. *Yule*: Bk. II. Ch. XXXV. *Benedetto*: Ch. CVI]

Marcus Paulus *was made the Emperoures Embaſſador.*

THe great CANE ſent me MARCUS PAULUS as his Embaſſador towards the Occident or Weſtwarde, in the which meſſage I was fourtéene moneths, from the time that I went from *Cambalu*. And héere I will declare to you of the meruailous things that I ſaw with mine own eies, aſwel at my going outwards, as at my commyng homewardes, as that at my going frõ *Cambalu*, and taking my iourney towards the Occident or Weſtwarde. And after that I had gone tenne dayes iorney, I founde a very great riuer which is called *Poluiſanguis*, and runneth his courſe into the Occean ſea. Vppon this riuer there is a bridge, the fayreſt in the worlde, it hath thrée hundred paces of length, and eighte paces of breadth, ſo that there may goe tenne menne in a rancke on horſebacke. This Bridge hathe foure and twentie arches of Marble, very artificially wroughte, at the heade of this Bridge at the one ſide ſtandeth a Piller being verye greate of Marble, hauing a Lion ſtanding on the toppe, and an other Lion at the neather ende, being very liuely made, and a pace and a halfe diſtant, from that ſtandeth an other like vnto it, and ſo orderly ſtandeth one by another, til you come vnto the further ende of the bridge, ſo there is on eche ſide of the bridge two hundred pillers, and in the middes of euery piller, there is made Images of men very artificially.

The riuer Poluiſanguis.

A goodly Bridge and long.

Of the Citie named *Goygu*, and of many meruellous things

CHAPTER 64

[*Marsden*: Bk. II. Ch. XXVIII. *Pauthier*: Bk. II. Ch. CV. *Yule*: Bk. II. Ch. XXXVI. *Benedetto*: Ch. CVII]

FRom this Bridge you ſhall goe tenne miles throughe fields full of Vines, & very faire palaces: at ẏ ten miles end, there is a Citie named *Goygu*, it is very great & faire, in it there ſtãdeth a gret Abby of Idolatry. The people of this Country liue vppon merchaundize, and be artificers, for they do make great plentie of cloth of golde and ſilke. Alſo there is plentie of lodgings for thoſe that do trauaile, and come thither out of other places.

Here is plentie of cloth of Golde.

Of the way that goeth vnto the Countrey of the Magos

CHAPTER 65

[*Marsden*: Bk. II. Ch. XXVIII (cont.). *Pauthier*: Bk. II. Ch. CV (cont.). *Yule*: Bk. II. Ch. XXXVI (cont.). *Benedetto*: Ch. CVII (cont.)]

Oyng from this Citie almoſte a myle, there parteth twoo wayes, the one goeth vnto the Occident or Weaſt, and the other goeth towardes the *Siroco*. The waye whiche goeth vnto the Occident or Weaſte, leadeth vnto the Occean Sea towards the high

Countrey of the MAGOS, and you may trauaile throughe the prouince of *Cataya* tenne dayes iourney, in the whiche waye there is many Cities and Townes.

Of the Citie named *Taraſu*

CHAPTER 66

[*Marsden*: Bk. II. Chs. XXIX, XXX. *Pauthier*: Bk. II. Ch. CVI. *Yule*: Bk. II. Ch. XXXVII. *Benedetto*: Ch. CVIII]

AFter you do goe from the Citie of *Goygu* trauailing ten dayes iourney, you come vnto a Citie named *Taraſu,* whiche is the heade Citie of that countrie or prouince, where there is plentie of vines & muche wine, and there they doe make all kinde of armoure for the greate CANES Court. In the Countrie of *Cataya*, there is no wine, for they prouide themſelues of wine out of this region.

Here is much armor made.

Of the Citie named *Paimphu*

CHAPTER 67

[*Marsden*: Bk. II. Chs. XXIX, XXX (cont.). *Pauthier*: Bk. II. Ch. CVI (cont.). *Yule*: Bk. II. Ch. XXXVII (cont.). *Benedetto*: Ch. CVIII (cont.)]

TRaueling from thẽce towards the Occident or Weaſt eighte dayes iourney throughe fayre Cities and Townes, wherein they doe traffike Merchandizes, at the eyght dayes iorney you ſhal come vnto a very gret and fayre Citie whiche is named *Paymphu*, and going twoo dayes iorney beyonde it, you ſhall come vnto a fayre Towne named *Caychin,* whiche was made by their King.

Of a King named BUR

CHAPTER 68

[*Marsden*: Bk. II. Ch. XXXI. *Pauthier*: Bk. II. Chs. CVII, CVIII (abridged). *Yule*: Bk. II. Chs. XXXVIII, XXXIX (abridged). *Benedetto*: Chs. CIX, CX (abridged)]

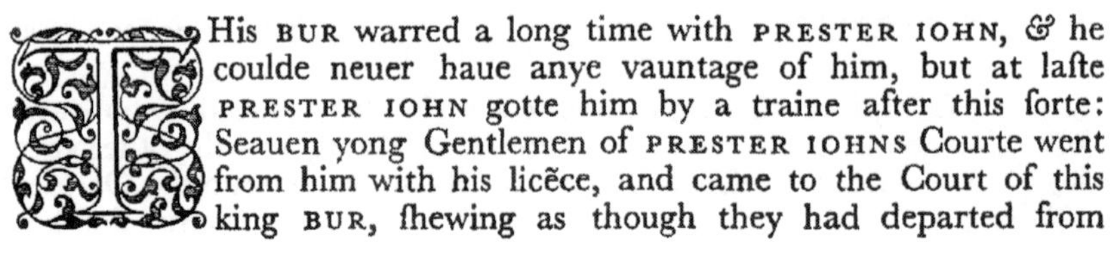

THis BUR warred a long time with PRESTER IOHN, & he coulde neuer haue anye vauntage of him, but at laſte PRESTER IOHN gotte him by a traine after this ſorte: Seauen yong Gentlemen of PRESTER IOHNS Courte went from him with his licẽce, and came to the Court of this king BUR, ſhewing as though they had departed from

PRESTER IOHN in great diſpleaſure, & ſo offered themſelues to ſerue the ſaid King BUR, who retayned them as ſquires and pages in his Courte, and after they had bin with hym two yeares, hauing greate confidence and truſte in them, thys King BUR on a tyme roade abroade for his pleaſure, and taking with him the ſaide ſeauen Gentlemen, and being the diſtaunce of a myle from his Caſtell, perceyuyng they had him now at aduantage to execute their purpoſe, tooke him, and carryed him to PRESTER IOHN, and PRESTER IOHN made him his ſhéepehearde, and kept his ſhéepe two yeares, and afterwardes gaue him horſes and menne, and ſent him to his Caſtell as his ſhéepehearde.

A King was made a ſheep-hearde by Preſter Iohn.

Of the Citie named *Caſiomphur*

CHAPTER 69

[*Marsden*: Bk. II. Ch. XXXII. *Pauthier*: Bk. II. Ch. CIX (abridged). *Yule*: Bk. II. Ch. XL (abridged). *Benedetto*: Ch. CXI (abridged)]

BEyond this caſtel twentie miles towardes the Occident, there ſtandeth a great Citie named *Caſiomphur*, and the people of it worſhip Idolles. The like doe all thoſe of the Countrey of *Cataya*. In this Citie there is made muche cloth of golde and of ſilke.

Cloth of gold and cloth of ſilke made.

Of the Citie named *Bengomphu*, and of many things that there is found in thoſe parties

CHAPTER 70

[*Marsden*: Bk. II. Ch. XXXIV. *Pauthier*: Bk. II. Ch. CX (abridged). *Yule*: Bk. II. Ch. XLI (abridged). *Benedetto*: Ch. CXII (abridged)]

GOing from *Caſiomphur* eight dayes iourney towards the Occident, you ſhal goe alwayes by greate Cities and faire Townes, and excellente places, with goodlie and faire Gardens, with principal houſes: there is great plentie of wilde beaſts and foules, for hunting and hauking, and at the ende of theſe eight dayes iourney, there ſtandeth a faire Citie whiche is called *Bengomphu*, and is the head Citie of that realme. There is in this Citie as king, one of the great CANES ſonnes, who is called MAGALA. The people of this Realme are Idolatours. This Citie hath plentie of all things, and without this Citie ſtandeth the pallace royall of the king, the which with the Wal of the Citie is tenne myle compaſſe. In this Citie there is a lake made of many fountaines, that runneth and ſerueth the Citie. The Walles of this Citie haue very faire battlementes, and on the inſide of the Wall of that Pallace it is layde on with gold, like playſter, and without this Pallace, round about that lake, there is very faire and delectable ground and fields.

The inſide of the pallace wall is layde on with gold.

Of the prouince named *Chinchy*

CHAPTER 71

[*Marsden*: Bk. II. Ch. xxxv. *Pauthier*: Bk. II. Ch. cxI. *Yule*: Bk. II. Ch. XLII. *Benedetto*: Ch. cxIII]

GOing from thys pallace towards the Occident thrée dayes iourney, you come vnto a playne full of faire Cities and townes, and at this thrée dayes iourneys ende, there bée greate mountaines and valleis belonging to the prouince of *Chinchy*, in theſe mountaines and valleys there be many Cities and townes, and all the people there are Idolaters, huſbandmen, and hunters. This iorney endureth twentie dayes, there be in it manye Lions, and plentie of other wilde beaſtes, and in all theſe twentie dayes iourney there is plentie of lodging for thoſe that doe trauell.

Of the Countrey and Citie called *Cineleth Mangi*, and many other things which be founde there

Mangi *a citie.*

CHAPTER 72

[*Marsden*: Bk. II. Ch. xxxv (cont.). *Pauthier*: Bk. II. Ch. cxII. *Yule*: Bk. II. Ch. XLIII. *Benedetto*: Ch. cxIv]

AT the end of twentie dayes iourney ſtandeth a Citie named *Cyneleth*, a noble and a greate Citie, and vnder the obedience of this Citie there be many Cities & townes toward the Occident. The people of thys Countrey are Idolatours, they haue great trade of Merchandiſe. In this countrey there is plentie of Ginger, and from thence the Merchaunts do carrie it vnto *Cataya*. Alſo there is aboundance of wheate and other graine. Thys countrey is called *Cyneleth Mangi*, and it hath two dayes iourney of plaine countrey. Beyond this countrey, there be great playnes and valleys & mountaines, being greatly inhabited, with Cities and townes, for the ſpace of twentie days iourney, where there be many Lions and beares, beſides other wilde beaſtes. Alſo there is greate plentie of Muſkcats, and other noble and faire beaſtes.

Great trade of Merchandiſe.

Mangi.

Here be many Muſke cattes.

Of the countrey and Citie named *Cindarifa*, and of a maruellous bridge

CHAPTER 73

[*Marsden*: Bk. II. Ch. XXXVI (in part). *Pauthier*: Bk. II. Ch. CXIII (in part). *Yule*: Bk. II. Ch. XLIV (in part). *Benedetto*: Ch. CXV (in part)]

AFter you haue gone these twentye dayes iourney, you come vnto a great plain, being of the countrey named *Cindarifa*, whiche is twenty miles compasse, and the great CANE before he died, diuided it into thrée partes, & al thrée parts be strongly walled rounde about. Through the middest of this countrey runneth a great riuer, which is called *Champhu*, half a mile brode. There is in this riuer plentie of fish, and there is scituated vpon this riuer many Cities and townes: also by shipping vpon this riuer they sayle from Citie to Citie, with all kind of Merchaundises. From the beginning and heade of this riuer, vntill the entring into the maine sea, there is thirtie days iourney, and the chiefe Citie of this countrey is named *Sindarifa*. From this citie ouer the riuer, there is a bridge of a mile long, and eight paces brode, made of marble stone, and couered with timber of Pineaple trée, verye fayre. On the sides of this bridge, there be houses and shops for Merchauntes, and of diuerse occupations, and at the foote of this bridge there standeth a custome house, verye faire made, where they do gather their Lords customes, and euery daye they receiue tenne thousande Bisancios of God. The people of this countrey are Idolatours.

A bridge of a myle long, and eight paces brode of marble and housen on it.

Of the prouince named *Cheleth*

CHAPTER 74

[*Marsden*: Bk. II. Chs. XXXVI (last 7 lines), XXXVII (in part). *Pauthier*: Bk. II. Chs. CXIII (last 7 lines), CXIV. *Yule*: Bk. II. Chs. XLIV (last 5 lines), XLV. *Benedetto*: Chs. CXV (last 5 lines), CXVI]

GOing from this countrey, you shal trauell through a faire plaine country, ful of many townes and Cities, it indureth fiue dayes iourney, and then you shal come vnto a prouince, whiche is called *Cheleth*, which was destroyed by the great CANE. In this prouince there bée Canes which are called *Berganegas* of fiftéene paces long, and tenne spannes in compasse euerye one of them, and they haue from the one knot to the other thrée spans. The trauellours make fire with these Canes, for they haue this propertie, that as soone as they féele the heate of the fire, they giue such a great cracke, that

Here be Canes of fifteen paces long and ten spans about.

the ſound is harde many miles off, and the Lyons and wilde beaſtes that are thereabouts, be ſo fearefull of that noyſe, that they do run away, and do no hurt vnto thoſe that trauell, and the horſes that the trauellours doe ride on, haue ſo much feare of that noyſe, being not vſed vnto it, that they breake theyr brydles and haulters, and runne away, ſo that ſometimes they cannot finde them againe, therefore thoſe that trauell, doe tye their horſes and Aſſes in certaine holes or Caues that they finde in the Mountaines. This countrey is twentie dayes iourney long, where they finde nothing to eate, nor yet to drinke, nor no habitation, therfore thoſe that trauell that way do carrie prouiſion for thoſe twenty daies iourney, whiche they do paſſe with great feare and trauell.

Of the Prouince named *Thebet*, and of the maruellous beaſtlineſſe and filthie liuing of the people there

CHAPTER 75

[*Marsden*: Bk. II. Chs. XXXVI (last 7 lines), XXXVII (in part). *Pauthier*: Bk. II. Chs. CXIII (last 7 lines), CXIV. *Yule*: Bk. II. Chs. XLIV (last 5 lines), XLV. *Benedetto*: Chs. CXV (last 5 lines), CXVI (All continued)]

AT theſe twenty dayes iourneys end, you come vnto a Prouince or Countrey, that is full of Cities and Townes. And the cuſtome in this Countrey is, that none dothe marrie with maydes nor virgins, but that firſt ſhe muſt be knowen carnally of many men, and ſpecially of ſtrangers. And for this occaſion, when the mothers meane to marrie anye of their damſels, the mother dothe carrie them néere the high way ſide, and with mirth and chéere procureth thoſe that do trauell, to ſléepe with hir, and ſometimes there lyeth with hir ten, and with ſome other twenty. And when the ſtranger or traueller goeth his wayes from any ſuche Damſell, hée muſt leaue vnto hir ſome iewell, the whiche iewell, the ſaide damſels or wenches do hang at their neckes, in token and ſigne that they haue loſt their virginitie wyth ſtrangers. And ſhe that hathe vſed hir ſelfe with moſte ſtrangers, it ſhall be knowen by the moſt quantitie of iewels that ſhe weareth aboute hir necke, and ſhe moſt ſooneſt ſhall finde a mariage, and ſhall be moſt prayſed and loued of hir huſband. And thoſe of this prouince are Idolaters, euill men, cruell, and robbers. In this Countrey there be manye wilde beaſtes, and ſpecially of Muſkettes. All thoſe of this Countrey doe weare Canuas, and Cowhydes, and the ſkinnes of wilde beaſtes, whych they do take in hunting. This Countrey is named *Thebethe*, and is adioyning vnto the Prouince of *Maugy*.

No maydens may marrie in this Countrey.

For lacke of wollen cloth, they do wear Canuas, and wilde beaſtes Skynnes.

Maugy.

Of the Prouince and Countrey named *Maugi*

CHAPTER 76

[*Marsden*: Bk. II. Ch. XXXVII (in part). *Pauthier*: Bk. II. Ch. CXV (abridged). *Yule*: Bk. II. Ch. XLVI (abridged). *Benedetto*: Ch. CXVII (abridged)]

Maugy. *Heere is found plenty of golde. Their money is made of Corrall. Here is cloth of gold, cloth of silke, and Chamlets made. Heere groweth spices. Masties as bigge as Asses.*

M*Augi* is a great prouince and Countrey, and it hathe vnder it eyghte Kingdomes and Riuers, and in the same there is found much gold of *Payulsa*. And they doe vse money made of Currall, and the Currall is there very déere, for that the women do vse to weare it about their neckes, and doe decke their Idols with it. In this Countrey they doe worke cloth of gold and silke, and of Chamlet great plenty. Also, there groweth much spice. Also, there be manye Negromancers, Astronomers, Inchanters, and euill dispofed men. Also, there be in this Countrey Masties as bigge as Asses, and the people be subjects to the great CANE.

Of the Prouince and Countrey named *Candon*, and of the iewels that grow there, and of the beastly conditions of the people

CHAPTER 77

[*Marsden*: Bk. II. Ch. XXXVIII. *Pauthier*: Bk. II. Ch. CXVI. *Yule*: Bk. II. Ch. XLVII. *Benedetto*: Ch. CXVIII]

Plenty of Pearles and precious stones.

C*Andew* is a Countrey that lyeth towards the Occident, and it hathe vnder it seauen Kingdomes of Idolaters, subiectes vnder the greate CANE. In this Countrey there be many Cities, Townes, and Villages. And in one place of this Countrey, there is greate plenty of Pearles and precious stones, but the great CANE dothe not suffer them to be had out. And in the Mountaynes in this Countrey there be foũd many Turquesses, and they may not be had out of the Countrey, without expresse licence of the greate CANE. Also, the custome of the people in this Countrey is, that as soone as there commeth a stranger to lodge in his house, the good man goeth out, commaunding his wife, children, and seruantes to obey that Stranger, as his owne proper person, and hée neuer commeth home vnto his owne house, vntill he know that the Stranger is gone from his house, and he knoweth it by a signe and a token that the Stranger dothe leaue at his going at the dore. And when the good man spyeth the signe or token, he entreth into hys house. This vse they doe

Heere they haue an ill custome.

kéepe thorough all that Countrey, and take it for no ſhame, although the Strangers do vſe their wiues. But rather they doe take it in greate honor and eſtimation, that they do ſo well enterteyne the Strangers. And theyr Idols tell them, for that they doe honoure the Strangers, their Gods do encreaſe their ſubſtance. The people of this Coũtrey do vſe money made of gold, that euery péece is worth .7. Duckets. In this prouince and Countrey there is great plenty of all kinds of ſpice and muſke, and great plentye of fiſhe, by reaſon of the greate lakes and pooles that be there.

Heere is money of golde.
Here is great plenty of Spices.

Of another Prouince, where there is found gold and other things

CHAPTER 78

[*Marsden*: Bk. II. Ch. XXXVIII (cont.). *Pauthier*: Bk. II. Ch. CXVI (cont.). *Yule*: Bk. II. Ch. XLVII (cont.). *Benedetto*: Ch. CXVIII (cont.)]

GOing out of the foreſaid prouince, and trauelling tenne dayes iourney through a Countrey full of Cities and Townes, and verye much people, ſeming much in their vſe and cuſtome, vnto thoſe of the laſt rehearſed Countrey. And at the tenne dayes iourneys end, you come vnto a greate Riuer, whiche is named *Brus*, at the which endeth the Countrey and prouince named *Candèw*. In this Riuer there is founde great plentie of gold. And faſt by this riuer groweth very much Ginger. And thys Riuer falleth into the Occean Sea.

Heere is found greate plentye of gold.
A Riuer into the Seas.

Of the Prouince named *Caraya*

CHAPTER 79

[*Marsden*: Bk. II. Ch. XXXIX. *Pauthier*: Bk. II. Ch. CXVII. *Yule*: Bk. II. Ch. XLVIII. *Benedetto*: Ch. CXIX]

BEyonde this Riuer you come vnto a Prouince named *Caraia* towards the occident. In this Countrey there be ſeauen Kingdomes, ſubiectes vnder the greate CANE. Héere raigneth one of the greate CANES ſonnes, named ESENTEMUR, being rich, wiſe, and a valiant man, and gouerneth his ſubiects with great prudence and iuſtice. Theſe people be Idolaters. And after that you haue paſſed the ſaide Riuer, and trauelling fiue dayes iourney, there be many Cities and Townes, and there is brought vp and bredde great plentie of Horſes.

Heere be many Horſes bredde.

Of the Prouince named *Ioci*, and of their beaſtly cuſtomes

CHAPTER 80

[*Marsden*: Bk. II. Ch. XXXIX (cont.). *Pauthier*: Bk. II. Ch. CXVII (cont.). *Yule*: Bk. II. Ch. XLVIII (cont.). *Benedetto*: Ch. CXIX (cont.)]

AT fiue dayes iourneys end, you come vnto a Citie which is named *Ioci*, and is verye great and full of people Idolaters, ſauing that there be ſome Chriſtian people Heretikes Neſtorians. They do vſe for their money fine ſhelles white, whiche are founde in the Sea, and foureſcore of them are worth a Sazo of gold, whyche is worth two grotes of golde. And eyght Sazos of ſiluer, which is an ounce, and is worth a Sazo of golde. There they do make Sault of the water of Welles great plẽty. And in this Countrey no man careth though another man haue to do with his wife. There is a Lake in thys Prouince, hauing in compaſſe a hundred miles. Therein is plentie of excellent good fiſh. The people of this Countrey do eate rawe fleſhe after this manner. They cut it in ſmall péeces, and ſauce it with Garlike and ſpices, which giueth them a good taſt vnto the fleſh.

A Sazo of gold is worth eyght of ſiluer, which is an ounce.

Of the Prouince named *Chariar*, and of the ſtrange Serpents that be there

CHAPTER 81

[*Marsden*: Bk. II. Ch. XL. *Pauthier*: Bk. II. Ch. CXVIII (abridged). *Yule*: Bk. II. Ch. XLIX (abridged). *Benedetto*: Ch. CXX (abridged)]

GOing from this Prouince *Ioci*, and trauelling tenne dayes iourney, you come vnto another Prouince named *Chariar*, ſubiect vnto the greate CANE, and it is full of people of Idolaters, and one of ẏ great CANES ſonnes named CHOCAYO, ruleth and gouerneth them. And in this Countrey there is found great plenty of gold. And a Sazo of gold goeth there for ſixe of ſiluer. And they doe vſe in this Countrey little white ſhelles of the Sea, in ſtead of money, which is broughte from *India*. In this Prouince there be certayne Serpents of tenne paces in length, and their gaule is ſolde very déere, for they do vſe it in manye medicines: for if a man ſhoulde be bitte with a madde Dogge, laying vppon the ſore ſo muche quantitie of that gaule as will lye vpon a farthing, it healeth it immediately. Alſo, it eaſeth a woman of hir pangs, that is in trauell.

Heere is great plenty of golde.

The men of this Countrey are peruerſe people, and cruell, for if they do ſée anye trauellers that are prudente and faire, they do marke where the night doth take them, and thither they come and kill them, ſaying, that the faireneſſe and prudence of the dead, doth paſſe vnto them, and therefore they do kill them, and not for to rob them. This peruerſe cuſtome was among them before they became vnder the great CANE. But .95. yeares hitherto that they were vnder the greate CANE, they dare not doe anye ſuch thing, and therefore become a greate deale better people, and of a better diſpoſition.

Of the Prouince named *Cingui*, and of many things that be there, and of the Citie named *Caucaſu*

CHAPTER 82

[*Marsden*: Bk. II. Ch. XLI. *Pauthier*: Bk. II. Ch. CXIX. *Yule*: Bk. II. Ch. L. *Benedetto*: Ch. CXXI]

AFter that a man departeth from *Chariar*, he goeth fiue dayes iourney towards the Occident, and commeth to another Prouince named *Noctèam*, and alſo the Citie named *Nociam*, whiche is the head of thys Prouince, and it is vnder the great CANE. All the men of this Prouince haue their téeth couered with golde. And the women do dreſſe their Horſes. The men doe no other thing, but goe on Hunting, paſſing the time in the fields, and goe vnto the warre. The women doe buy and ſell, and do all things neceſſarie belonging to the houſe, and gouerne all the goodes, and their men and women Seruantes. Ouer and aboue this, the women of this Countrey haue this cuſtome, that as ſhe is deliuered of childe, ſhe riſeth and wrappeth the childe, and dothe all things belonging to the houſe, and receyueth no more payne, than though ſhe had not bin deliuered of childe, but in giuing the childe ſucke, and as ſoone as ſhe is deliuered, the huſband lyeth in the bedde, laying the childe by hym, as though he had borne it himſelfe, for the ſpace of fortye dayes, and the woman dothe ſerue him. He is viſited of the kinſmen and friends & neyghbours, as though he had bin deliuered himſelfe, making great feaſtes for the ſpace of thirtie dayes. In this Countrey they doe giue a Sazo of golde, which is an ounce, for fyue Sazos of ſiluer, being fyue ounces. Alſo, they doe vſe Perſiuolas, béeyng little ſhelles of the Sea, whiche come from *India*, in ſtead of móney. Theſe people haue no Idols, but euery houſeholde worſhippeth theyr Superiour and Mayſter. None

The men of this countrey haue theyr teeth couered with gold.

Heere is a cuſtome, that the good man is much made of, after hys wife is broughte a bed.

A Sazo of gold is an ounce, and is worth fyue of Siluer.

of them can write nor reade, for that they dwell among the moyſt Mountaynes, corrupted with euill ayres. In thys Prouince, and in the other two afore ſpecifyed, there be no Phiſitions, but when they doe fall ſicke, they cauſe to come vnto their houſes certayne Miniſters, which vſe inchantmentes by the power of the Diuell, and declare the ſickneſſe that the diſeaſed hathe, and theſe Miniſters ſounde their inſtrumentes in honor of theyr Idols, in ſo muche that the Deuill entereth into one of thoſe Miniſters, Inchanters, or Idols, and falleth downe as though hée were dead, and thoſe Miniſters, or Mayſters of the Idols, demaunde of hym that lyeth inchanted, or in a trance, wherefore that man fell ſicke, and hée aunſwereth, for that he hathe angered ſuche or ſuche an Idoll, and then thoſe Mayſters or Miniſters of the Idols ſaye vnto him that is inchanted, we requeſt thée to pray vnto that Idoll that is angrie wyth the ſicke bodye, to pardon hym, and wyll make hym Sacrifice with hys owne bloud. And if hée that is in thys trance, doe beléeue that the diſeaſe is mortall, hée aunſwereth, thys ſicke man hathe ſo diſpleaſed the Idoll, that I knowe not whether he will pardon hym or not, for that hée hathe determined that hée ſhoulde dye, and if he thynketh that hée ſhall eſcape hée ſayeth, if hée wyll lyue, it behoueth hym to gyue vnto the Idoll ſo manye Shéepe that haue blacke neckes, and to dreſſe ſo many ſortes of meates dreſſed with ſpices, ſufficient to make the ſacrifices vnto the Idoll that is angry with him, and for the miniſters that ſerue him, and for the women that ſerue in his temple, whiche is all fraude and guile of the inchanters for to gette victuals, by this meanes all are damned vnto Hell. To this banket there is conuited the maiſters and miniſters of the Idols, the inchanters and women that ſerue in ẏ temple of that Idoll. And before they ſitte downe to the Table, they doe ſprincle the broath aboute the houſe, ſinging and daunſing in the honor of that Idoll. And they doe aſke the Idoll, if he haue forgiuen the ſicke man. And ſometimes the Féende aunſwereth, that there lacketh ſuch or ſuche a thing, whiche immediately they do prouide: and when he anſwereth that he is pardoned, then they do ſitte downe to eate and to drinke that ſacrifice which is dreſt with ſpices, and this done, they go vnto his houſe with great ioy. If the paciente heale, it is good for him, but if he dye, it is an euerlaſting payne for him, and if he recouer, they do beléeue that the diueliſhe Idol hath healed him, and if he die, they ſay that the cauſe of his deathe was for the greate offence that he had done vnto him, and ſo they be loſt as brute beaſts in all that Countrey.

A ſtrange kind of Phyſicke

Of another Prouince named *Machay* where there be Vnicornes, Elephants, and wilde Beaftes, with many other ftrange things

Vnicornes.

CHAPTER 83

[*Marsden*: Bk. II. Ch. XLIII and a few lines in the middle of Ch. XLIX. *Pauthier*: Bk. II. Ch. CXXIII and a few lines in the middle of Ch. CXXIX. *Yule*: Bk. II. Ch. LIII and a few lines in the middle of Ch. LIX. *Benedetto*: Ch. CXXV and a few lines in the middle of Ch. CXXXI]

GOing from the Prouince of *Charian*, you go downe a greate penet or hill, whiche endureth two dayes iourney, without any habitation, fauing one towne, where they doe kéepe holyday thrée dayes in the wéeke. There they doe take a Sazo of golde for fyue of filuer. And paft thefe two dayes iourney, you doe come vnto the prouince named *Machay* whyche lyeth towardes the midde daye or South, adioyning vnto the *Indias*, and through this prouince you trauell fiftéene dayes iourney, through deferte mountaines, where there be many Elephants, and other wilde beaftes, for that the countrey is not inhabited. Alfo there is found Vnicornes. When they wil take any Elephant, they do compaffe him with dogges, and fo they do hunt him, that they make him wearie, and fo he is faine to reft for wearineffe, and his refting is, leaning vnto a great trée, for that he hath no ioyntes in hys knées, fo that he can not lye downe nor rife vp. The Mafties dare not come neare him, but barke at him aloofe, & the Elephante hath neuer his eye off thofe Mafties, and then thofe that be expert and hunt him, hurle Dartes, and fo kil him. In this countrey is much gold and filke.

Of a prouince named *Cinguy*, and of the Citie named *Cancafu*

CHAPTER 84

[*Marsden*: Bk. II. Ch. XLIX (in part). *Pauthier*: Bk. II. Chs. CXXIX (last 6 lines), CXXX. *Yule*: Bk. II. Chs. LIX (last 7 lines), LX (in part). *Benedetto*: Chs. CXXXI (last 5 lines), CXXXII]

BEyond this prouince *Machay*, there is another prouince named *Cinguy*, and trauelling foure dayes iourney in it, you paffe manye Cities and townes, and at thefe four daies iournyes ende, ftandeth a greate Citie named *Cancafu*, being verye noble, fituated towards the

reat plenty of loth of Gold nd Silke.

mydday or South, and this is of the ſtreight of *Cataya.* In thys Citie there is wroughte cloth of Golde, and ſilke greate plentie.

Of the Citie named *Cianglu*

CHAPTER 85

[*Marsden*: Bk. II. Ch. L. *Pauthier*: Bk. II. Ch. CXXXI. *Yule*: Bk. II. Ch. LX (in part). *Benedetto*: Ch. CXXXIII]

FRom this Citie trauelling fiue dayes iourney, you come vnto another Citie named *Cianglu*, which is very noble and great, ſituated towards the midday, or ſouth, and it is of the ſtreight of *Cataya*, here is made greate plentie of ſalte: and there runneth through this countrey a very great riuer, that vp and down this riuer there trauell many ſhips with merchaundiſe.

Of the Citie named *Candrafra*, and of the Citie named *Singuymata*

CHAPTER 86

[*Marsden*: Bk. II. Chs. LI, LII, LIII (abridged). *Pauthier*: Bk. II. Chs. CXXXII, CXXXIII, CXXXIV (much abridged). *Yule*: Bk. II. Chs. LXI, LXII (much abridged). *Benedetto*: Chs. CXXXIV, CXXXV, CXXXVI (much abridged)]

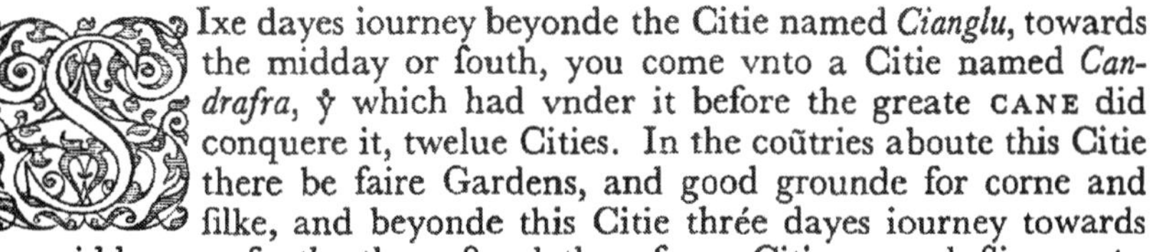

SIxe dayes iourney beyonde the Citie named *Cianglu*, towards the midday or ſouth, you come vnto a Citie named *Candrafra*, ẏ which had vnder it before the greate CANE did conquere it, twelue Cities. In the coũtries aboute this Citie there be faire Gardens, and good grounde for corne and ſilke, and beyonde this Citie thrée dayes iourney towards the midday, or ſouth, there ſtandeth a fayre Citie named *Singuymata*, which hath a great riuer that the Citizens made in two parts, the one way runneth towards the eaſt, and the other towardes the Occident, or Weaſt through *Cataya*, and vppon this riuer there ſayle ſhippes with Merchaundiſes in number incredible.

Of the Riuer *Coromoran*, and of the Citie *Choygamum*, and of another Citie named *Cayni*

CHAPTER 87

[*Marsden*: Bk. II. Ch. LIV. *Pauthier*: Bk. II. Ch. CXXXVII (in part). *Yule*: Bk. II. Ch. LXIV (in part). *Benedetto*: Ch. CXXXIX (in part)]

GOing from *Singuymata* ſeuentéen dayes iourney towards the midday or ſouth, you paſſe throughe manye Cities and townes, in the whiche there is greate traffique of Merchaundiſe. The people of this countrey are ſubiectes vnder the greate CANE. Their language is *Perſian*, and they do honour Idols. At the ſeauentéen dayes iourneys ende, there is a greate riuer that commeth from the Countrey of PRESTER IOHN, which is named *Coromoran*, hauing a myle in bredth, and it is ſo déepe, that there may ſayle any great veſſel laden with Merchandiſe. Vpon this riuer the great CANE hath fiftéene great ſhips for to paſſe his people vnto his Idols, that are in the Occean ſeas, euery ſhippe of theſe hath fiftéene horſes, and fiftéene mariners, and al victuals neceſſarie. Vpon this riuer there ſtãdeth two Cities, one on the one ſide, and the other one the other. The biggeſt of them is named *Choyganguy*, and the other *Caycu* and they be both a dayes iourney from the ſea.

Of the noble prouince named *Mangi*, and of many maruellous things that were there, and how it was brought vnder the great CANES gouernaunce

CHAPTER 80 [88]

[*Marsden*: Bk. II. Ch. LV (abridged). *Pauthier*: Bk. II. Ch. CXXXVIII (abridged). *Yule*: Bk. II. Ch. LXV (abridged). *Benedetto*: Ch. CXL (abridged)]

Mangi.

PAſſing the ſaide riuer, you enter into ẏ prouince of *Mangi*, where raigneth a king named FUCUSUR, of more power and riches than any King in ẏ worlde ſauing the great CANE. In this realme there be no men of warre, nor horſes for the wars, for it is ſituated ſtrongly, in a place compaſſed rounde about with many waters. And rounde about his Cities and townes, there be verye déepe ditches and caues, being brode and full of water. The people of this countrey are giuen to féebleneſſe, they do liue delicately: if they were giuen to warres, and feats of armes, all the worlde could not conquere the prouince of *Mangi*. This king of *Mangi* was very leacherous, but hée had in himſelfe two good properties, the one was, that he maintayned his realme in great iuſtice and peace, that euery one remayned in his place, and both day and nighte you myght traffique and trauell ſurely: the other propertie was, that he was verye pitifull, and did greate almes vnto the poore, and euerie yeare he brought vppe twentye poore ſtriplings, and he gaue them as ſonnes and heires vnto his Barrons and knightes. In his Courte he hadde alwayes tenne thouſande Squires that ſerued hym. It fortuned that in the yeare of our Lord .1267. CUBLAY CANE got perforce the countrey of *Mangi*, and the ſayde king of this prouince fledde with .1000. ſhippes vnto his Ilandes that were in the Occean Seas. He lefte the principall Citie of his prouince *Mangi* named *Gaiſſay* vnder the guiding of his Quéene, and when ſhe knew that there was entred into hir land BAYLAYNCON CAN a Tartarous name, which is as much to ſay in Engliſhe, as a hundreth eyes, a Captaine belonging to the greate CANE with a greate hoſte, and ſo without any reſiſtance, ſhe ſubmitted hir ſelfe with all hir country, and al the cities ſauing one named *Sinphu*, whiche kepte it ſelfe thrée yeares before it yéelded. Thys Quéene was carryed vnto the greate CANES Courte and kepte like a Quéene, and the King FUCUSUR came not out of thoſe Ilandes vntill he died, being out of his ſeigniorie.

Mangi.

Mangi.

Of the Citie named *Coygangui*, and many other thyngs

CHAPTER 91 [89]

[*Marsden*: Bk. II. Ch. LVI. *Pauthier*: Bk. II. Ch. CXXXIX. *Yule*: Bk. II. Ch. LXVI. *Benedetto*: Ch. CXLI]

HEre I will tell you of the faſhion and condition of this ſaide prouince *Mangi*. The firſt Citie at the entring is named *Coygangui*, whiche is a greate and a noble Citie ſcituated towards the wind *Syroco* or Eaſt ſoutheaſt. The people of this Citie doe worſhip the Idolles, and haue the Perſian tongue. They haue many ſhippes, and burne their dead bodies. This citie ſtandeth vppon the riuer *Coromoran*. In this Citie they make ſo muche ſalte as woulde ſuffice for fortie great cities, and of the abundaunce of thys ſalte, there groweth greate profites vnto the greate CANE.

Of the noble Citie named *Panguy*, and of another Citie named *Cayn*

CHAPTER 92 [90]

[*Marsden*: Bk. II. Chs. LVII, LVIII. *Pauthier*: Bk. II. Chs. CXL, CXLI. *Yule*: Bk. II. Ch. LXVII. *Benedetto*: Chs. CXLII, CXLIII]

PAſſyng from *Coygangui* towardes the winde *Siroco*, which bloweth betwéene *Leuant* and the midday, which we call Eaſte Southeaſte, you trauaile vpon a fayre ſtonye Cawſey well made. It beginneth at the entring of *Mangi*, and there be very déepe waters on ech ſide of the cawſey. In this country of *Mangi* there is a citie named *Pangui*, very faire, and of greate magnificence. In this prouince they doe vſe that money that the greate CANE doeth vſe in his countrie, and here is greate ſcarcitie of corne, and of al things elſe that ſuſteineth the body. And at another iorneys end towards *Siroco* there ſtandeth another noble and greate citie named *Cayn*, and all the inhabitants are Idolators, and there is abundaunce of fiſhe and beaſts, and wildfoule, ſo that there is boughte thrée good Feſants for the value of ſixe pence.

Of the Citie named *Tinguy*

CHAPTER 93 [91]

[*Marsden*: Bk. II. Ch. LIX. *Pauthier*: Bk. II. Ch. CXLII. *Yule*: Bk. II. Ch. LXVIII (first half). *Benedetto*: Ch. CXLIV]

A Dayes iorney beyond *Cayn* you ſhal find fayre villages, and eared grounde, and ſo you come vnto the grounde of *Tinguy*, plentiful of Wheate, and of al things neceſſary for ſhipping. The people of thys countrey doe honour the Idolles, and thrée dayes iourney from this Citie you come vnto the Occean Sea: and at the ſea ſide there is greate plentie of ſalte.

Of the Citie named *Mangui*, whiche haue vnder their Lordſhip ſeuentéene Cities, and of an other Citie named *Saimphu* which hath vnder it twelue Cities

CHAPTER 94 [92]

[*Marsden*: Bk. II. Ch. LX. *Pauthier*: Bk. II. Ch. CXLIII. *Yule*: Bk. II. Ch. LXVIII (second half). *Benedetto*: Ch. CXLV]

BEyond *Tinguy* a dayes iorney towards the winde *Siroco* you come vnto a faire Countrie, and at the ende of it ſtandeth a Citie named *Manguy* very fayre and greate, and there they honour the Idolles, and ſpeake the Perſian tong. This Citie hathe vnder it ſeauentéene Cities, and I MARCUS PAULUS did gouerne this vnder the great CANE thrée yeares. Toward the Occident or Weſt ſtandeth a prouince or Citie named *Manguy*, where they doe make greate plentie of cloth of Golde and ſilke. Alſo there is greate plentie of corne, and of all manner of victualles. And beyonde this Citie ſtandeth the Citie of *Saimphu* whiche hathe vnder it twelue Cities, whiche is the Citie that reſiſted it ſelfe agaynſte the power of the greate CANE the ſpace of thrée yeares.

Howe this prouince was wonne by the great CANE

CHAPTER 93

[*Marsden*: Bk. II. Ch. LXII (abridged). *Pauthier*: Bk. II. Ch. CXLV (abridged). *Yule*: Bk. II. Ch. LXX (abridged). *Benedetto*: Ch. CXLVII (abridged)]

AFter that the great CANE had wonne the prouince of *Mangi*, conqueſted by induſtry and councell of NICHOLAO and MATHIO and MARCUS PAULUS, as nowe you ſhall perceiue in this preſent chapter: From the hoſte of the greate CANE I write vnto the greate CANE, that that prouince by no manner of way coulde be wonne or taken, of the whiche newes the greate CANE was ſore abaſhed, and we perceyuing his heauineſſe, wée went vnto hym and ſayde: *Potentiſſimo* and mightie Lord, receiue you no conceite nor heauineſſe, for wée wil haue ſuche means, that this prouince ſhall come into youre hands: who béeing comforted with oure promiſe, gaue vs full power and libertie to doe all thoſe things that vnto vs ſhould ſéeme beſte, and that we ſhoulde be obeyed as to his owne proper perſon. And then I MARCUS PAULUS tooke vppon mée this charge, and gathered togither certaine VENETIANS that I founde in thoſe Countries, being diſcréete menne, and exerciſed in feates of armes, and I cauſed to be made thrée greate Trabuco or greate péeces of ordinaunce, whiche ſhotte a pellet of a thouſande pounde waighte, and hadde them vnto the campe, and planted them where they ſhould be ſhotte off, and this done, by the meanes of theſe péeces I ſhotte into the Citie greate pellettes, and when thoſe of the Citie ſaw their houſes fall about their eares, by ſuche meanes as they neuer ſaw nor hearde of before, they receyued great feare, and immediately they yéelded themſelues vnto the great CANE.

Of the Citie named *Singuy*, and of many other things

CHAPTER 94

[*Marsden*: Bk. II. Ch. LXIII (abridged). *Pauthier*: Bk. II. Ch. CXLVI (abridged). *Yule*: Bk. II. Ch. LXXI (abridged). *Benedetto*: Ch. CXLVIII (abridged)]

GOyng from *Siamphu*, and trauelling fiftéene dayes iourney towardes *Syroco*, or to the Eaſte ſoutheaſt, you come vnto the Citie named *Singuy*, wherevnto belongeth a greate number of ſhips: and this Citie is ſcituated vpon the greteſt riuer of the world named *Tuognrou* which is .17. miles in breadth, and one hundred dayes iorney in length, and there is neuer a riuer in the worlde, where there ſayleth ſo manye ſhippes with Merchaundizes, as there. And I MARCUS PAULUS was in this Citie, and did tell ſtanding vpon a bridge at one time fiue thouſande ſhippes or barkes that ſailed vppon this riuer, and vppon this riuer there ſtandeth two hundred Cities, being greater than this that we haue ſpoken of. Thys riuer paſſeth throughe ſixetéene prouinces.

The riuer Tnoguron *the greateſt riuer in the world.*

Fiue thouſãd veſſels on this riuer.

Of the Citie named *Cianguy*

CHAPTER 95

[*Marsden*: Bk. II. Ch. LXIV. *Pauthier*: Bk. II. Ch. CXLVII. *Yule*: Bk. II. Ch. LXXII. *Benedetto*: Ch. CXLIX]

C*Ianguy* is a ſmall Citie ſtanding vpon the ſaide riuer, it hath nothing vnder it but good ground, where they do gather plentie of corne, and rice, which is caried vnto *Cambalu*, that the great CANE may haue greate plentie of victualles in his Courte. This Citie ſtandeth towardes the *Siroco*, and they doe carry this prouiſion vnto *Cambalu* vpon this riuer, and not by ſea. Therefore there commeth through this riuer greate profite vnto *Cambalu*, for it is better prouided with barkes than with cartes, or horſes.

Of the Citie named *Pingramphu*, and of many other things that be in that Countrey

CHAPTER 98 [96]

[*Marsden*: Bk. II. Chs. LXV, LXVI. *Pauthier*: Bk. II. Chs. CXLVIII, CXLIX. *Yule*: Bk. II. Chs. LXXIII, LXXIV. *Benedetto*: Chs. CL, CLI]

P*Ingramphu* is a Citie of the prouince *Mangi*, in the which there is two churches of Chriſtians Neſtorians, edified by MARSAR CONOSTOR, which was Lord of that Citie vnder the greate CANE, and it was in the yeare of oure Lord .1288. Whẽ you do go from *Pingramphu*, you goe thrée dayes iorney againſte *Solano*, whiche is Eaſte and by South, throughe many Cities and Towns, where there is traffiqued muche merchaundizes, and many artes. At theſe thrée dayes iourneys ende ſtandeth the citie of *Tigningui*, greate, riche, and abundant of all things to liue vpon, and alſo of Wine. On a time certaine Chriſtian men named ALANOS tooke this citie, and that nighte they drunke ſo much wine, that they were all drunke, and ſlepte like dogges al that nighte, and the Citizens perceyuing that they were all aſléepe, killed them, and BARAYN King of theſe ALANOS, aſſoone as he knewe this, gathered a great hoſte, and went againſt thys citie, and tooke it perforce, and cauſed to be killed all thoſe that he found in the citie, men, women and children, ſmall and gret, in the reuenging of his Chriſtians.

Of the Citie named *Singuy*, and of many other things there

CHAPTER 99 [97]

[*Marsden*: Bk. II. Ch. LXVII (in part). *Pauthier*: Bk. II. Ch. CL (in part). *Yule*: Bk. II. Ch. LXXV (in part). *Benedetto*: Ch. CLII (in part)]

S*Inguy* is a very great and a noble citie whiche is .40. miles in compaſſe. There is in this citie people innumerable, where you may beléeue, that if the people of *Mangi* were exerciſed in the feate of warre, all the worlde coulde not winne it, but they be all Philoſophers, Phiſitions, Merchaunts and Artificers, very cunning in all artes. There be in this Citie .7000. bridges of ſtone, very faire wroughte, and vnder any of theſe bridges there may rowe a Galley, and vnder ſome twoo Galleys maye rowe togither. In the mountaines of this Citie groweth Rewbarbe greate plentie, and ſo muche Ginger, that for ſixe pence they doe giue more than fiue pound of Ginger. Vnder this Citie there be .17. Cities greate and fayre. In this Citie they do worke greate plentie of cloth of golde & ſilke, for that the Citizens there delighte muche to weare ſuche cloth, and of many coloures.

Seauen thouſand bridges of ſtone. Plentie of Rewbarbe. Fiue pounde of Ginger for ſixe pence.

Of the Citie named *Quinſay*, that is to ſay, the Citie of Heauen, which is a hundred miles in compaſſe, hauing twelue thouſand Bridges, and fourteene Bathes, and many other thinges of wonder

Quinſay

CHAPTER 97 [98]

[*Marsden*: Bk. II. Ch. LXVIII. Sects. I, VI, VII, VIII (in part). *Pauthier*: Bk. II. Ch. CLI. *Yule*: Bk. II. Ch. LXXVI. *Benedetto*: Ch. CLIII]

GOing from *Singuy*, and traueling fiue dayes iorney, you come vnto a noble and famous Citie named *Quinſay*, that is to ſay, the citie of Heauen. This is the nobleſt Citie of the worlde, and the heade Citie of the prouince of *Mangi*. And I MARCUS PAULUS was in this citie, and did learne the cuſtomes of it, and it was declared vnto me, that it was one hundred miles in compaſſe, and .12000. bridges of ſtone with vaultes and

The nobleſt Citie of the worlde, it is an hundred miles cōpaſſe. Twelue thouſand bridges of ſtone.

arches ſo highe, that a greate ſhippe mighte paſſe vnder, and this Citie ſtandeth vppon the water as *Venice* doth, and the people of this citie euery one of them muſt vſe the ſcience of his fathers, and of his predeceſſors. In this Cittie there ſtandeth a lake whiche is in compaſſe thyrtie myles, and in this lake there is builte the faireſt Pallaces that euer I ſaw: And in the mids of this lake ſtandeth two Pallaces wherein they do celebrate all the weddings of that Citie, and euer there remayneth within them all the things neceſſary whiche belong vnto the weddings. Alſo there is rounde aboute this Citie other Cities, but they be ſmall ones. In this Citie they doe vſe money of *Tartaria*, to wit of a Mulbery trée, as it is vſed in the great CANES Court, and as it is afore mentioned. Vppon euerye one of theſe .12000. bridges of ſtone, continually there ſtandeth watch and warde, bycauſe there ſhall be no euill done, and that the Citie doe not rebell. In this citie there is an highe mountaine, and vppon it there ſtandeth a very highe Tower, and vppon it there is a thing to ſounde vppon, and it is ſounded when there is anye fyre or anye rumour in the Countrey. There is in this citie fourtéene Bathes: and the great CANE hath great watch and ward in this Citie.

Of the Citie named *Ganſu*

CHAPTER 99

[*Marsden*: Bk. II. Ch. LXVIII. Sects I, VI, VII, VIII (in part). *Pauthier*: Bk. II. Ch. CLI. *Yule*. Bk. II. Ch. LXXVI. *Benedetto*: Ch. CLIII. (All continued)]

BEyonde *Quinſay* fiftéene myles, bordereth the *Occean* ſea betwéen eaſt and North, and there ſtandeth a Citie named *Ganſu*, which hath a fayre porte or hauen, and thyther come many ſhips out of the *Indias*: betwéen the Citie and the Sea, runneth a great riuer, that paſſeth through many countries, and out that way there go many ſhips vnto the ſea.

Of the diuiſion which the great CANE made of the prouince *Mangi*

CHAPTER 100

[*Marsden*: Bk. II. Ch. LXVIII. Sect. X. *Pauthier*: Bk. II. Ch. CLI (cont.). *Yule*: Bk. II. Ch. LXXVI (cont.). *Benedetto*: Ch. CLIII (cont.)]

THe prouince *Mangi* was diuided into .8. kingdomes, by the greate CANE, and of euery kingdome there is aboute .140. Cities vnder a king. There is in all the prouince of *Mangi* .1202. Cities al ſubiect vnto the great CANE, and al thoſe whiche be borne in this prouince of *Mangi*, are written by dayes and houres, that the prouince may knowe the number of ẏ people, and that they may not rebel. When they do goe on any iourney, they conſult with the Aſtrologers, and when any dieth, the parents do cloth the deade in Canuas, and burne the bodies with papers, wherevpon is paynted, mony, horſes, ſlaues, beaſtes for their houſes, apparell, wyth all other things, for they doe ſaye that the deade vſeth all this in the other worlde, and that with the ſmoke of the deade bodie, and of thoſe papers, whereon there is paynted all thoſe things rehearſed, beléeuing, that it goeth all with him, into the other world, and when they burne thoſe bodies, they ſing and playe vpon al kinde of inſtrumentes and muſicke that they can finde, and ſaye, that in that order and pleaſure, theyr Gods doe receyue them in the other worlde. In this Citie ſtandeth the greate Pallace of *Eſtnoſogi*, which was Lorde and King of that prouince of *Mangi*.

¶ This Pallace is made after this wiſe, it is ſquare and ſtrongly walled, tenne myles in compaſſe. It is high and fayre, with faire chambers, Hals, Gardens, fruites, fountaines, and a lake with many fiſhes. In this Pallace there is twentie Halles, wherin there may ſitte downe at meales, twentie thouſand perſons: by this it may be comprehended how bigge this Citie is. In this Citie there is a famous Churche or Temple of Chriſtians Neſtorians, and euerye one that dwelleth in this Citie hath written his name, and of his wife, Children, menne ſeruauntes, and women ſeruauntes, and horſes that he hath in hys houſe, ouer the Porch of his doore. Alſo when there is anye that goeth to another Citie, it behoueth that the Inholders that lodge ſtraungers, doe bryng a Regiſter vnto the officers appoynted, giuyng relation howe long they doe remayne, and when they goe away.

Of the rent which the great CANE hath of the prouince of *Quynſay*

CHAPTER 101

[*Marsden*: Bk. II. Ch. LXIX (abridged). *Pauthier*: Bk. II. Ch. CLII (abridged). *Yule*: Bk. II. Ch. LXXVIII (abridged). *Benedetto*: Ch. CLIV (abridged)]

SEing I haue declared vnto you of the City and prouince of *Quinſay*, now I wil declare you what rent the greate CANE hath yearely, out of this prouince only, of the ſalt euery yere .4500. Hanegs or buſhels of Gold, and to euery meaſure goeth .18000. Sazos, and euery Sazo of Gold is worth ſeauen Duckets, and of the other rentes ouer and aboue the ſalte he hath euerye yeare .10000. hanegs of gold.

Of the Citie named *Thampinguy*, and of many other maruellous things

CHAPTER 102

[*Marsden*: Bk. II. Chs. LXX, LXXI, LXXII (in part). *Pauthier*: Bk. II. Ch. CLIII (in part). *Yule*: Bk. II. Ch. LXXIX (in part). *Benedetto*: Ch. CLV (in part)]

GOing from *Quinſay*, trauelling towardes *Solano* a dayes iourney, you do goe by Cities and townes, and manye Gardens, and at the ende you come vnto the Citie named *Thampinguy*, which is faire and gret hauing abundaunce of all things, and it is vnder the Seigniorie of ỳ greate CANE: the people are Idolaters, and paſſing other .3. days iourney, you come vnto an other citie named *Vguy*, & going two days iourney beyond, towards *Salano*, or eaſt and by South, there is ſo many Cities & townes y he that trauelleth, thinketh that he neuer goeth out of townes, & there is great plentie of all prouiſion, here is Canes great and thicke of foure ſpannes in compaſſe, and fiftéene in length. At two iourneys ende ſtandeth the Citie named *Greguy* verye noble and greate, hauing aboundance of all things néedeful. The people are Idolatours, and vnder the greate CANE. And going from this Citie thrée dayes iourney towarde *Solano*, you ſhall finde many Cities and townes, and many Lyons. The people do kill them in this manner, the man doth put of his hoſen, and apparell, and putteth on a wéede of Canuas, carriyng a certaine thing pitched, vpon his ſhoulders, and carrieth a ſharpe kniſe in his handes with a pointe, and in this manner he goeth vnto the Lions denne, and as the Lion ſéeth him come, he maketh towards him, and the man when he is neare caſteth vnto him the pitched thyng whyche hée hath vpon his ſhoulders. The Lyon taketh it in hys mouthe, thinking that he hath the manne, and then the man doth wounde him with the ſharpe poynted knife, and as ſoone as the Lyon féeleth hymſelfe hurt he runneth away, and as ſoone as the colde entereth into the wounde he dyeth. In this maner they do kill many Lyons in that countrey, whych is of the prouince of *Mangi.*

The manner how they do kil the Lions.

Of the Citie named *Cinaugnary*, and of many other noble Cities, and of the cruelty of the people that inhabit there, and of other things

CHAPTER 101 [103]

[*Marsden*: Bk. II. Chs. LXXII, LXXIII, LXXIV, LXXV (all in part). *Pauthier*: Bk. II. Chs. CLIII (in part), CLIV. *Yule*: Bk. II. Chs. LXXIX (in part), LXXX. *Benedetto*: Chs. CLV (in part), CLVI]

TRauelling forward foure dayes iourney you come vnto a citie named *Cinaugnary*, a great and a famous Citie ſtanding vppon a Mountayne, which parteth a riuer into two partes, and trauelling foure dayes iourney forwarde, you come vnto a Citie named *Signy*, whiche is vnder the ſegniorie of *Quinſay*. And after you enter into the Realme of *Fuguy*, and trauelling forward ſixe dayes iourney towardes *Solano*, or Eaſt, and by South, through mountaynes and valleys, you ſhall finde many Cities and Townes, hauing plenty of all victuals, and ſingular for Hunting and Hawking, and plenty of ſpices, and ſuger ſo plenty, that you may buy forty pound of Suger for a Venice groate. There groweth a certayne ſwéete fruite like vnto Saffron, and they vſe it in ſtead of Saffron. The people of this Countrey eate mans fleſh, ſo that he dye not of naturall death. When the people of this Countrey go vnto the warres, they doe make certayne ſignes in their forheads, to be the better knowen: and they go all on foote, except their Lorde, who rideth on Horſebacke. They are very cruell people, and vſe the ſpeare and ſword. They do eate the fleſhe of thoſe men that they kill, and drinke their bloud. In the middes of theſe ſixe dayes iourney, ſtandeth the Citie named *Belimpha*, whiche hath foure bridges of marble, with very fayre pillers of marble. Euery bridge of theſe

Good cheape Suger.

is a mile in length, & nine paces in breadth. Vnto this Citie there commeth great plenty of Spices. Alſo, there is in thys Citie very faire men, and more fayre women, and there be blacke Hennes, and fatte without feathers, and verye perfect to eate. In this countrey there be Lions, and other wilde & perillous beaſts, ſo ẏ they trauel in this cuntrey in great feare. At theſe ſixe dayes iourneys ende, ſtandeth the Citie named *Vguca*, where there is made great plentye of ſuger, which is all carried vnto the great CANES court.

Faire men and women heere. Blacke Hens and fatte without feathers.

Of the Citie named *Friguy*, and of manie other maruellous things which be there

CHAPTER 104

[*Marsden*: Bk. II. Ch. LXXVI. *Pauthier*: Bk. II. Ch. CLV. *Yule*: Bk. II. Ch. LXXXI. *Benedetto*: Ch. CLVII]

PAſſing out of the Citie of *Vgucu*, and trauelling fiftéene miles, you come vnto the Citie named *Friguy*, which is the head of ẏ Realme of *Tonca*, which is one of the nyne Kingdomes of *Mangi*. Through the middeſt of this Citie runneth a Riuer of ſeauen miles in breadth. And in this Citie there be made manye Ships, and is laden greate plentie of Spices, and diuers other Merchandizes that is gathered néere to that Riuer, and Precious ſtones whiche be broughte out of *India maior*. This Citie ſtandeth very néere vnto the Occean Seas, and hath abundance of all kind of victuals, or any thyng elſe néedefull.

A Riuer of ſeauen mile broad. There be many Ships made.

Of the Citie named *Iaython*, and of many other things

CHAPTER 105

[*Marsden*: Bk. II. Ch. LXXVII (abridged). *Pauthier*: Bk. II. Ch. CLVI (abridged). *Yule*: Bk. II. Ch. LXXXII (abridged). *Benedetto*: Ch. CLVIII (abridged)]

GOing from *Quinſay*, and paſſing the ſayd Riuer, trauelling fyue dayes iourney towardes *Solano*, or Eaſt, and by South, you find many Cities and Townes, hauing abundance of all victuals. And at the ende of theſe fyue dayes iourney, ſtandeth a great and a faire City named *Iaython*, whiche hath a good Hauen, and thither come many Shippes from the *Indyes*, with many Merchandiſes, and this is one of the beſt Hauens

This Citie hath the beſt Hauen in the world.

that is in the world, and there commeth Shippes vnto it in ſuch quantitie, that for one Shippe that commeth vnto *Alexandria*, there commeth .100. vnto it. The great CANE hathe great cuſtome for Merchandiſes, in and out of that Hauen, for the Ship that commeth thither, payeth tenne in the hundred for cuſtome, and of Precious ſtones and ſpices, and of any other kind of fine wares, they pay thirtie in the hundred: and of Pepper .44. of the hundred, ſo that the Merchants in freight, tribute, and cuſtomes, pay the one halfe of their goodes. In this Countrey and Citie there is great abundance of victuals.

For one Shippe that commeth to Alexandria, *there commeth hither a hundreth.*

Great cuſtome is payd heere.

Of the Ilande named *Ciampagu*, and of things which be found there, and how the great CANE would conquer it

CHAPTER 106

[*Marsden*: Bk. III. Chs. II, III, IV (abridged). *Pauthier*: Bk. III. Chs. CLVIII, CLIX, CLX (abridged). *Yule*: Bk. III. Chs. II, III, IV (abridged). *Benedetto*: Chs. CLX, CLXI, CLXII (abridged)]

I Will paſſe from hence vnto the Countreys of *India*, where I MARCUS PAULUS dwelte a long time: and although the things which I will declare, ſéeme not to be beléeued of them that ſhall heare it, but haue it in a certaynetie and of a truth, for that I ſawe it all with mine owne eyes. And now I will beginne of the Iland named *Ciampagu*, whiche ſtandeth in the high Sea towardes the Orient, and it is ſeparated from the mayne land .1500. miles. The people of this Countrey are fayre, and of good maners, although they be all Idolaters. There is in thys Iland a King franke and frée, for he payeth no tribute at all to any Prince. The people of this Countrey ſpeake the Perſian tong. And there is found in this Iland great plenty of golde, and they neuer haue it forthe vnto anye place out of the Ilande, for that there commeth thyther fewe Shyppes, and little Merchandiſe. The Kyng of thys Ilande hathe a maruellous fayre and great Pallace, all couered with golde in paſte, of the thickneſſe of a péece of two Ryals of plate. And the windowes and pillers of this Pallace bée all of golde. Alſo there is greate plenty of precious ſtones. And the great CANE knowing of the greate fame and riches of this Iland, determined to conquere it, and cauſed to be made great prouiſion of munition and vittayles, and a greate number of Shippes, and in them he put many Horſemen and footemen, and ſent them vnder the gouernance of two of his Captaynes, the one was named ABATAN, and the other VONSAUCIN,

The Iland of Ciampagu *is fifteene hundred miles from the mayne lande.*

In this Iland is great plenty of golde.

The Kings Pallace is couered wyth cleane golde. The windowes and pillers thereof is golde. Great plenty of Precious ſtones.

and theſe two went with this great armie from the Hauen of *Iaython* and of *Glunſay*, and they went vnto the Iland *Ciampagu*, where they went alande, and hauing done great hurt in Mountaynes and valleys, there entred ſuche enuie and hatred betwéene theſe two Captaynes, and ſo much diſcord, that loke what the one would haue done, the other did againe ſay it, and through this meanes they toke neyther Citie nor Towne, but only one, and they killed all them that they founde therein, for that they would not yéeld, ſauing eyght men, whyche could not be killed with any iron, for that eache of them had a precious ſtone enchanted in his righte arme, betwéene the fleſhe and the ſkynne, and theſe ſtones did defend thẽ from death to be killed with yron, and knowing of it, theſe two Captaynes procured to kill theſe eyghte men with clubs of wodde, and toke thoſe ſtones for them ſelues, and in that inſtant there aroſe ſuche a tempeſt of wind of Septentrion or North ſo terrible, and doubting that their Shippes would breake, they hoyſed vp Sayle, and went vnto another Iland, tenne miles diſtant off frõ this, and the wind was ſo terrible, that it opened many of their Ships, and manye were forced to make backe towards their owne Countrey againe, and about .30000. of them fledde by land, of theſe they thought that they were all killed. And as ſoone as it was caulme on the Sea, the King of this Iland which had bin ſo ſpoyled, wente with a great armie of Shippes vnto the other Ilande, where as they were gone to haue taken them that were fledde, and as ſoone as hée was on lande with his men, the TARTARES like wiſe and politike men, retired backe by the Ilande, and went vnto the Shyppes of this King whiche they had lefte without ſtrength, entred in, hoyſed vp the Sayles, with the Auncientes and Flagges of that King, whiche they left behinde in the Ilande, and ſayled vnto the firſt Iland, where they were receyued, and the gates opened, thinking it had bin their owne King. And in thys manner the TARTARES tooke that Citie, wherein the King had his habitation, and ranſacked it. And as ſoone as the King of this Ilande knewe of it, he cauſed many other Shippes to be prepared, and with the men that hée had, and many other that hée tooke of new, enuironed his proper Citie, hauyng it beſéeged ſeauen moneths. And finallye the aboueſayd TARTARES hopyng for no ſuccoure, delyuered vp the Citie vnto the right King, conditionally to let them go with their liues, bagge and baggage. Thys hapned in the yeare of our Lorde .1248. In this Ilande there bée Idols, that ſome haue heads like Wolues, ſome heads like Hogges, ſome like Shéepe, ſome like Dogs, ſome haue one head and foure faces, ſome thrée heads, hauing one only necke, and onely one right hande, ſome haue onely one lefte hande, ſome haue foure handes, and ſome tenne,

Men hauing ſtones that were inchanted, could not be ſlaine with weapons of iron, but with clubbes

The Citie taken by a prettie meanes

and the Idoll that hath moſt handes, is taken to be the moſt beautifull: and to him that demaundeth of them, wherefore they haue ſo many Idols, they doe gyue no other reaſon, ſauing that ſo did their predeceſſors. Whẽ the people of this Iland do take in battell any ſtranger, if he doe not raunſome himſelfe for money, they kill him, drinke his bloud, and eate his fleſh. This Ilande is enuironed round about with the Occean ſea. The portes are frée for themſelues. The Marriners which vſe that Sea, ſay, that there is in it .7448. Ilandes. There is no trée there, but he is of a ſwéete odoure, frutefull, and of greate profite. In this Iland groweth the white Pepper. From the Prouince of *Mangi* vnto the *India* and home, is a yeares ſayling, the reaſon is, for that there raygneth two ſtedfaſt windes, the one in the winter, and the other in the Sommer, contrary the one vnto the other.

In this Sea is .7448. Ilands, whiche be verye frutefull and pleaſant. Heere groweth whyte Pepper.

Of the Prouince named *Ciabane*, and of that King, who hath .325. ſonnes and daughters of his owne. There be many Elephants and much ſpices

CHAPTER 107

[*Marsden*: Bk. III. Ch. VI. *Pauthier*: Bk. III. Ch. CLXI. *Yule*: Bk. III. Ch. V. *Benedetto*: Ch. CLXIII]

WHen you do go from *Iaython*, whych is vnder the ſegniorie of the greate CANE towardes the Occidente, and ſomewhat declining towarde the midday fiue dayes iourney, you come vnto a Countrey named *Cyaban*, wherein there is a Citie riche, great, and famous, ſubiect vnto a King that he and his ſubiects ſpeake the Perſian tong. And in the yeare of our Lord .1248. the greate CANE ſente thither a great Baron, named SAGATO, with a greate armie, to conquere that Prouince, and hée coulde do nothing, but deſtroy muche of that Countrey, and for that he ſhould do no more hurt, that King became tributarie vnto the greate CANE, and euery yeare he ſente him his tribute. And I MARCUS PAULUS was in this countrey in the yeare of our Lord .1275. and I found this King very olde. He had many wiues, and amongſt ſonnes and daughters he had .325. Among his ſonnes he hadde .25. of them that were very valiante men of armes. In thys Countrey there be many Elephants and Lyons greate plenty, and great Mountaynes of blacke Ebbanie.

This King had .325. children.

Of the great Iland named *Iaua*, and of many Spices that grow there

CHAPTER 108

[*Marsden*: Bk. III. Ch. VII. *Pauthier*: Bk. III. Ch. CLXII. *Yule*: Bk. III. Ch. VI. *Benedetto*: Ch. CLXIV]

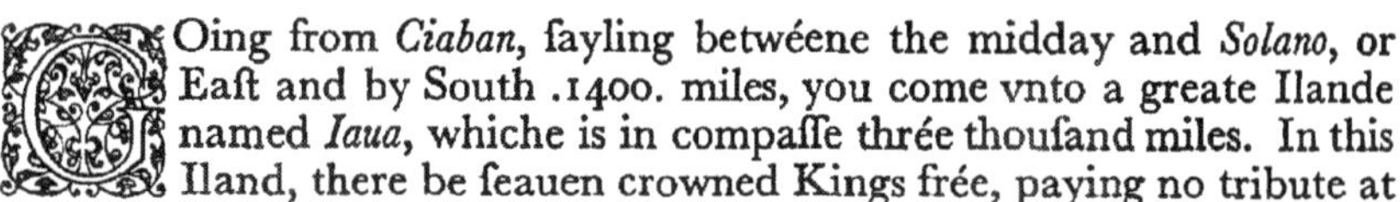

GOing from *Ciaban*, ſayling betwéene the midday and *Solano*, or Eaſt and by South .1400. miles, you come vnto a greate Ilande named *Iaua*, whiche is in compaſſe thrée thouſand miles. In this Iland, there be ſeauen crowned Kings frée, paying no tribute at all. In this Ilande there is great abundance of victuals, and greate riches, hauing very muche Pepper, Cinamon, Cloues, and many other ſingular Spices in great quantitie. The people do honour the Idols. The great CANE could neuer make himſelfe Lord of it.

A very riche Iland of ſpices and golde in great plentye.

Of the Iland named *Iocath*, and of other two Ilands, their conditions and properties

CHAPTER 109

[*Marsden*: Bk. III. Ch. VIII. *Pauthier*: Bk. III. Ch. CLXIII. *Yule*: Bk. III. Ch. VII. *Benedetto*: Ch. CLXV]

SAyling ſeauentéene myles from *Iaua*, betwéene the midday and *Solano*, or Eaſt and by South, you come vnto two Ilands, the one is named *Sondure*, and the other *Condur*. And beyond theſe two Ilands almoſt two hundreth miles, ſtandeth the Countrey named *Iocathe*, great and rich. They ſpeake the Perſian tong, and worſhip Idols. They pay no kinde of tribute to any man, for there is no man that can do them hurt. There is found greate plentye of gold, and a greate number of the ſmall white ſhels of the Sea, whyche is vſed in ſome places in ſtead of money, as before it is rehearſed. Alſo, there be many Elephantes.

Heere is found entie of golde.

¶ Vnto this Ilande there commeth very fewe Strangers, for that it ſtandeth out of the way.

Of the Kingdome named *Malenir*, and of the Ilande named *Pentera*, and of *Iaua* the leſſe, and of their cuſtomes

CHAPTER 110

[*Marsden*: Bk. III. Chs. IX, X. *Pauthier*: Bk. III. Chs. CLXIV, CLXV (in part). *Yule*: Bk. III. Chs. VIII, IX (in part). *Benedetto*: Chs. CLXVI, CLXVII (in part)]

SAyling beyond *Iocath* fiue miles towardes the midday, you come vnto the Iland named *Penthera*, full of Mountaynes. And in the middes of this Iland, about forty miles, there is but foure paſſes of water, therefore the great Shippes do take off their Rudders: and being paſt theſe fiue miles towards the midday, you come vnto a Realme named *Malenir*. The Citie and the Iland is named *Pepethan*, where there is plentie of Spices. And going forwarde, ſayling by *Solano*, or Eaſt, and by South a hundred miles, you come vnto the Ilande named *Iaua* the leſſe, which is in compaſſe two hundred miles. In this Iland there is eyghte Kings, euery one hauing his Kyngdome by himſelfe. They doe all ſpeake the Perſian tong, and honour Idols. They haue ſcant of victuals. From this Ilande you can not ſée the North Starre little nor muche. Beyonde it ſtandeth the Realme of *Ferlech*. The people are Moores. They do honor MARTIN PINIOLO, which is Mahomet. There dwell others in the Mountaynes that haue no kind of law. They doe liue as beaſtes, honouring the firſt thing that they do ſée in the morning, as their God. They doe eate all kinde of dead fleſhe, and the fleſh of man, caring not howe, nor yet after what ſorte it dyeth.

Of the realme named *Baſſina*, and of the Vnicornes, and other wilde beaſtes

CHAPTER 111

[*Marsden*: Bk. III. Chs. XI, XII. *Pauthier*: Bk. III. Ch. CLXV (in part). *Yule*: Bk. III. Ch. IX (in part). *Benedetto*: Ch. CLXVII (in part)]

GOing from *Ferlech* you come vnto ẏ realme of *Baſſyna*, wher the people are without law, liuing as beaſtes, being ſubiect at their will vnder the gret CANE, although they do giue him no tribute, ſauing, that at ſometimes when it pleaſeth them they do ſende vnto him ſome ſtrãge thing. In this realme there be Apes of diuerſe ſorts, and Unicornes, little leſſe than Elephants, hauing a head like vnto a ſwyne, and alwayes hanging it downward to the grounde, and ſtandeth with a good will in Cieno or miery puddel. They haue but one horne in their forehead, wherby only they are called Unicornes, theyr horne is large and blacke, their tong is rough and full of prickles long and thicke. The Apes of this country are ſmall, hauing a face like vnto a childe, and thoſe in that countrey do flaye them, ſo that they looke like vnto a naked childe. They ſéeth it, and dreſſe it with ſwéete ſpices, ſo that they haue no euil ayre nor ſtrong ſent, and ſo ſodden, they doe ſende them aboute in the worlde to ſell, ſaying they be ſodden children. In this countrey there be haukes as blacke as Rauens, very ſtrong and good to hauke with.

Of the realme named *Samara*, and of many ſtraunge things that are founde in the ſayd countrey

CHAPTER 112

[*Marsden*: Bk. III. Chs. XIII, XIV. *Pauthier*: Bk. III. Ch. CLXV (in part). *Yule*: Bk. III. Ch. X. *Benedetto*: Chs. CLXVIII, CLXIX]

GOing from the Realme of *Baxina*, you enter into the realme of *Samara* beyng in this ſame Iland, where I MARCUS PAULUS was fiue moneths, by fortune of weather, and for feare of the euill people of that countrey, for the moſt parte there liueth vppon mans fleſhe. From hence, you ſée not the North ſtarre, nor yet the other ſtars that rule the principal winde, the people there are ruſtical and worſhippe Idols, there is ſingular good fiſh, they haue no wine, but they get it in this wiſe. They haue manye trées like vnto the paulme trée, they breake the braunches and from them commeth water, as it commeth from the vyne. This licour is white and redde like vnto Wine, béeing very perfect to drinke, there is great plentie of it. Another realme there is in this Iland, which is named *Deragoya*, the people are ruſticall, and worſhip Idols. They haue no king, and ſpeake the *Perſian* ſpeach. In this Iland there groweth great plentie of the *Indian* nuts. They haue this cuſtome in this Iland, that when any falleth ſicke, his kinſfolke demaunde of them if the patient ſhall liue or dy. Then theſe maiſters make Diuelliſh inchauntments, if they ſay that he ſhall eſcape, they let him lye, and if they ſay that he ſhall dye, they ſende for the Butchers, whiche ſtoppe his breath till he dye, and when he is deade, they ſéeth the bodie, and the parents eate the fleſh, and kepe his bones in a cheſt. Thys they do, ſaying, if the wormes had eaten the fleſhe they ſhould die for hunger, and the ſoule of the deade bodie ſhoulde ſuffer greate penurie in the other world. They do hide this cheſt with the bones, in a caue of the mountaines, ſo that it maye not be founde. All the ſtraungers that they doe finde, they kil and eate them, if they be not ranſomed for money as ſoone as they take them.

Of the Kingdome named *Lambry*, and of the ſtraunge things there founde, and of the realme *Samphur*, and of the things founde there

CHAPTER 113

[*Marsden*: Bk. III. Chs. XV, XVI. *Pauthier*: Bk. III. Ch. CLXV (in part). *Yule*: Bk. III. Ch. XI. *Benedetto*: Chs. CLXX, CLXXI]

L*Ambry* is another realme in this Ilande, where there is great plentie of ſpices. The people are Idolaters. In this realme there be men that haue feathers about their priuities, great and bigge, and of the length of a gooſe quill. The fift realme of this Iland *Iaua* is named *Samphur*, where there is found the beſt *Camphore* that is in the world, and it is ſolde for the waight of gold: here they do vſe the Wine of trées. In this prouince there is a kinde of great trée, and it hath a very thinne ryne, and vnder the ryne it is full of ſingular meale, and of thys meale they do make perfect meats, of the which I MARCUS PAULUS did eat many times.

Of two Ilandes, and of the euill liuing and beaſtlyneſſe of the people

CHAPTER 114

[*Marsden*: Bk. III. Chs. XVII, XVIII. *Pauthier*: Bk. III. Chs. CLXVI, CLXVII. *Yule*: Bk. III. Chs. XII, XIII. *Benedetto*: Chs. CLXXII, CLXXIII]

GOing from *Lambry* ſayling .140. myles towardes the North, you come vnto two Ilands, the one is named *Necumea*, and the other *Nangania*. The people of *Necumea*, liue like beaſtes, the men and women go naked, couering no part of their ſecrets: they do vſe carnallye like beaſts or dogs in the ſtréets, or whereſoeuer they doe finde, without any ſhame at all, hauing no difference, nor regard, the father vnto the daughter, nor the ſonne vnto ẏ mother, more than vnto another woman, but euery one doth as he lusteth or may. Here there be mountaines of *Sandolos* or Sauders, and of nuts of *India*, and of *Gardamonia*, and many other spyces. *Nangama* is the other Ilande, it is fayre and great. The people therof are Idolaters, they liue beastly, and eate mens flesh, they are very cruel, they haue heades lyke great Mastie dogges, and the men and women haue téeth like dogs. In this Ilande there is great plentie of spices.

Of the Iland *Saylan*, and many noble things which be founde there

CHAPTER 115

[*Marsden*: Bk. III. Ch. XIX. *Pauthier*: Bk. III. Ch. CLXVIII (first part only). *Yule*: Bk. III. Ch. XIV (abridged). *Benedetto*: Ch. CLXXIV (abridged)]

AFter that you go from *Nangana*, you go towarde the Occident, and declynyng against *Arbyno* about ten hundred myles, you come vnto the Iland of *Saylan*, whiche is the beste and the greatest Iland in the world, being in compasse thirtie thousand myles. In this Iland there is a very rich king, the people are Idolatours, and they goe all naked in this Ilande, sauing that they do weare a linnen cloth before their secretes. There is great plentie of Rice and of cattel, and of the Wyne of trées. In this Iland are founde the best Rubies, that bée in the worlde, and they be founde in no other place than here. And here there be founde manye precious stones, as Topases, Amatistes, and of diuerse other kindes. Thys king hath the fairest Rubie in the world, the length of a spanne, and is as thicke as ones arme, as redde as fire, glistering without any blemmish. The men of this countrey are wonderfull leacherous, and they are worth nothing for the warres.

Of the prouince named *Moabar*, wherin there be fiue kingdomes, and of the noble things that be founde there

CHAPTER 116

[*Marsden*: Bk. III. Ch. XX. Sects. I, II, III, IV (in part). *Pauthier*: Bk. III. Ch. CLXIX (abridged). *Yule*: Bk. III. Chs. XVI (in part), XVII. *Benedetto*: Ch. CLXXV (in part)]

PAſſing from thys ſayde place, and trauelling towardes the Occident fortie myles, you come vnto a greate prouince named *Moabar* in the great *India*. This is the greateſt and the beſt prouince that is in ẏ world, ſtanding in the firme land, being an excellent region. There is in thys prouince, Margarites verye ſayre and great. This prouince is diuided into fiue kingdomes, wherevpon raigneth fiue brethren legitimate. In the firſt beginning of this prouince ſtandeth the firſte kingdome gouerned by one of thoſe fiue brethren, named SENDARBA, and is entituled as king of *Nor*, here is fine great pearles, in great number. This king hath the tenth of all ẏ pearls whych are founde in his kingdome. The fiſhermen do fiſh theſe pearles, from the beginning of April, vntill the middeſt of May, in a gulfe of the Sea, where there is greate plentie of them, they are founde in the Oyſters. The men and women of this realme goe all naked, ſauing that they do weare a certaine cloth to couer theyr priuities. Alſo the king goeth naked, and to be knowen, he weareth about his necke a lace full of precious ſtones, whyche are in number a hundereth & foure, in the remembrance of a hundreth & foure prayers, that he vſeth to ſay in the honour of his gods morning and euening, and on his armes, legges, féete, and téeth, he weareth ſo manye precious Stones, that tenne riche Cities be not able to paye for them.

¶ This king hath fiue hundreth wiues, and one of them he toke from his brother. In this realme there be verye faire women of themſelues: alſo they do vſe paynting, ſetting more beautie vnto their faces and on their bodies. Thys king hath alwayes a greate companie with him, to ſerue him: when the king dyeth they burne his bodye, and with him of their owne voluntarie willes, all thoſe that accompanied and ſerued him in his life time, leape into the fire, and burne themſelues with him, ſaying, that they do go to beare their king companie in ẏ other world, and liue as they did here in this worlde. Yerely this King buyeth tenne thouſande horſes of the countrey named *Cormos*, at the price of fiue ounces of gold euery horſe, ſome more, ſome leſſe, according vnto the goodneſſe and beautie of the horſe. The merchaunts of *Quinſay*, of *Suffer*, and of *Beden*, ſell thoſe horſes vnto the merchauntes of this realme. Theſe horſes lyue not in this prouince aboue one yeare: by this meanes that king conſumeth a greate part of his treaſure in horſes. In this countrey they doe vſe this cuſtome, that is, when a man is condemned to dy, he is begged of the Prince that he maye kill himſelfe, and when they haue obtayned the kings good will, he killeth himſelfe, in the loue and honour of his Idols. After thys wiſe, hauing obtayned the kings grace and fauour, the wife of this malefactour and kinred, taketh him, tying about his necke twelue kniues, and in this manner he is carried by them vnto a place of iuſtice, where he crieth as lowde as he may, ſaying, I doe kill my ſelfe in the honour and for the loue of ſuche an Idoll, and with one of thoſe kniues ſtriketh himſelfe, and then with another, vntyll ſuch time as he falleth downe deade: this done, hys parents with great ioye and gladneſſe burne the dead body, thinking that he is happy. In this countrey euerye man hath as many wiues as he is able to maintain: whẽ ẏ huſbande dyeth, according vnto their cuſtome, his bodie is burnt, and his wiues of their owne frée willes burne themſelues with him, and ſhée that leapeth firſte into the fire, the beholders take hir to be the beſt. They are all Idolaturs, and for the more part of them, worſhip the Oxe, ſaying, he is a Sainct, for that he laboureth and tilleth the grounde, where the corne growth, and ſo by no manner of meanes they will eate anye kinde of Oxe fleſhe, nor yet for all the golde in the world, will they kill an Oxe, and when any Oxe dyeth, with his tallow they do rubbe al the inſides of their houſes.

¶ Theſe people deſcende of thoſe that killed SAINT THOMAS the Apoſtle, and none of them can enter into SAINCT THOMAS Churche, whiche he edified in that countrey: beſides this, if one will preſume to enter into the Temple, he falleth ſtreight deade. It hath bene proued oftentimes, that ſome of them would enter perforce into the Church, and it hath not bin poſſible for them to doe it. The king and thoſe of this prouince eate alwayes vpon the ground, and if it be demaunded of them by queſtion why they doe ſo, they doe aunſweare, for that they doe come of the earth, and to the earth they muſte, and they cannot doe ſo much honour vnto the Earth as is worthy. In thys prouince there groweth nothing elſe but Rice. theſe people go naked vnto the warres, hauing no other weapon but ſpeare and ſhield, and they kill no wilde beaſtes at all for their eating, but they cauſe ſome other that is not of their lawe to kill them. All the men and women do waſh themſelues twice aday, morning, and euening, for otherwiſe they dare neyther eate nor drinke, and he that ſhould not kepe this vſe among them, ſhoulde be reputed to bée an Hereticke: and they do waſhe themſelues in thys manner, as we haue rehearſed: they goe all naked, and ſo they go vnto the riuer, and take of the water, and powre it vpon their heads, and then one doth helpe to waſhe another. They are good men of warre, and verye fewe of them drinke wine, and thoſe that doe drinke it, are not taken to be as a witneſſe, nor yet thoſe that go vnto the Sea, ſaying, that the Marriners are dronkards. They are deſperate men, and eſtéeme lecherie to be no ſinne. This countrey is intollerable hote, and the boyes go altogither naked. It neuer rayneth in that Countrey, ſauing in Iune, Iuly, and Auguſt. In this Region there be many Philoſophers, and many that vſe Negromancie, and verye manie of them that tell fortunes. There be Hawkes as blacke as Rauens, bigger than ours, and good to kill the game. Alſo, there be Owles as bigge as Hennes, that flye in the ayre all night. Many of thoſe men doe offer their children vnto thoſe Idols that they haue moſt reſpect vnto, and when they worſhip and feaſt thoſe Idols, they do cauſe to come before them, all the yong men and maydes, whiche are offered vnto them, and they doe ſing and daunce before the Idols, and this done, they do cauſe their meate to be broughte thither, and they doe eate the fleſh, ſaying, that the ſmell of the fleſh filleth the Idols.

Of the Realme named *Muſuly*, where there be found Adamants, and many Serpents, and of the manners of thoſe in that Countrey

CHAPTER 117

[*Marsden*: Bk. III. Chs. XX. Sect. IV (in part), XXI (abridged). *Pauthier*: Bk. III. Chs. CLXX, CLXXI (abridged). *Yule*: Bk. III. Ch. XVIII and few lines of Ch. XIX. *Benedetto*: Chs. CLXXVI, CLXXVII (a few lines only)]

MVſuly is a Region that ſtandeth beyonde *Moabar*, trauelling towards Septentrion which is the North .1000. miles. The people of this Realme worſhip Idols. And in the Mountaynes of this Countrey, there be found fine Adamants. And after they haue had muche rayne, the men goe to ſéeke them in the ſtreames that runne from the Mountaynes, and ſo they do find the Adamants, whiche are brought from the Mountaynes in Sommer when the dayes are long. Alſo, there be ſtrong Serpents and great, very venemous, ſéeming that they were ſette there to kéepe the Adamantes that they might not be taken away, and in no parte of the world there is found fine Adamants but there. There be in this Countrey the biggeſt Shéepe in the worlde. And in the Prouince of *Moabar* afore-named, lyeth the body of the Apoſtle SAINCT THOMAS, buryed in a ſmall Citie, whither there goeth but few Merchants, for that it ſtandeth farre from the Sea. There dwell manye Chriſtians and Moores, hauing great reuerence vnto the body of SAINCT THOMAS, for they doe beléeue and ſay, that he was a Moore, and a great Prophet, and they do call him THOMAS DAUANA, which is to ſay, a holy man. The Chriſtians that go on Pilgrimage to viſit the body of SAINCTE THOMAS, take of that earth where he was martired, and when any falleth ſicke, they doe giue him of it to drinke, with wine and water. In the yeare of our Lorde .1297. it chanced there to be a miracle in this wiſe· A Knight gathered ſo much Rice, that he had no place to put it in, but put it into a houſe of SAINCT THOMAS, and the Chriſtian men deſired him not to peſter the holy Apoſtles houſe with his Rice, where the Pilgrims did lodge, yet the Knighte would not heare them, and the ſame night, the ſpirite of SAINCT THOMAS appeared with a Gallowes of iron in his hande, putting it aboute the Knightes necke, and ſayde, If thou cauſe not thy Rice to be taken out of the houſe of SAINCT THOMAS, I will hang thée. This miracle the Knight told with his owne mouth, vnto all the people of that Countrey, and forthwith the Chriſtians rendred hartie thankes to the holy Apoſtle, who dothe many miracles on the Chriſtians that committe themſelues deuoutely vnto him: All the people of this Countrey be blacke, not bycauſe they be ſo borne, but for that they woulde be blacke, they annoynt themſelues with a kind of oyle, called oyle of *Aiomolly*, for the blackeſt are eſtéemed moſt fayre Alſo, the people of this Countrey cauſe their Idols to be paynted blacke, and the Diuels to be painted white, ſaying, that God and his Sainctes are blacke, and the Diuels white. When they of this Countrey go on warfare, they weare hattes vpon their heads, made of the hides of wild Oxen, and vpon their ſhieldes. And to the féete of their Horſes, they faſten the heares of an Oxe, ſaying, that Oxen heares be holy, and haue thys vertue, that whoſoeuer carieth of them aboute him, can receyue no hurt nor danger.

Of the Prouince *Lahe*, and of the vertue that is in the people

CHAPTER 118

[*Marsden*: Bk. III. Ch. XXII. *Pauthier*: Bk. III. Ch. CLXXII. *Yule*: Bk. III. Ch. XX. *Benedetto*: Ch. CLXXVIII]

GOing from that Towne of SAINCT THOMAS towardes the Occidente, you come vnto a Prouince named *Lahe*, and there dwell the men named *Bragmanos*, which are the trueſt men in the world. They will not lye for all the worlde, nor yet conſent vnto any falſehoode for all the world. They are very chaſt people, being contented only with one woman or wife. They neuer drinke wine, and by no manner of meanes they will take another mans goodes, nor will eate fleſhe, nor kill any kinde of beaſt for all the world. They do honour the Idols, and haue much vnderſtanding in the arte of Fortunes. Before they doe conclude anye greate bargayne, and before they doe anye thing of importance, firſte they doe conſider theyr ſhadowe agaynſte the Sunne, whereby they iudge the thyng that they muſte doe by certayne rules which they haue deputed for it. They doe eate and drinke temperately. They are neuer let bloud, therefore they be very wiſe. In this Countrey there be many religious men, which are named *Cingnos*, and liue a hundred and fiftie yeares, for their greate abſtinence and good liuing. In this Countrey there be alſo certayne religious men Idolaters, who goe altogither naked, couering no part of their body, ſaying, that of themſelues they be pure and cleane from all ſinne. Theſe doe worſhip the Oxe. Theſe religious men weare eache of them vppon his forhead an Oxe made in mettall. They do oynt all their bodie with an oyntment, which they make with great reuerence of the marou of an Oxe. They do neyther eate in diſhes, nor vppon trenchers, but vppon the leaues of the Apple trée of Paradiſe, and other drye leaues, and not gréene by no manner of meanes, for they ſaye, that the gréene leafe hath life and ſoule. They do ſléepe naked vppon the ground.

Of the Kingdome named *Orbay*, and of many things and ſtrange beaſtes found there, and of their beaſtly liuing

CHAPTER 119

[*Marsden*: Bk. III. Ch. XXV. *Pauthier*: Bk. III. Ch. CLXXIV. *Yule*: Bk. III. Ch. XXII. *Benedetto*: Ch. CLXXXI]

Rbay is a Kingdome that ſtandeth towards the Orient, or the Eaſt, beyond *Marbar* fiue miles. In this Kingdome there be Chriſtians, Iewes, and Moores. The King of *Orbay* payeth no tribute. Héere groweth more Pepper, than in any place of the world. There is a thyng in couloure redde, which they do call *Indyaco*, there is plẽtie, and it is good to dye withall, and is made of hearbes. A man can ſcarce kéepe himſelfe in health, for the greate heate that is there, whiche is ſo vehemente, that if you ſhould put an Egge in the water of the riuer at ſuch time as the Sunne hathe his ſtrength, it woulde ſéeth it as though it were put in ſéething or ſcalding water. There is greate trade of merchandiſe in this Countrey, by reaſon of the greate gaynes. There is very muche Pepper, and very good cheape. In thys Countrey there be manye and ſtrange Beaſtes to beholde. There groweth no other kynde of grayne for ſuſtenance, but Rice. There bée many Phiſitions and Aſtrologers. The men and women are blacke, and go naked, ſauing that they do couer theyr priuities. Héere they do marrie the Couſen with the couſen, and the ſonne in lawe with the mother in lawe, and throughout all *India* they do kéepe this manner of wedding.

Great trade for Pepper by reaſon of the quantity.

Of the Prouince named *Comate*, and of the people and ſtrange Beaſtes that be there

CHAPTER 120

[*Marsden*: Bk. III. Ch. XXVI. *Pauthier*: Bk. III. Ch. CLXXV. *Yule*: Bk. III. Ch. XXIII. *Benedetto*: Ch. CLXXXII]

Omate is a Countrey of *India*, from whence you can not ſée the North Starre, nor yet it can not be ſéene from the Ilande named *Iaua* to this place. But going from hence, ſayling vppon the Sea thirtie miles, you ſhall diſcouer the North Starre ſtreight. In this Countrey there are verye ſtrange people, and verye ſtrange Beaſtes, but ſpecially Apes that are like men.

Of the Kingdome named *Hely*, and of the ſtrange beaſtes found there

CHAPTER 121

[*Marsden*: Bk. III. Ch. XXVII. *Pauthier*: Bk. III. Ch. CLXXVI. *Yule*: Bk. III. Ch. XXIV. *Benedetto*: Ch. CLXXXIII]

GOing from *Comate* agaynſte the Occident, or the Weaſt thirtie miles, you ſhall playnely ſée the North Starre, and come to the Region of *Hely*, where they are all Idolaters. The King of this place is very rich of treaſure, but he is weake of people. Thys Countrey is ſo ſtrong, that no man can enter into it perforce. And when any Shippe commeth thither by force of weather, or otherwiſe, thoſe of the Countrey robbe hym, ſaying, that thoſe Shyps come not thither, but to robbe them, and therefore they do earneſtly beléeue that it is no ſinne to robbe them. Héere be Lyons, and other wylde beaſtes a great number.

Of the Kingdome named *Melibar*, and of the things found there

CHAPTER 122

[*Marsden*: Bk. III. Ch. XXVIII. *Pauthier*: Bk. III. Ch. CLXXVII. *Yule*: Bk. III. Ch. XXV. *Benedetto*: Ch. CLXXXIV]

Melibar.

M*Elibar* is a greate Kingdome in *India*, towards the Occidente, and the King payeth no tribute. All the people of this Countrey be Idolaters. Out of thys Realme and the nexte, there goe manye Shippes vnto the Sea a rouing, whiche robbe all kind of people. They do carrie with them their wiues and chyldren, and they ſayle in all the Sommer a hundred Shippes togither, and when they doe come to the ſhore, they roue into the Countrey a hundred miles, taking all that they can finde, doyng no hurte vnto the people, ſaying vnto them, Go, and gette more, for peraduenture you ſhall come againe into our hands. In this Countrey there is plentye of Pepper, of Ginger, and of Turbit, which is certayne rootes for medicines. Of thys Countrey, and their conditions, I will not rehearſe, for it would be very tedious, therefore I will paſſe vnto the Realme of *Gieſurath*.

Plenty of Pepper and Ginger, and Turbit.

Of the Kingdome named *Giesurath*, of their euill conditions

CHAPTER 123

[*Marsden*: Bk. III. Ch. XXIX. *Pauthier*: Bk. III. Ch. CLXXVIII (abridged). *Yule*: Bk. III. Ch. XXVI (abridged). *Benedetto*: Ch. CLXXXV (abridged)]

G*iesurath*, is a Kingdome, in lawe, faith, and tong of the Persians, standing towards the Occidente. All the people are Idolaters. Frõ hence you maye playnely sée the North Starre. In this kingdome be the worst and cruellest Rouers in the worlde, they doe take the Merchantes, not onely taking their goodes, but setting a price of their ransome for their bodyes, and if they do not pay it in a short time, they giue them so great tormentes, that many dye of it. Héere they worke good Leather of all maner of coloures.

Of the Kingdome named *Thoma*, and of the Kingdome *Sembelech*, which stand in *India* the great

CHAPTER 124

[*Marsden*: Bk. III. Chs. XXX, XXXII (in part). *Pauthier*: Bk. III. Chs. CLXXIX (in part), CLXXXI. *Yule*: Bk. III. Chs. XXVII (in part), XXIX. *Benedetto*: Chs. CLXXXVI (in part), CLXXXVIII]

GOing from *Giesurath* towardes the Occidente, you come vnto the Kingdomes of *Thoma* & vnto *Sembelech*. In these Realmes there is al kind of Merchandizes. And these Realmes haue the language and fayth of *Persia*, and in none of them both there groweth anye other sustenance than Rice. They are Realmes and Prouinces of *India* the great.

Of the things already declared

CHAPTER 125

[*Marsden*: Bk. III. Ch. XXXIII (last few lines). *Pauthier*: Bk. III. Ch. CLXXXII (last few lines). *Yule*: Bk. III. Ch. XXX (last few lines). *Benedetto*: Ch. CLXXXIX (last few lines)]

I Haue onely declared of the Prouinces and Kingdomes of *India*, which stande only vpon the Sea coast, and haue declared nothyng vnto you of the Prouinces and Kingdomes within the land, for then this treatise would be very long and tedious vnto the Readers, but yet something of those partes, I will not let to declare.

Of two Ilands, the one of men, and the other of women, Chriſtians, and how there is much Amber

CHAPTER 126

[*Marsden*: Bk. III. Ch. XXXIV. *Pauthier*: Bk. III. Ch. CLXXXIII. *Yule*: Bk. III. Ch. XXXI. *Benedetto*: Ch. CXC]

WHen you go from *Beſmaceian*, ſayling thorough the meane ſea towards the midday or South .25. miles, you come vnto two Ilandes of Chriſtians, the one thirtie miles diſtant from the other. The Iland where there is all men, is named *Maſculine*, and the other where there is all women, is named *Feminine*. The people of thoſe Ilands are as one. The men go not vnto the women, nor the women vnto the men, but thrée monethes in the yeare, as to witte, Auguſt, September, and Octóber, and theſe thrée moneths, the men and women are togither, and at the third moneths end, they returne vnto their owne houſes, doing the reſt of their buſineſſe by thẽſelues. The children Males tarrie with their mothers vntill they be ſeauen yeares of age, and then they goe vnto their fathers. In this Ilande there is greate plentye of Amber, by reaſon of the greate number of Whales that they do take. In thys Iland they are good fiſhers, and take greate plentie of fiſhe, and drye it at the ſunne, hauing great trade with it. Here they liue wyth fleſhe, milke, fiſhe, and rice, and there increaſeth no other ſuſtenaunce. Here ruleth, and gouerneth a Biſhop ſuffragane of the Archbiſhop of *Diſcorſia*.

Of the Iland named *Diſcorſia*, whiche are Chriſtians, and of the things that be founde there

CHAPTER 127

[*Marsden*: Bk. III. Ch. XXXV. *Pauthier*: Bk. III. Ch. CLXXXIV. *Yule*: Bk. III. Ch. XXXII. *Benedetto*: Ch. CXCI]

GOing from theſe two Ilandes, and ſayling towards the middaye .500. myles, you come vnto an Ilande named *Diſcorſia*, wherein are Chriſtians, and haue an Archebiſhoppe. Here is great abundãce of Amber. Alſo they do make very faire clothes of Cottenwooll, the people goe all naked without any clothing. Here is the ſtall of Rouers and Pirates, and the Chriſtians buy with a good wil the goods whiche they bring, & haue

8-2

robbed, for that thefe Pyrates do not robbe but only the Moores and Paynims, and meddle not with the Chriftians. When a fhip fayleth vnder fayle with a profperous winde, a whole day, the day following the Pyrates, with inchauntmentes of the Diuel, caufe the fhippe to haue a contrarie winde, and fo take it.

Of the Ilande named *Maydeygaftar*, where Elephantes be founde, and other ftrange things, and the foule named *Nichas*, which hath quils on his wings twelue paces in length, and of many other conditions

CHAPTER 128

[*Marsden*: Bk. III. Ch. XXXVI (abridged). *Pauthier*: Bk. III. Ch. CLXXXV (abridged). *Yule*: Bk. III. Ch. XXXIII (abridged). *Benedetto*: Ch. CXCII (abridged)]

M*Aydeygaftar* is an Ilande ftanding towardes the midday, diftaunt from *Difcorfia* about a thoufande myles. This Ilande is gouerned by foure Moores, and hath in compaffe a thoufand four hundred myles. Here is greate trade of Merchaundife for Elephantes téeth, for that there is great plentie: they eate no other flefh in this Iland but of Elephants, and of Cammels. Here be many mountaines of redde *Sandalos* or Saunders trées, alfo there is founde greate plentie of Amber. Here is good hunting of wilde beafts, and hauking of foules, and hither come many fhippes with Merchaundife. Alfo there is very great plentie of wilde Boares. There was fente from hence vnto the greate CANE the Iawe of a wilde Boare which wayghed twentie fiue poundes. In fome times of the yeare, there is founde in this Ilande a certaine foule named NICHAS, which is fo big, that the quill of his wings is of twelue paces long, and he is of fuche bigneffe and ftrength, that he with his talents [*sic*] taketh an Elephante, and carrieth him vp into the ayre, and fo killeth him, and the Elephant fo being dead, he letteth him fal, and leapeth vpon him, and fo féedeth at his pleafure.

Of the Iland named *Tanguybar*, where there be men like Gyants

CHAPTER 129

[*Marsden*: Bk. III. Chs. XXXVII (in part), XXXVIII. *Pauthier*: Bk. III. Ch. CLXXXVI (abridged). *Yule*: Bk. III. Ch. XXXIV (abridged). *Benedetto*: Ch. CXCIII (abridged)]

T*Anguybar* is an Ilande of great nobility, being tenne thouſand myles in compaſſe, and the people of this countrey are Idolatours, and ſo bigge and groſſe, that they ſéeme like Giants. One of them wil bear a burthen as waightie as ſixe of our men may beare. They are all black, and go naked without any couer. Theſe men are fearefull to beholde, hauing greate mouthes, and a great redde noſe, great eares, and bygge eyes, horrible in ſight. The women are filthy and euil fauoured. There is great trade of Merchandiſe. Theſe people are bigge of their bodies, ſtrong, and great fighters, and eſtéeme not their liues. The wilde beaſtes of thys Iland differ much from other wilde beaſtes of other Ilãds and countries.

Of the things rehearſed

CHAPTER 130

[*Marsden*: Bk. III. Chs. XXXVII (in part), XXXVIII. *Pauthier*: Bk. III. Ch. CLXXXVI (abridged). *Yule*: Bk. III. Ch. XXXIV (abridged). *Benedetto*: Ch. CXCIII (abridged). (All continued)]

YOu ſhall vnderſtande that all whyche I haue declared of *India*, is only of the noble and great prouinces bordering vppon the ſea coaſtes, and I doe beléeue that there was neuer man, Chriſtian, nor Iew, nor Paynim, that hath ſéene ſo much of the leuaunt parties as I MARCUS PAULUS haue ſéene, for I haue ſéene *India* bothe the greate and the leſſe, & *Tartaria*, wyth other prouinces & Ilands, which are ſo many, ẏ the age of one man, yea peraduenture of .ij. men, would not ſuffice to trauel them all. And now I will declare vnto you of *India* the great.

Of *Abaſhya*

CHAPTER 131

[*Marsden*: Bk. III. Ch. XXXIX (first part only). *Pauthier*: Bk. III. Ch. CLXXXVII (first part only). *Yule*: Bk. III. Ch. XXXV (first part only). *Benedetto*: Ch. CXCIV (first part only)]

IN *India* the greate, there is a greate prouince named *Abaſhia*, whych is to ſay the middle *India*, for it ſtandeth betwéene *India* the greate, and *India* the leſſe. The king of the prouince is a Chriſtian, and the Chriſtians that be vnder hym carrye two tokens made with a burning yron, from the forheade vnto the pointe of their noſe. The great King dwelleth in the middeſt of the prouince, the Moores dwel towardes the prouince of *Cadamy*.

¶ The holye Apoſtle SAINT THOMAS did conuerte muche people vnto the Chriſtian faith in this prouince, and afterwards went from thence vnto the prouince of *Moaber*, where he was martyred. In this prouince there be many valiant knights, and mẽ of armes, and they do euer make war againſt the SOULDAN of *Aden*. The people of thys countrey liue vpon fleſhe, milke, and Rice, and of no other thing. There they vſe muche vſurie, and in this prouince there be many Cities and townes.

Of the prouince of *Adem* or *Ades*, and of the things found there

CHAPTER 132

[*Marsden*: Bk. III. Ch. XL. *Pauthier*: Bk. III. Ch. CLXXXVIII. *Yule*: Bk. III. Ch. XXXVI. *Benedetto*: Ch. CXCV]

THe prouince of *Adem* hathe a King, and he is named the *Sowdan* of *Adem*. There be in this prouince many Cities and Townes, and the people are Moores, and haue greate ſtrife with the Chriſtians. There be in this prouince Ports and Hauens, whither many ſhippes come with merchaundize, and the moſte of this prouince liue vppon Rice, for that they haue little fleſhe, and leſſe milke. This country is very dry and without fruite, and there groweth no graſſe, and therefore the beaſtes of this prouince liue vppon drie fiſhe, ſalte and rawe, which they doe eate in ſteade of ſtrawe and barley.

Of a mightie King of the Orient parties

CHAPTER 133

[*Marsden*: Bk. III. Ch. XLIV (in part). *Pauthier*: Bk. IV. Ch. CCXVI (in part). *Yule*: Bk. IV. Ch. XX (in part). *Benedetto*: Ch. CCXVIII (in part)]

NOwe I haue tolde you of *India* the greate, *India* the leſſe, and of middle *India*, and nowe I haue remayning to tell you of the Countries whiche are towards Septentrion or the North, where there raygneth a King of the imperiall houſe of the greate CANE. Theſe people do worſhippe the ſame Idoll that the TARTARIANS doe worſhippe, whiche they name NAZIGAY. This prouince hathe plaines and mountaines. There groweth no kinde of ſuſtenaunce neither corne nor Rice, and the people liue onelye vppon fleſhe and milke of Mares, and no man maketh warre againſte them, nor they againſte no manne, Here bée manye Camelles and other beaſtes, but they are deade. Vppon the Seigniorie of this Kyng there is a Countrey ſo ſtrong, that no manne maye enter into it, nor yet beaſte being bigge, by reaſon of the ſtraites, lakes, and fountaynes whyche bée there, and for that alwayes there is ſuche feruent colde, that it is alwayes frozen, and vnto them there can come no ſhipping. This Countrey is in compaſſe twelue dayes iorney.

Howe Armines are boughte, and of other beaſtes

CHAPTER 134

[*Marsden*: Bk. III. Chs. XLIV (in part), XLV. *Pauthier*: Bk. IV. Chs. CCXVI (in part), CCXVII. *Yule*: Bk. IV. Chs. XX (in part), XXI. *Benedetto*: Chs. CCXVIII (in part), CCXIX]

IWill declare vnto you howe in theſe twelue dayes iourney they doe buy the wilde beaſts for to haue theyr ſkinnes. In euerye place of theſe twelue dayes iourney there is plentie of habitations, and there be maſties or dogges little leſſe than Aſſes. Theſe maſties doe drawe after them a certaine thing made of Woodde, whiche is called *Slioiala*, whiche is a ſleade, as the Oxen or Horſes doe drawe a Carte, ſauing it hathe no whéeles as oure Cartes haue, and theſe Slyoialas or ſleddes, are as bigge as twoo menne maye be in it, that is to ſaye, the Mayſter of the mayſties or carte, and the

Merchaunt that goeth to buy the ſkinnes. And theſe maſties ceaſe not drawing, excepte it be in ſome myry place, they ſette foure or ſixe maſties to drawe, as among vs wée doe ſette Oxen or Horſes, & when they do come to their iourneys end, the Merchaunt hyreth an other carter with his ſlead and maſties, for that the firſte coulde not endure ſo muche labor, and ſo he maketh his twelue dayes iourney, till he come to the mountains where the Armins and ſkinnes are ſold, where they buy them, and afterwardes they retourne as they came. At the ende of this Countrey there ſtandeth a Kingdome whiche is named the *Darkland*, for it is there euer darke, as wée call the Twylight, for the Sunne ſhyneth not there, and is not ſéene. The people of this Countrey haue no King, but liue as beaſtes without lawe. In this Countrie the men and women are well made of their bodies, although they be ſomewhat yellowe of coloure. The TARTARIANS that border vppon them, doe ſpoyle them very muche, and when the TARTARIANS doe goe to robbe in that darke valley, they ride vppon mares that haue horſe or mare coltes following them, for they doubte to come oute that wayes that they were in, by reaſon of the darkneſſe and wooddes, and when they come neare vnto the place where they meane to robbe, they doe tye their horſe or mare coltes vnto the trées, and ride vppon the mare, and doe their feate, and as they haue done it, they lette their mares goe whither they liſte, and the mares goe ſtraighte vnto their horſe or mare coltes, where they lefte them tyed vnto the trées. Thoſe in that Countrie, wyth certaine deuiſes doe take many Armines, and diuers other wilde beaſts, and take the ſkinnes and dreſſe them, & make merchaundize. This obſcure and darke Countrey, ioyneth one parte with *Rouſelande.*

A darke land.

Of *Rouſeland*, and of other thinges which be founde there

CHAPTER 135

[*Marsden*: Bk. III. Ch. XLVI. *Pauthier*: Bk. IV. Ch. CCXVIII (abridged). *Yule*: Bk. IV. Ch. XXII (abridged). *Benedetto*: Ch. CCXX (abridged)]

ROuſelande is a greate Prouince towardes *Traſmontana* whiche is the North. The people of *Ruſſi* are Chriſtians, according to the vſe of the Gréekes. Touching the things of the holy Church, they are verye ſimple. *Rouſeland* is a ſtrong Countrey, and hathe very ſtrong paſſages. There be very fayre menne and women, and vnto no man they giue tribute, ſauing vnto the King of *Tartarie* of the Occident. There is made greate merchaundize of noble furres for apparell. In *Rouſeland* there be founde many mines of ſiluer, alſo there is ſuch feruent colde, that the people can ſcarce liue. This prouince reacheth vnto the Occean Seas towardes the Septentrion, in which Seas there be many Ilandes wherein bréedeth many Gerfaulcons, and ſingular Hawkes.

FINIS

THE TRAVELS OF NICOLÒ DE' CONTI IN THE EAST

The Introduction

FOr that this treatiſe which I found in the ſecõd Booke towards the end, that Maiſter *Pogio Florētine*, Secretary vnto POPE *Eugenius* the fourth wryteth of the varietie or chaunge of fortune, it maketh muche vnto the confirmation and proofe of the things that Maiſter *Marcus Paulus* writeth in his Booke, for that by the mouth of two or three (as our Redeemer ſayth) there is proued the truth I thoughte good to tranſlate it out of Eloquent Latine, whiche hée did write it in, and to communicate it into my rude Caſtilian and naturall tongue, for that ioyntly ſuche twoo witneſſes in thys preſent worke may make a full, or almoſte a ſure proofe of ſome things, for that it hath not bin ſeene in our *Europa*, or that in any auntient writing appeareth, it may be thoughte harde or difficile credence. And the ſaide *Pogio* followeth in this manner, in the ende of his ſeconde Booke.

IT ſéemeth not vnto me a thing ſtraunge from reaſon, if I decline from the ſtile that hitherto I haue vſed in this Booke, declaring of the harde fortune making an ende, counting the diuerſities of thinges, wherein the heartes of the Readers finde more taſte, and amiable gladneſſe, than in thoſe that already I haue written. Notwithſtanding that alſo in the cauſe I will declare, appeareth plainelye the force of Fortune, in retourning a man vnto *Italy* oute of the extreame partes of the worlde of the Orient, after that he had ſuffered and paſſed fiue and twentie yeres ſuch greate fortunes, aſwell by ſea as by lande. The olde Authors do write many things of the INDIANS with the common fame, of the whiche the certaine knowlege that ſince we haue hadde, ſheweth them to be rather fables than of truth, as it appeareth by the referring of one NICHOLAS a VENETIAN, that after he had trauailed the intrailes of the *Indias*, he came vnto EUGENIUS the fourth Pope of that name, who then was in *Florence*, to reconcile himſelfe, and to haue pardon, for that comming oute of *India*, and néere vnto *Egipt* towards the redde Sea, hée was conſtrayned to renounce and forſake the faith, for feare of death, more of his wife and children, than of hymſelfe. And for that I hearde by manye, that he declared of manye ſingular things, I deſired muche to heare hym, and not onelye to demaunde of him concerning the things whiche hée hadde ſéene, in the preſence of wiſe Barons, and of greate authoritie, but alſo to enforme my ſelfe wyth hym in myne owne houſe, and to take a note of his relation, for that there mighte remaine a remembraunce of it vnto thoſe that hereafter ſhoulde come after mée. And of a trueth hée tolde ſo certaynelye, ſo wiſely, and ſo attentiuely all hys trauaile made amongeſt people of ſo farre Countries, the vſe, manners, and cuſtome of the INDIANS, the diuerſitie of wilde beaſtes, trées, the lynages of Spices, and in what place it groweth, that it appeared well, hée dydde not declare a fained tale, but the trueth of that whiche hée hadde ſéene. And as it ſéemeth, this man went ſo farre, as none of the olde tyme hadde béene, for he paſſed the riuer *Gangy*, and wente beyonde the Ilande *Taprobana*, where we reade there came none, excepte one Captaine of ALEXANDERS fléete named ONESYCRITO, and a Citizen of *Rome*, that by fortune of tẽpeſt arriued in thoſe parties in the time of TIBERIUS CESAR. This NICHOLAS VENETIAN being yong, was as a Merchaunt in the Citie of *Damaſco* in *Syria*, and hauing learned the ARABIAN tongue, he departed from the ſayde Citie in the company of .600. Merchantes, the whiche company they do call CAROUANA, or CARAUANA, & trauailing with his merchaundize through the deſerts of *Arabia*, otherwiſe named *Petrea*, and from thence thoroughe *Chaldie*, he came vnto the greate riuer *Euphrates*.

¶ Hée ſaide, that at the going out of the Deſerte, hée ſawe a meruailous thing, that aboute midnighte, being all at reſte, he heard a great noiſe and ſound, that they thoughte it hadde bin companies of ALARABES wild naked menne, or robbers, and that they were comming to doe them ſome hurte, and all the whole company aroſe and were al ready with the feare, and they ſawe manye battels of horſemen whiche paſſed harde by their tents much like an hoſte, dooing vnto them no hurte at all, and thoſe that hadde vſed that way, ſaid it was certaine companies of fiends which did ouerrun in that ſorte the Deſerts.

¶ There ſtandeth aboue *Euphrates* a noble Citie that the walles of it be of fouretéene thouſande paces. And this Citie was a parte of the olde *Babilon*, and thoſe of that Countrie name it by a newe name *Baldachia*, and *Euphrates* runneth in the middes of it, and they doe paſſe ouer a bridge that hath fourtéene arches of ech ſide, where appeareth many remembraunces of the olde *Babilon*, and manye edifications throwne downe. It hath a ſtrong and a greate Pallaice royall ſtanding vpon a mountaine. The King of this prouince is of a mightie power. From hence vp the riuer twentie dayes ſayling, he ſawe manye noble and populous, and earable groundes of Ilandes, and ſo he trauailed eight dayes iourney by lande vnto the Cittie *Balſera*, and from thence in foure dayes he came vnto the Sea of *Perſia*, whiche ebbeth and floweth as ours doeth, and ſo there ſaylyng fyue dayes, he arryued at a Hauen called *Chalcou*, and from thence hée wente vnto an Iland named *Omerſia*, whych is a ſmall Iland, & diſtant from the firme lãd about .12000. paces, & frõ thẽce he paſſed toward *India* a hundred myles, and came vnto a Citie named *Calabatia*, which is a noble Citie of the PERSIANS, where merchaunts vſe to traffique, and here he was a certaine time, and learned the Perſian tongue, and made him apparell as the PERSIANS had, and ſo he paſſed from thence forward al his time and trauell. And here he tooke ſhipping in a ſhippe with company of the PERSIANS, and of the MOORES, & among them they kéepe muche their promiſe, lawes, and othes made in company, and ſo ſaylıng a moneth, he came vnto a noble Citie named *Cambayta*, ſituated at the ſecond entraunce that the riuer of *India* maketh in the lande. In this Countrie there is founde the pretious ſtones whiche are called SARDINS or SARDONICAS: and here when the huſband dieth, they do burne his wife or wiues that he hath, with his body, and ſhe that he moſt loued, layeth hir neck vppon hir huſbandes arme, and in this wiſe being in hir huſbandes armes, they burne them: and the other wiues they burne in an other fire whiche is made for that purpoſe, and of this vſe it ſhall be rehearſed hereafter. And paſſing on twentie dayes iourney, he founde two Cities, the one named *Pacamunria*, and the other *Hely*. In this Countrey there groweth Ginger whiche is called in that countrie BELLYEDY, GEBELLY, and BELLY, and it is the roote of trées of two cubites in height, the leaues are great, and after the faſhion of a kettle, ỹ bark is hard like ỹ barke of Canes, & it couereth his fruit: out of it procéedeth the ginger, which mingled with aſhes, & layd againſt the Sun, it drieth in thrée dayes. From hence he went trauailing frõ the ſea coaſte thrée hundreth myles, and he came vnto the greate Citie named *Berengalia*, whych is in compaſſe thrée ſcore myles, being enuironed on the one ſide with harde and highe rockes, and on the other ſide towards the valleys and playne grounde with ſtrong adarues and boughes. They ſaye héere is .900000. menne that may weare armoure. The men of that country take as manye wiues as they liſte, and are burnte with them when they dye. In this their King hath ouer them greate vantage, for he taketh twelue thouſãd wiues, and of theſe there goeth on foote after him whereſoeuer he goeth foure thouſande, whych do only prepare and dreſſe his victuals: and there rideth foure thouſand on horſebacke, well apparelled, and of more eſtimation than the firſte. The other foure thouſand ryde in carts and wagons, and of theſe at the leaſte there be two thouſand or thrée thouſand of them that he taketh with condition, that when the king is dead, they of their owne frée willes muſte be burnte wyth him: vnto theſe they do great ſeruice and obedience. This king hath another very noble Citie, which hath ten thouſande paces in compaſſe, being eight dayes iourney from *Berengalia*, from whence in twentie dayes iourney by lande, hée came vnto a Citie vppon the ſea coſte, with a good hauen called *Pedifetaman*, and in theſe twentie dayes iourney hée went through two Cities, the one named *Odes Chyria*, and the other *Conteri Chyrian*, where there groweth the redde *Sandolos* or Saunders. From hence he paſſed vnto a Citie named *Malpurya*, whiche ſtandeth beyonde the ſeconde entring, that the riuer *India* maketh in ẏ end, wher the body of SAINT THOMAS the Apoſtle lyeth honourably in a fayre and famous Church, where he is greatlye honoured and worſhipped by the Heretickes Neſtorians: and there liue almoſt a thouſande men of them in this Citie. Theſe doe liue throughout all *India* ſcattered as the Iewes doe among vs. All this prouince is named *Mahabaria*, beyonde there ſtandeth a Citie named *Cayla*, where there be plenty of peares, and many trées that beare no fruite, of ſixe Cubits high, and as muche in compaſſe: the leaues of theſe trées are ſo thinne, that being playted or foulded vp, you may put one of them in the palme of your hand. They doe vſe theſe leaues in ſteade of Paper to write vpon, and for to couer their heades with when it rayneth, for one leafe will couer thrée or foure men, when they doe trauell. In the middeſt of this ſea there ſtandeth a noble Ilande named *Zaylan*, whyche is thrée thouſande myles in compaſſe, where there be many precious ſtones, as Rubies, Saphires, Granates, and thoſe that are named Cattes eyes, whyche are muche eſtéemed there. Alſo there is plentye of Synamon, whiche is a trée muche like vnto oures of the greateſt Hawthornes, ſauing that the braunches runne not vpwarde, but open and ſtreight ſlopewiſe: the leaues be muche like vnto our Bay leaues, ſauing that thoſe of ẏ Synamon are bigger: the rine or barke of the braunches is beſt and thinneſt, and the rine of the bodye and roote is thickeſt, and of leſſe taſte: the fruite is like vnto the Baye berries, out of whych there commeth a very ſwéete Oyle, and the people vſe to make oyntment of it, wherewith the *Indians* do annoint themſelues: they burne the wood of the trée, when the rine is taken away. There is in this countrey a lake, and in the middeſt of it ſtandeth a royall Citie of thrée myles compaſſe. The Lords of this Iland are of the lynage of the *Bragmanos*, and are taken to be of more witte than the others. The *Bragmanos* Studie Phyloſophy all their life, and alſo Aſtrologie, and liue honeſtly. From hence he paſſed vnto the famous Ilande named *Taprobana*, which the

Indians call *Scyamucera*, where is a noble Citie, and there he was a twelue month· it is ſixe myles in compaſſe, and is a famous Citie, hauing greate trade of Merchaundiſe there, and in al that Iland. From hence he ſayled with a proſperous winde, leauing on the right hand the Iland *Adamania*, which is as much to ſay, as the Ilande of Golde, whyche is .800 myles compaſſe, wherein the *Euitrofagitas* doe liue, and no ſtraungers goe thyther, except it be for neceſſity of weather, and immediately thoſe barbarous people hewe them in péeces, and eate them. He ſayde that *Taprobana* is .1600000 paces in compaſſe, the men are verye cruell, and of ſtubberne conditions, and the men and women haue very bigge eares, laden with Hoopes of golde, and with precious ſtones. They do weare linnen and cloth of ſilke or cruell downe vnto their knées: they take many wiues: their houſes are lowe, by reaſon of the greate heate that the ſunne hath there. They are Idolatours, and haue muche Pepper named the greateſt, and of the long Pepper, and greate plentie of Camphore and golde. The trée that maketh the pepper is like the *Yedra*, or Iuie trée, the berries are gréen lyke vnto the Iuniper berries, and redde, and being mingled wyth aſhes, they harden with the ſunne: there is a gréene fruite named Duriano, of the bigneſſe of Cucumbers. And there be ſome of them lyke long Orengies or Lemans, of diuerſe ſauours and taſte, as like butter, lyke milke, and like curdes In that part of this lande, whiche is named *Bateth*, ÿ *Antropophagos* dwel, and haue continuall warre with their neyghbours, and eate the fleſhe of their enimies that they doe take, and kéepe their heades for treaſure, and vſe them in ſteade of money, when they do buy anye thing, in giuing moſte heades for the thing that is moſt worth, and he that hath moſte heades of the deade men in his kéeping, is eſtéemed to be moſt rich.

¶ Hauing the Iland *Taprobana*, and ſayling fiftéene days, he arriued by tempeſt of weather, vnto the entring of a riuer called *Tenaſerim*, and in this region there be manye Elephants, and there groweth much Braſill And goyng from thence trauelling many dayes iourney by land, and by ſea, he entred at the mouth of the Ryuer *Gangey*, and ſayled fiftéene dayes vp the riuer, and came vnto a Citie named *Cernomen*, very noble and plentiful.

¶ Thys Riuer *Gangey* is of ſuche breadth, that Saylyng in the middeſt, you ſhall ſée no lande on neyther ſide, and hée affyrmeth that it is in ſome places fiftéene myles in breadth. In the armes and braunches of this ryuer there be Canes of ſuche a maruellous lẽgth, and ſo bigge, that ſcarce a manne maye compaſſe one of them wyth both his armes. and of the hollowneſſe or pith of them, they do make things to fiſhe with, and of ÿ wood which is more than a ſpanne thick, they do make boates to trauell with vpon the riuer, and from knot to knotte of theſe Canes it hath of hollowneſſe the length of a man.

¶ There be in this riuer certaine beaſts, hauing four féete, named *Crocodiles*, which liue in the day time vpon the lãd, and in the night in the water: and there be many kindes of fiſhe whiche are not founde among vs, and vppon the braunches of this riuer be manye fayre Gardens, habitations, and delectable grounde. On eche ſide there groweth a kinde of fruite muche like vnto a figge, whych is named *Muſa*, and it is verye pleaſaunte, and more ſwéete than honnye. Alſo there is another fruite, whyche we call Nuttes of *India*, and manye other diuerſe fruites. Going from hence vppe the ryuer thrée moneths, leauing behinde him foure famous Cities, he came to a goodlye famous Citie named *Maarazia*, where there is great plenty of the trées called Alloes, and plentie of golde, and ſiluer, Pearles, and precious ſtones. And going from hence he directed hys waye vnto the mountaines of the Orient, for to haue Carbuncles, and trauelling thirtéene dayes, he returned firſte to *Cermon* and afterwardes vnto *Buſſetanya*.

¶ And after that, ſayling a whole moneth by ſea, he came vnto the entring of the riuer *Nican*, and ſayling vppon it ſixe dayes, he came vnto the Citie alſo named *Nican*, and he went from thence ſeauentéene dayes iourney throughe deſerte mountaynes, and plaine countrey, the fiftéene days of plaine countrey, vntil he came to a riuer greater than the riuer *Gange*, which the people of that countrey cal *Claua*, and ſayling vp this riuer a month, he came vnto a famous great Citie called *Aua*, being .15 miles in compaſſe

¶ This prouince is named of the inhabitauntes *Marcino* They haue greate plenty of Elephantes, for their Kyng dothe kéepe tenne thouſande of them for the warres, and ſetteth vpon euery Elephantes backe a Caſtell, whyche maye carrie eyghte or tenne men with Speares and Shields, or Bowes, or Croſſebowes. He rehearſed that they toke the Elephantes in this manner, PLINIE agréeth vnto the like. They let the tame Elephants females goe vnto the mountaynes, vntill ſuche time as the wilde bée acquainted with them, for the male commonly doth content himſelfe with one female, and when they haue once acquaintance, the female bringeth the wild, by little and little, graſing, vnto a ſmall yard ſtrongly walled, hauyng two dores, one to come in at, and another to goe out at. The female when ſhe is in at the firſt gate, ſhe goeth out at the ſeconde, and the male following hir, the two dores be locked againſte him, and then hauing him within, by certayne loupe holes made for the purpoſe, there commeth in to the number of a thousand men, euery one with his snare in his hande, and one of those men presenteth himselfe before the Elephant, which runneth, thinking to kill the man, and then all those men runne vnto the Elephant, fastning those snares on his féete, and whẽ they be fastned, with great diligence, they do tye the snares vnto a great post, which is set there for that purpose, and they let him alone so thrée or four dayes, till he be more féeble, and after the space of fiftéenė dayes, they giue him a little grasse, in the whiche time he waxeth tame, and then they do tye him among other tame Elephants, and carrie him aboute the Citie, and in tenne dayes he becommeth as gentle as one of the others. Also he sayde, they did tame them in this other wise, that they had and draue them vnto a valley compassed round about, where they did put vnto them the females that were tame, and being somewhat féeble with hunger, they draue them into strayter places made for the nonce, where they be made tame, and these the Kings do buy for their owne vse. Some are fedde with Rice, and Butter, and some with grasse. The wilde Elephantes féede vpon grasse, and vpon the trées of the fields. He that hathe charge of them, ruleth them with a rodde of yron, or a ring whiche he putteth round about his head. The Elephants haue so much prouidence, that manye with their féete, pull away the Speares from their enimies, for that they shoulde not hurt those that be vpon their backes The King rideth vpon a white Elephant, which hath a chayne of golde about his necke, being long vnto his féete, set full of many precious stones. The men of this Countrey haue but one wife a péece. Both men and women of this Countrey pricke themselues, making diuers markes, and of diuers couloures, on theyr bodyes. They be all Idolaters, and assoone as they do rise in the morning, they looke into the Orient, holding their hãds togither, and worship. There is in that Countrey a certayne kinde of fruite, like vnto the Orenge, whiche they doe call *Cyeno*, full of iuice and swéetenesse. Also, there is a trée whiche they doe call *Tall*, whereon they do write, for in all *India*, except it bée in the Citie of *Combahita*, they doe vse no paper, and it beareth a fruite like vnto the Turnep, but they are greate and tender like vnto Gelly It is pleasant in eating, but the ryne is more pleasant. There be in that Countrey daungerous Serpents, of sixe cubites in length, and as thicke as a man, hauing no féete. The people of that Country haue great delight in eating of those Serpẽts rosted. Also they do eate a certayne redde Ante as bigge as a crabbe, estéeming it much drest with Pepper. Also, there is a certaine Beast, hauing a head like vnto a Hogge, the tayle lyke vnto an Oxe, and a horne in his forehead, like vnto a Unicorne, but smaller by a cubite. He is in couloure and bignesse like vnto the Elephante. He is an enimie to tne Elephant. The vtter part of his hornes is good for medicines against poyson, and for this cause he is had in great price and estimation. At the end of this Region towards *Catay*, there be Oxen both blacke and white, had in great estimation They haue a mane and a tayle lyke vnto a Horse, but more hearie, and reacheth vnto their féete. The heares of their tailes be very fine, and like vnto feathers, and they be sold by weight, and therof they do make Moscaderos or Table clothes, for the Altares of their Gods, or for to couer the Table of then King, or for to trimme them with gold and siluer, to couer ẏ buttocks or breasts of their Horses, for beautyfulnesse, & they estéeme thẽ for principall ornaments. Also, the Knightes hang of these heares fast by the yron of their Speares, in token among them of singular nobilitie.

Cataya. *The great* Cane.

¶ Beyond the sayde *Marcino*, there is another Prouince more principall than the others, which is named *Cataya*, and he is Lord of it that is named the great CANE, whych is as muche to saye in their tong, as Emperoure, and the City royall, which is 28. miles in compasse, four square, is named *Cymbalechya*. There standeth in the middest thereof, a very faire and strong Pallace, that serueth for the King. At euery corner standeth a round fortresse of .4 miles compasse, whiche serue for houses of all manner of armoure, and necessarie engines for the warre, and combat against any Citie And from the Pallace royall there runneth a wall with arches vnto euery one of these fortresses, whereon the King may go vnto any of them, if in case they would rise against him in the Citie. From thys Citie fiftéene dayes iourney, there standeth another Citie newly edifyed by the great CANE, and is named *Nentay* It is in compasse thirtie miles, and is most populous of all the rest And this NICHOLAS affirmeth, that the houses and Pallaces, and all other policies of these two Cities, séemed much like vnto those of *Italy*, the men béeing modest and curteous, and of more riches thã the other be.

¶ Going from *Aua* vpon a small riuer seauentene dayes iourney, he came vnto a Hauen Citie, being very greate, named *Zeitano*, and from thence he entred into another Riuer. and in tenne dayes, he came vnto another greate and populous Citie, whiche is in compasse .12000. paces, whiche is called *Paconya*, where he remayned foure monethes In this Citie he founde Uines though they were few, for all *India* lacketh Uines and Wine, nor they make no wine of the Grapes. This Grape groweth among the trées, and after the Grape is cut, the first thing of all, if they do not sacrifice with it vnto their Gods, it is by and by auoyded out of their sight. Also, there be in this Countrey Pines, Aberrycocks, Chestnuttes, and

Mellons, although they be ſmall and gréene Héere is whyte Sandalos or Saunders, and Camphora, or Camphire

¶ There is in *India* farre within, almoſt at the furtheſt end of the world, two Ilandes, and both of them are named *Laua*, the one is of two miles in length, and the other of thrée, towards the Orient, and they are knowen in the name, for the one is called the greate, and the other the leſſe. And turning vnto the Sea, he went vnto them, béeing diſtant from the mayne land a monethes ſayling, and the one is a hundred miles diſtãt from the other. He was in theſe with his wife and children nine moneths, for in all his pilgrimage he had them euer with him The dwellers in theſe Ilands are the moſt cruell and vncharitable people in the world They eate Rattes, Cattes, Dogges, and other viler beaſtes. They eſtéeme it nothing to kill a man, and he that doth any crime, hathe no penaltie, and the debters be giuen to be as ſlaues vnto the creditors, and ſome debters will rather dye than ſerue, and take a Sword, and kill thoſe that are weaker than they, till they find one that is of more ſtrength than themſelfe, who killeth them, & then they carrie the creditor of that murtherer before the Iudge, and cauſe him to pay the debtes of the debter. If any of them do buy a new Sword or knife, he proueth it vpon the body of the firſte that he méeteth, and there is no penaltie for it. Thoſe that come by looke vpon the wound, and prayſe the hardineſſe of him that did it, if it be a great wound. They take as many wiues as they liſt. They do vſe much the game of Cockfighting, and they that bring them as well as the lokers on, lay wagers whiche Cocke ſhall ouercome, and winne the game. In *Laua* the great, there is a Fowle like vnto a Doue, which hath no féete, his feathers light, and a long tayle: he reſteth alwayes on the trées, hys fleſh is not eaten, the ſkinne and tayle are eſtéemed, for they do vſe to weare them on their heads.

¶ Sayling fiftéene dayes beyond theſe two Ilandes towards the Orient, you come vnto two other Ilands, the one is named *Sanday*, where there is Nutmegges and *Al maxiga* or Maſticke. The other is called *Bandan*, where Cloues grow, and from thence it is caryed vnto the Ilands named *Clauas*. In *Bandan* there be thrée kinds of Popiniayes or Parrets, with redde feathers, and yellowe billes, and others of diuers couloures, whiche are called *Noros*, that is to ſay, cleare. They are as bigge as doues. There be other white ones as bigge as Hennes, named *Cachos*, that is to ſay, better, for they excéede the others, and they ſpeake like men, in ſo muche, that they doe aunſwere vnto the things that they are aſked of. The people of theſe two Ilandes are blacke, by reaſon of the greate heate. Beyond theſe Ilands there is a mayne Sea, but the contrary winds will not ſuffer men to trauell on it.

¶ Leauing these sayde Ilands, and hauing done his Merchandise, he toke his waye towards the Occidente or Weste, and came vnto a Citie named *Cyampa*, hauing abundaunce of Aloes and of *Camphora*, or Camphire, and of golde, and in so muche time as he came hither, whiche was a moneth, he came vnto a Citie named *Coloen*, whiche is a noble Citie of thrée miles compasse, where there is Ginger named *Conbobo*, and Pepper, and Uergino, and Sinamon, which is named *Gruessa* Thys Prouince is named *Melibarya*. Also, there be Serpents of sixe cubites in length, and fearefull to behold, but they do no hurt, except they receyue hurt. They do delight muche to sée children, and for to sée them, they come where men be. Their heads when they be layde, séeme like to Celes heads, and when he lifteth vp his head, it seemeth bigger. It hath at the hinder partes a face like to a man, and as though it were paynted of diuers couloures They doe take them by inchantments, which the people vse muche there, and carrie them to be séene. and doe no hurt to anye body Also, there is in this Prouince, and in the nexte adioyning named *Susynaria*, another kind of Serpẽts, which hath foure féete, and a long tayle lyke mastyes They doe take them hunting, and eate them, for they doe no hurte, and are to eate as amongst vs the Hinde or wilde Goate The people say they are good meate. Their Skinnes be of diuers couloures, and those people vse them for diuers couertings, for it is very fayre to behold. Also, there be other Serpentes of a maruellous figure in that Countrey, of ẏ length of one cubite, with wings like vnto Battes. They haue seauen heads, ordinarily sette of the length of his bodye. They dwell among the trées, and are of a swifte flighte. They are more venomous than the other, that onely with their breath they kill a man. Also, there be Cattes of the Mountayne, that flye, for they haue a small skinne from the backe vnto the bellie, ouer all theyr body and féete, whyche is gathered vp when they are still: and when they will flye, they spredde it, and moue it lyke wings, leaping from one trée vnto another. The Hunters do follow them, till they be wéerie with flying, that they fall downe, and so are taken. Also, there is in this Countrey a trée named *Cachy*, that of the troncheon there groweth a fruite lyke vnto a Pyne, but it is so great, that a man can scarce beare it The hull is gréene and harde, but it is of suche a sorte, that if you thrust it with youre finger, it gyueth place It hath within it two hundred and fiftie, or thrée hundred Apples, like vnto Figges. They are of a pleasante tast, and are separated with a very thinne rine The hull within is like vnto the Chestnut in hardnesse and sauoure, and in like maner they are rosted. They are windie, so that if they be putte into the fire, except they be cut, they will start out They do giue the vtter rine vnto the Oxen to eate. Sometimes they fynde this fruite vnder grounde in the rootes of the trées, and thofe be of a pleafanter taft, therefore they doe vfe to prefente them vnto the Kings and Nobles The fruite within hath no rine. This trée is muche like vnto a great Figge trée: the leaues are like vnto the leaues of *Platanos*, or ragged. The wodde is like vnto Boxe, therefore it is hadde in eftimation, and is vfed aboute manye things Alfo. there is another fruite named *Amba*, verye gréene, like vnto a Walnut, but bigger than a Peache. The rine is bitter, and within, it hath the fauour of hony. They lay them in water before they ripe, and dreffe them as we doe the gréene Olyues for to eate.

¶ From *Coloen* he wente thrée dayes iourney vnto a Citie named *Cochin*, it is fiue myles in compaffe, fcituated at the entring of a Riuer, of the whiche it hath the name, and fayling a certayne time vpon the Riuer: he faw manye fiers and nettes faft by the Riuer, and thought there had bin fifhermen, and he demaunded what thofe fifhermen did with thofe fires euery nighte, and thofe of that Coũtrey gaue him anfwere *ycepe, ycepe*, that is to fay, they were fifhes or monfters, hauing humane forme, that on the daye time liued in the water, and in the night they doe come out of the water, and gather wodde togither, and make a fyre, ftriking one ftone agaynfte another, whiche Monfters did take and eate fifhe, for there woulde come manye vnto the lighte of the fire, and fometimes there is taken fome of them, and there is found no difference in them from other men and women. In this Region, the frutes are like vnto thofe of *Coloen*. Beyond this, there ftandeth another Citie named *Calongunia*, ftanding at the entring of another Riuer into the Sea, and beyond, there ftandeth *Paluria*, and *Malyancora*, and this name among them fignifyeth a great Citie, it is nine miles in compaffe. He wente through all thofe, and came vnto *Colychachia*, a City ftanding vpon the Sea coafte, it is eyght miles in compaffe, it is the moft noble in trade of Merchandife, that is in all *India*.

¶ There is héere very much Pepper, Laccar, Ginger, groffe Sinamon, and other fpices Aromatike, and of a fwéete fauoure. Only in this region, the woman taketh as many hufbands as fhe lifteth, and the hufbands agrée among themfelues what eache fhall giue towardes the mayntenance of the wife. Euery hufband is in his owne houfe, and when he goeth vnto his wife, he fetteth a figne at the dore, and when another of them commeth, and féeth the figne, he goeth another way. The children are the hufbands that the wife lifteth to giue them vnto. The fonne dothe not inherit his fathers lande, but hys fonnes fonne.

¶ From hence he trauelled fiftéene dayes, tyll hée came to a Citie called *Cambayta*, ftanding néere the Sea. It is twelue miles in compaffe towardes the Occidente. There is plentye of Eſpico, Nardo, or Lacca Indico, or Gome Laka, Myrabolanos, & Crewill.

¶ There is héere a certayne kind of Prieſtes, whiche are named *Bachales*, hauing but one wife a péece, and ſhe (by their law) is burnt with hir huſband. This kind of people eateth no fleſh, but onely fruites of the grounde, and Rice, milke and hearbes.

¶ Here be many wilde Oxen, they haue manes like vnto Horſes, but longer, and his hornes are ſo long, that when he turneth his heade they reache vnto his tayle, and for that they be ſo bigge, they doe vſe them in ſteade of bottels to drinke in by the waye. Returning to *Colicuchia*, hée paſſed vnto an Iland named *Secutera*, whiche ſtandeth towards the Occident, diſtant from the mayne lande a hundreth myles. It is ſixe hundreth myles in compaſſe, and it is repleniſhed for the moſt parte with Chriſtians Neſtorians Heretickes. Right againſt this Iland no more thā fiue myles, there ſtandeth two Ilands, a thouſande myles diſtant the one from the other, the one is of men, the other of women, ſometimes the men paſſe vnto the women, and ſometimes the women go ouer vnto the men, and they returne backe vnto their Ilande before ſixe moneths, for if they ſhoulde tarrie any longer, they thinke they ſhoulde dye.

¶ From hence he paſſed by ſea, vnto a Citie named *Adena* in fiue days, which hath many edifications, and from thence in ſeauen dayes he wente vnto *Ethiopia*, vnto a hauen named *Barbara*, and from thence in a monthes ſayling he came vnto the redde ſea, vnto a hauen called *Byonda*, and from thence he ſayled two monthes with great difficultie, and landed in a countrey neare vnto mounte *Sinay*, & from thence trauelling through the deſerts, he came vnto *Carras*, a Citie in *Egipt* with his wife, foure ſonnes, and as many ſeruaunts. In this Citie his wife, two ſonnes, and his ſeruauntes died of the plague, and finallye after long perilous and daungerous pilgrimages, he came vnto *Venice*, his own countrey.

Pogio

I Demaunding him of the life and cuſtomes of the *Indians*, he gaue me aunſweare that all *India* was diuided into thrée parts, the one from *Perſia* vnto the riuer *Indo*, another from the riuer *Indo*, vnto the riuer *Gange*, and the other ſtādeth beyond theſe, and excéedeth the others in riches, humanitie, and pollicie, and are equal vnto us in cuſtomes, life, and pollicie, for they haue ſumptuous and neate houſes, and all their veſſels

and houſholde ſtuffe very cleane: they eſtéeme to liue as noble people, auoyded of all villanie and crueltie, being courteous people & riche Merchauntes, in ſo muche that there is one merchaunte hauing fortie ſhippes for his owne trade, and euery one of them is eſtéemed in .50000. Duckets. Theſe only vſe as we do, tables couered with table clothes, and haue theyr Cupboardes of plate, for the other *Indians* eate vppon a thing layde vppon the grounde. The *Indians* haue neyther vines nor Wine, they doe make their drinke of grounde Rice mingled with water, putting vnto it a certaine redde coloure all tempered with the iuyce of a certayne trée.
¶ Alſo they make their pottage like vnto their Wine. In the Ilande named *Taprobana* they doe cutte the braunches of a certaine trée, whiche is named *Tall*, and leaue them hanging, and out of them there runneth a ſwéete licour whiche they vſe to drinke Alſo there is a lake betwéene the riuers *Indo* and *Gange*, of a maruellous ſauerie and pleaſaunt water to drinke, and al thoſe that dwell there about drink of it, and alſo farre off, for they haue ſet horſe from place to place, for the purpoſe, ſo that they haue it brought freſh euery daye: they haue all want of breade: they liue vppon Farro or Rice, fleſhe, milke, and chéeſe. They haue gret plentie of Hennes, Capons, Partridges, Feyſauntes, and manye other wildefoules. They doe vſe much fowling and hunting. They ſhaue their beardes, and nouriſhe a Heare tayle· and ſome tye their haire wyth a ſilken lace, behinde their ſhoulders, like a tayle, and ſo they weare them vnto the warres They haue Barbars as we haue, they are tall of bodye as we be, and alſo in their time of life, they doe lye in ſumptuous beddes, and couered with quilles of Cotten. Their apparell is diuerſe according vnto the diuerſitie of the countrey. They haue all ſcante of woollen cloth, they do vſe cloth of lyne and of cruell and make apparell of it. As well the men as the women couer their ſecréetes vnto their hammes, with a péece of linnen, & vpon it they put a veſture of linnen, or of ſilke, for the greate heate will not ſuffer them to weare more apparell, and therefore they doe goe ſo ſingle tyed with Crimſon lace, and of gold tyed as we do ſée the painters make on the auntient pictures. The women vſe certaine thinne ſhoes of leather, trimmed wyth Golde and cruell.
¶ Alſo they doe weare for gallauntneſſe Hoopes of golde on their armes, and about their neckes, about their breaſtes, and on their legges, the waight of thrée pounde ſet with precious ſtones: the common women kepe theyr houſes as baudes: there be manye and eaſie to finde, for they are almoſt in euery ſtréet, the which with perfumes and ſoft oyntmentes, with their tender age and beautie prouoke muche the menne, for in that countrey they are muche inclyned vnto thoſe women, and for thys cauſe

the *Indians* knowe not what thyng is that abhominable ſinne. Of manye wayes they doe dreſſe theyr heades, but commonlye moſte of them vſe to couer their heades with fine lawnes wreathed, and their haire laced with a ſilken lace. in ſome other places they binde theyr haire vp to their heades, in manner like vnto a peare, and on the knot aboue on their haire they ſet a pinne of golde, whereby they do hang certaine cordes of golde, being of diuerſe colours, hanging betwéene the haires. Some women vſe commonly blacke haire, and among them it is moſt eſtéemed. Some women couer theyr heades wyth certaine painted leaues of trées, and they doe not paynte their faces, but thoſe inhabiting the prouince named *Cataya* doe.

¶ In the *India* within, they do not conſent to a man to haue but one wife. In the others they haue as manye as their carnall luſt wil, ſauing the Chriſtian Hereticke Neſtorians, which dwel ſcattered throughout all the *Indias*, for they take but one woman The maner of their tombes is not as one in all the *Indias*, for the moſte *India* excéedeth other, in diligence and ſumptuouſneſſe, for they doe make caues vnder grounde, in trimming it with a fine wal, and laye in the deade body in a precious bedde, trimmed wyth Ornaments of Golde, ſetting certaine baſkettes round about wyth his moſte precious apparell, and put on rings, as though the deade bodye ſhoulde enioye thoſe things in Hell. They cloſe the mouth of the caue very ſtrongly, that none may enter, and vpon it they do make a ſumptuous and rich tombe ſtrong to abide rayne, and to be the more durable: but in the middle *India* they doe burne the deade bodies, and moſt commonly they do burne their wiues alyue with the deade body, one or manye, according as hée had.

¶ They doe by law burn the firſt wife with him, although it be but one. Alſo they doe take other wiues on this condition, for to honoure him in his death, burning hir ſelfe with him: and this among them is no little honour. They do laye the deade bodie in a bedde trimmed with the beſte apparell that he hath. They do make a fyre rounde about with ſwéete wood, and when it burneth, his wife is trimly dreſt with hir beſte aray, and comming with Trumpets and Shawmes and ſongs merily, as thoughe ſhe did ſing, ſhe goeth rounde about the fire. At this there is preſente the Prieſte, whiche they name BACHALE in a Pulpit, preaching vnto hir howe ſhe muſte not eſtéeme the life nor death, ſaying, that ſhe ſhall haue in the other worlde with hir huſband muche pleaſure, and ſhall poſſeſſe greate riches, honour and apparell: ſhe inflamed with thoſe words that he telleth hir, after that ſhe hath gone a certaine time rounde aboute the fire, ſhée ſtandeth nigher the Prieſtes Chayre or Pulpit, and putteth off all hir apparell, & putteth on a white linnen ſhéete, and leapeth into the fyre. If ſome of them be fearefull (for they haue ſéene the lyke of ſome) that lamenteth and ſtriueth with death, after that ſhe hath leapte in, then the ſtanders by doe throwe hir in wheather ſhée will or no. After they be burnte, they gather the aſhes, and putte them into pottes, and ſome into the graue.

¶ They doe wéepe for the deade after diuers manners The inner INDIANS couer theyr heades with a ſacke, and ſome putte boughes of trées in the highe wayes, and doe hang from the toppe to the grounde painted verſes, playing thrée daies vpon certain inſtrumēts of Copper. They do giue vnto the poore for Gods ſake. Other do wéep thrée dayes for the deade, and all the kinſfolkes and neighbors goe vnto the deade bodies houſe, and they doe carry victualles, but it is not dreſte in the dead mans houſe. In theſe thrée dayes, thoſe that haue buried their father or mother, do carry a bitter leafe in their mouth, and in a whole yere after they doe not chaunge their apparell, nor eate not but once a day, nor yet cutte theyr nailes, nor haire of their heade or bearde. The women which wéepe for the dead, are many, they ſtande neare vnto the deade bodies bedde, being naked vnto the nauell, and ſtrike theyr breaſtes wyth a loude voyce, ſaying, alacke, alacke: and one of them beginneth to praiſe the vertues of the deade bodye, and all the reſte aunſwered vnto hir wordes, ſtriking theyr breaſts: ſome put in certaine veſſels of gold, and of ſiluer. The aſhes of their Prince they cauſe to be caſt into a lake that they haue, ſaying, it is hallowed by their Goddes, and that that waye they goe downe vnto their Gods. The Prieſtes whyche they doe call BACHALES, eate of no kinde of beaſtes, eſpecially not of the Oxe, for they will neither eate, nor kill him, ſaying, he is verye profitable vnto menne aboue al beaſtes. They doe eate Rice, hearbes, fruites, and ſuch like, and haue but one wife, whiche is borne with hir huſbande when he dyeth, laying hir armes aboute his necke, receyuing hir death with ſo good a wil, that ſhe ſheweth no ſigne of paine. Through out al *India* there is founde a lynage of Philoſophers named BRAMANOS, whiche ſtudye Aſtrologie, and prognoſticate things to come They are apparelled more honeſtly, and liue more holily than the others. NICHOLAS ſaide, that he hadde ſéene amongeſt theſe men, ſome of .300. yeares, and among them it was hadde for a miracle, for wherefoeuer that man wente, the boyes woulde followe hym, as a thing of noueltie: and among them is muche vſed the ſuperſtition whyche they doe call GEOMANCIA, by the whiche they tell thinges to come, as thoughe they were preſent. Alſo they are gyuen vnto inchauntementes, ſo that dyuers tymes they doe moue and cauſe tempeſtes to ceaſe, and for this

Men liue thrée hūdreth yeares.

caufe manye do eate in fecret, for that they fhould be enchaunted by thofe that looke vppon them

¶ The faide NICHOLAS dydde tell for a trueth, that hée béeyng patrone and owner of a Shyppe, hée hadde a calme feauen dayes, and hys marriners fearyng, they wente all vnto the mayne mafte, and fette vppe a Table, and after they had made their facrifices vppon it, they leapte and daunced rounde aboute, calling manye times the name of their Gods, whyche they name MUTIA: and among thefe there entred a Féend in a Alarabe or Moore, whyche was amongeft them, he beganne to fing maruelloufly, running aboute the Shippe lyke a madde man, and afterwardes he came vnto the Table, and dydde eate vppe all the meate vnto the bones and fire. Alfo hée didde demaunde a Cocke, and killed it, and drunke vp the bloude, and immediatelye hée demaunded of thofe of the Shippe, what they woulde haue that hée fhoulde doe, and they demaunded that he fhoulde gyue them wind, he promifed to giue it them within thrée dayes, and fuche, that they fhoulde come vnto harborowe: and he fhewed, fetting his handes behinde, from whence the wind fhould come, and willed them to prepare for the ftrength that the winde woulde bring: and when he hadde thus faide, the manne fell downe as halfe deade, without anye knowlege or remembraunce of anye thing that he hadde faide, and in fewe dayes after they were fette in harborowe. Commonly the INDIANS fayle by the guiding of the Starres of the Pole Antartique, for feldome times they doe fée oure North Starre. They vfe not the Loademans ftone as wée doe: they doe meafure their waye, and diftaunce of places, according as their Poale rifeth and falleth, and fo they doe knowe by this meanes, what place they are in. They doe make bigger Shippes than wée doe, that is to faye, of twoo thoufande Tunnes, wyth fyue fayles, and fo manye maftes: they builde their Shippes wyth thrée planckes one vppon another vnder water. that they maye the better refifte the tempeftes, for there chaunceth many These Shippes are made with Chambers, after fuche a forte, that if one of them fhoulde breake, the others maye goe and finifh the voyage. Throughout al *India* they doe worfhippe Idolles, and haue Churches muche like vnto oures, painted within with diuers pictures, which they doe decke with floures at their feafts. They haue within Idolles of ftone, and gold, of filuer, and of Iuorie, fome of .60 foote in height. They haue among themfelues diuers manners in worfhipping, and facrifizing. When they enter into the Church, they wafh themfelues in cleane water, and fo they go in the morning, and in the afternoone, they go in lying along vpon the ground, lifting vppe their féete and handes, and fo praye a whyle, then they doe kiffe the grounde, and fenfe their

Their Pole rifeth and falleth.

Idolles with the ſmoake of ſwéete woodde. On this ſide of *Gange* the INDIANS vſe no belles, but in ſteade of them, they doe ſtrike vppon a veſſell of Copper, and with an other veſſel they doe offer victualles vnto their Gods as the Gentiles did, and afterwarde doe imparte it to the poore, that they maye eate it.

¶ In the Cittie whiche they name *Cambayta* the Prieſtes preache vnto the people in preſence of the Idoll their God, declaring howe they ſhoulde worſhippe him, & howe much it pleaſeth their Gods, when they do kill themſelues for their loue: and there ſtands in preſence many that determine to kill themſelues for them. They haue a hoope of Iron aboute their neckes, the vtter parte of the hoope is rounde, and within, it is ſharpe like vnto a Raſar: alſo they doe hang vnto the fore parte of the hoope down theyr breaſte a chaine, and being ſette downe, they faſten theyr féete vnto it, and béeyng thus, as the Prieſte ſayeth certaine wordes, they ſtretch forth their legges, and lifte vp their heades, and thus with the ſharpeneſſe of the hoope, cutte off their heades in ſacrifice of their Idoll, yéelding vppe their liues And they that kill themſelues in this order, are eſtéemed as Saints. In the Citie of *Bizenegalia* in certaine time of the yeare, they doe carry about the Cittie in proceſſion their Idoll betwéene two cartes, in the company of muche people, and the Damoyſelles ride in cartes in trimme aray, ſinging in the praiſe of hym with muche ſolempnitie, and manye induced by the ſtrength of theyr faith, do lay themſelues vpon the ground, that the whéeles of the cartes may goe ouer them, to bruſe their bones, and ſo to dye, ſaying, that that death is acceptable vnto theyr God. Others there be, that for the better adorning of the carts, make holes throgh the ſides of their bodies, putting a rope throughe it, and tye themſelues vnto the carte, and ſo hanging dead in the proceſſion, accompany theyr Idoll, thinking that they cannot doe greater worſhippe nor ſacrifice vnto their Gods And they make their ſolempnity thrée times in a yeare. In one time there gather togither all the menne and women, and people of all ages, waſhing themſelues in the ſea, or in a riuer, hauing all newe apparell, doyng nothing elſe in thrée dayes but feaſte, daunce, and ſing. Another feaſte they celebrate in burning manye lampes within and withoute their Churches, burning with oyle of *Ionjolly*, and the light goeth not oute daye nor nighte In the thirde, they doe ſette vppe poales like ſmall maſtes through all the ſtréetes, and from the toppe vnto the grounde, they doe hang very faire clothes, wroughte with golde, belonging vnto their Gods and painted, and on the toppe of theſe poales, al the whole nine dayes that it endureth, they do ſette a religious man that hathe a benigne and méeke face, who ſuffereth all that paine for to receiue the grace of his God, and the people throwe vnto him Orrenges, Lemmons, and other like fruites, and he suffereth it all with patience. There bée other thrée solempne dayes, that they doe caste Saffron water vppon those that passe throughe the stréetes, and manye laughe at it. They doe celebrate their weddings wyth banquets, songs, trumpets, and instrumentes muche like vnto ours, sauing Organs whiche they haue not: they doe make very sumptuous feasts day and nighte, with instrumentes, daunces, and songs. They daunce rounde aboute as wée doe, following one after an other in order, and twoo of them carrying twoo painted wandes in their handes, and as they doe méete, they doe chaunge stickes or wandes.

¶ And NICHOLAS rehearseth, that this was a fayre sighte to beholde. They doe vse no Bathes, sauing the INDIANS beyonde *Gange*. The others doe washe themselues manye times of the daye with colde water: they haue scant of oile, and other fruites of ours, as Peaches, Peares, Cherries, Damsons, Apples, and of Grapes they haue but fewe, and (as aboue is rehearsed) onely in one place. And in *Puditsetamas*, a prouince, there groweth a certaine trée withoute fruite, it groweth thrée cubites aboue grounde, and they call it shamefulnesse, for when a man commeth vnto it, it incloseth the braunches, and when he goeth away, it spreadeth abroade his braunches.

¶ *Birengalia* is a Mountaine whiche standeth beyonde towards the Septentrion fiftéene dayes iorney. It is enuironed with many lakes, named *Birenegalias*, whiche are full of venomous beastes, and the mountaine standeth daungerous to bée entred, by reason of Serpentes. And thereon growe the Adamantes: and for that menne dare not goe vnto it, the pollicie of manne founde a way to enter, and to take the Adamantes, for there standeth adioyning vnto it an other mountaine, being a little higher, and in certaine times of the yeare menne goe vppe vnto the toppe of it, where they doe kill certaine Oxen that they carrye with them, and the péeces of fleshe being hotte and bloudy, with certaine Crossebowes for the purpose, doe shoote them vppon the toppe of that other mountaine, and with the fall, it cleaueth faste vnto the Adamantes, and then the Bitturs and Eagles that flye in the ayre, snatche vppe that fleshe with their clawes or tallants, and flye vnto other places, where they maye féede vppon it without feare of those Serpentes, and so the men finde the stones that fall from the fleshe: they doe fynde wyth more ease the pretious stones, for they doe digge in sundrye places, where they vse to finde suche stones, so déepe, til they fynde water mingled wyth grauell, and then they doe take a syue for that purpose, and putte in of the grauell, and the water runneth out, and kéepe the ſtones that remaine behind, and after this ſorte in al theſe parties they doe vſe to finde them: and the Maiſters that ſette to ſéeke them, haue greate care that their ſeruauntes doe not ſteale of thoſe ſtones, for they haue thoſe that ſearch all their apparell, yea, and ſo neare, that they leaue not vnſearched their priuie partes, to knowe if they haue hidden anye. They diuide the yeare into twelue moneths as we doe, and counte the moneths according to the twelue ſignes of the Elements. They accompte the yeare in diuers manners, and the moſte parte doe recken it from Auguſt, for that in the time of AUGUSTUS OCTAUIUS CÆSAR there was an vniuerſall peace throughoute all the world, and they recken from that time .1490. yeares. In ſome regions they haue no money, but vſe in ſteade of money a certayne ſmall ſtone whiche they name Cattes eye, and in ſome other places they do vſe péeces of Iron like néedles, ſomewhat bigger. In other places they do vſe the Kings name written in paper in ſteade of money. In ſome prouinces of *India* more within the lande, they doe vſe *Venice* duckets of golde, and alſo other mony of two duckettes in one. Alſo they do vſe money of ſiluer, and of copper, and in other places they doe make certaine péeces of golde, and vſe them in ſteade of money The firſte INDIANS in the warres vſe dartes, & ſwords, a defence for their armes like Almaine riuets, rounde Targes, and bowes. The other INDIANS vſe ſkulles, backes, and breaſte plates. The INDIANS which are beyond, vſe Croſſebowes and gunnes, & al other ingenious artillerie vſed against Cities. Theſe name thoſe of the Weaſte frée, and ſaye, that all other people are blinde, ſauing they, whiche haue twoo eyes, and ſaye that we haue but one, ſignifying, that in prudence they do excéede all the worlde. And onelye the *Cambaytas* write in paper, and all the reſte write vppon leaues of trées, and of them make Bookes of a good liking: nor yet they write not as we doe, nor as the Iewes from one ſide vnto an other, but begin aboue, and ſo write downewardes. There be among the INDIANS diuers languages. They haue gret abundaunce of ſlaues. The debtor that can not paye, they cauſe him to ſerue the creditor, & he that is accuſed of any crime, there being no certaine witneſſe againſte him, is quitte by his oth. they vſe thrée manner of othes There commeth the partie before hys Idoll, and ſweareth by that Idoll, that he is not faultie, and they haue readye a hotte burnyng Iron like vnto a fiſhe hooke, and cauſe hym that ſwore, to touch it with his tongue, and to licke it, and if it doe him no hurte, he is quitte. And others bring the partie before hys Idoll, and cauſe hym to take that ſame burning yron in his hande, and ſo to carry it certaine paces, and if it hurt him not, he is quitte, but if it doe, he is guiltie. The thirde manner of ſwearing, whiche is moſte vſed, is ſuche:

They doe ſette before his Idoll, a potte full of hote melted butter, and he that ſweareth not to be guiltie, dippeth in two of his fingers into the butter, and ſo wrappeth them with a clowte, and ſealeth it, that it ſhall not vnlooſe, and at thrée dayes ende they vndoe it, and if there be founde any ſigne of burning, hée is guiltie, if not, he is quytte. There is no peſtilence in the *Indias*, nor yet other of the diſeaſes that vſe to trouble oure regions, and for this cauſe there is more Townes and people than is to be beléeued There be manye that make hoſtes of a million of menne, whych is .1000000. NICHOLAS declared, that of one towne, there went out againſt another towne great hoſts, and had battayle, and when the one had ouercome the other, for a great triumph, they did bring twelue Cart loades of gold laces, and of ſilke, with the whych the men that remayned deade, had tyed theyr locke hayres, that hanged downe vpon their backes. He ſayd more, that ſometimes he had gone to their wars, only for to ſée both parties, and they dyd not hurt hym, for that they knew hée was a ſtraunger. ¶ In an Iland named *Laua* the great, is founde in a fewe places a trée, that hath in the middeſt of the harte a rodde of yron, very ſmall, but ſo long as the hart goeth, and hée that hath of this yron next vnto his fleſh, ſhall not periſhe by no kinde of yron, and for this cauſe there be many that cut their ſkinnes and put a péece of it betwéene the ſkinne and the fleſh, it is much eſtéemed.

¶ The things that of the byrde Phœnix be declared and written in verſes by LATANCIO, ſéeme not to be fables, for the ſayde NICHOLAS doeth ſay, that at the end of *India*, there is only one byrde named Seuienda, whoſe bil is like vnto Alboge, or togither with many hoales, and when the time of his death commeth, he gathereth togither dry woodde into his neſte, and ſitting vpon it, he ſingeth ſo ſwéetely wyth his bill, that he delighteth and pleaſeth muche thoſe that heare him, and then flittering with his wings vppon the wood, there cõmeth fire, and he letteth hymſelfe burne, & then there commeth a worme out of his neſte, and of hys aſhes, and of it bréedeth the birde, vnto the likeneſſe of that byrdes byl. Thoſe of that country made the Aluogue with the which they play very ſwéetly. And NICHOLAS maruelling much of it, they tolde him of what the making of it procéeded. Alſo there is in the firſt *India*, in an Iland called *Saylana*, a riuer named *Arotanie*, ſo full of fiſhe, that eaſily they maye take them vp with their handes, but as ſoone as a manne holdeth one of theſe fiſhes in his hande, there commeth vnto him a Feuar, and letting the fiſh go, the Feuar is gone from hym, the cauſe of it appeareth to be the nature of the fiſh, as among vs there is a fiſh which we call *Torpedo*, whych fiſh if a man do hold in his hand, it will be num, and grieue him: although the *Indians* ſaye, that it commeth by meanes of their Goddes, by a certaine tale that they do tell of it.

AFter, for an information to the reader, kéeping ẏ truth of the Hiſtorie, I did write thoſe things rehearſed, as the ſayd NICHOLAS gaue report, and then there came another out of the high *India*, which ſtandeth towardes Septentrion, or the North, and he came, ſente vnto the Pope for to ſée the things and manners of theſe parties, for in thoſe parties they had fame, that in the Occident or weſt there was another worlde, being Chriſtians And this mã declared that neare vnto *Cataya* there was a kingdome, which indured twentie dayes iourney, the which king and people were Chriſtians, but of the ſect of the Neſtorians. He declared hath the Patriach of the Neſtorians had ſent him for to bring him tydings certaine from theſe parties. He rehearſed that they had bigger, & more richer Churches than ours, being al vaulted, and that their Patriarch was very rich in golde and in ſiluer, that euery father of family did giue yearelye vnto him an ounce of ſiluer. I communed with this man, by an interpreter whych could the Turkiſh tong, and the Latin, and I demaunded of him by meanes of this, the wayes, & townes, houſes, cuſtoms, manners, and of other things that a man delighteth to heare, there was great difficultie to learne it, for lacke of the interpreter, and alſo of the *Indian*, but he affirmed the power of the great CANE, or Emperoure of al men, to be greate and mighty, for he had vnder him nyne mighty kings.
¶ Alſo he declared that he hadde trauelled many months through the high *Scithia*, is nowe *Tartaria*, and throughe *Perſia*, and that finallye he came vnto the riuer *Euphrates*, from whence he entered into the ſea, and ſayled vnto *Tripole*, and from thence to *Venice*, and from thence to *Florence*. He reported to haue ſéene manye Cities more faire than ours, both in publike edifications, and of Citizens, for he declared to haue ſéene many cities ten myles, and of twentie myles in compaſſe. And after that this man had ſpokẽ with EUGENIUS the fourth Pope of that name, he wẽt from *Florence* for to ſée *Rome* in deuotion: he demaunded neyther ſiluer nor gold, ſéeming, that he came not for gain, but only to fulfil the meſſage of hym that ſent him.

IN the ſame time there came vnto the Pope certaine men from *Ethiopia*, in deuotion of the faith, with whõ I had communication, by an interpreter, to knowe if they knew any thing of ẏ riuer of *Nilus*, and of his ſpringing. Two of them gaue anſwere, that they were of a countrey being very neare vnto two welſprings, from whence the riuer *Nilus* procéedeth: when I hearde this, I coueted to knowe the things that of this matter the olde auntiente Phyloſophers, namely PTOLOMEUS, did write: firſte of the fountaines of *Nilus*. It appeareth not that they knewe it, but only by coniecture, to appeare that they drew out ſome things of the Originall increaſe of the ſayde riuer. And as theſe witneſſes of ſighte, did tell me of theſe and of others worthy to remayne in memorie, it ſéemed vnto me verye good to write them.

¶ They declared that the Riuer *Nilus* hadde his heade and Welſpryng neare vnto the Region Equinoctiall at the foote of verye hyghe mountaynes, whyche are alwayes couered on the toppe with Miſtes, from thrée welſprings, two of them ſtanding .40. paces the one from the other, and in .500. paces they méete, and make the riuer ſo great, that no man may paſſe ouer but with boate. The thirde which is the biggeſt, ſtandeth a thouſand paces frõ the other two, and he commeth into the riuer of the others, ten myles off. Alſo they ſayde that more than .1000. riuers did enter into *Nilus*, and it increaſeth ſo muche in thoſe countryes, with the raine of March, April, and May, that it maketh *Nilus* to ſwell ouer ſo muche, that it made wonderfull great floudes Alſo they declared that the water of *Nilus* was verye ſwéete and ſauerie, before he entereth among the other Riuers, and it hath vertue to heale thoſe that haue the leaprie and ſcabs, if they waſhe themſelues in it. And beyonde the headſprings of *Nilus* fiftéene dayes iourney, there be verye fruitefull countries, ful of people, and well tilled, hauing very notable Cities, and alſo ſayde that beyonde that countrey there was the ſea, but they had not ſéene it, and that neare vnto the ſpring of *Nilus* there was a Citie, wherein they were borne, and it was fiue and twentie myles in compaſſe, full of people, and in the night had .1000. watches for to defende the Citie from daunger and alterations that might riſe. This region is temperate, and delectable, and plentifull of all thyngs, in ſo much as .3. times in the yere there ſpringeth new graſſe, and twice in the yeare it beareth corne. It hath abundaunce of breade and wine, although the moſt parte of *Ethiopia* vſe (in ſteade of wine) barley ſodden in water. They haue figges, Peaches, Orenges, and Cucumbers like vnto our Lemmons, Sytrons, and ſauing Almonds They haue al our kinde of fruites. Alſo they named diuers trées that they had, whiche we neuer ſawe nor hearde of in our parties, and they are difficult to write, for that the interpreter could not altogither vnderſtande the *Arabian* tongue. But of one of thoſe trées, I muſte néedes rehearſe, whiche is as thicke as a man maye compaſſe, and as highe as a man. It hath many rynes one within another, and betwéene thofe rines hath his fruite like vnto the Cheftnut, and being ground, it becommeth meale, and of it they do make pleafant white bread, which they do vfe in their bākets. The leafe of this trée is more than a cubit in breadth, and more than two cubits in length. They fayd alfo, that towardes the Ilande *Meroe*, the *Nilus* coulde not be fayled, by reafon of the number of Rockes that were there, and that from *Meroe* vnto *Egipt*, it was nauigable, but they tarrie fixe moneths in the Nauigation, for that the riuer giueth manye turnes. Thofe that dwell in that Countrey, haue the face of the Sunne towardes the North, as we haue it towardes the South, and in March they haue it right ouer their heads. All *Ethiopia* hath one manner of letters, although they haue diuers languages, according vnto the greatneffe of the prouinces. Some of them that dwell in the regions towards the Sea coaft, and in the hart of the *India*, there was very much Ginger, Cloues, Nutmegs, and Suger. Betwéene *Ethiopia* and *Egipt* there be defertes of .50. dayes iourney, and they trauell fo farre, hauing with them prouifion of meate and drinke vpon Cammels It hath dangerous paffages in many places, by reafon of the wilde men that go naked in thofe deferts, like wild beaftes riding vppon Cammels, whofe flefh and milke they do eate. They doe robbe the Cammels and prouifion that the trauellers carie, fo that many dye for hunger, and for this caufe there paffe fewe that way vnto vs. The *Ethiopians* moft commonly are of longer life than we, for many liue vntil .120. yeares, and .150. yeares, and in fome places they liue tyll .200. yeares. It is a Countrey much inhabited, and neuer hath the plage, nor other infirmities, fo with this, & with their long liues, their multitude is much encreafed. They haue diuers cuftomes, according to the diuerfitie of the Countrey. They haue no wooll, but weare linnen and filk both men and women. And in fome places, the women weare long traines, and a girdle of a fpanne broade, trimmed with gold and precious ftones. Some of them weare vpon their heads a Lawne, weaued with gold: and fome weare their heare loofe: and fome wound vp in a lace hanging downe at their backes. They haue more plentye of gold and precious ftones than we. The men vfe to weare rings, and the women brafelets wrought of gold and precious ftones. From Chriftmas vnto Lent, they feaft euerie day, eating and daunfing. They do vfe little Tables, fo that two or thrée may fitte at one of them, and do couer them with table clothes as we do. They haue but one King, whiche is entituled King of Kings, after or vnder God, and they faye, he hath many Kings vnder him, and that they haue diuers kinds of beaftes. The Oxen are crooke backed, like vnto Camels, with hornes of thrée cubites in length bending vpon their backes, ſo that vppon one of their hornes they do carrie a Rundlet of wine. Their dogges are of the bigneſſe of our Aſſes, and there is ſome of them that may do more than a Lion, and hunt with them. They haue very great Elephants, and bring vp ſome of them for their pleaſure & for hoſtilitie, & ſome for the warres. They bring them vp of yong ones, & tame them, and then kill the old. Their téeth are of ſixe cubites in length. Alſo, they do tame and bring vp Lions, and to ſhew them for a magnificẽce and oſtentation. Alſo, there is a kind of beaſtes of diuers couloures like vnto the Elephant, but they haue not ſuche a tronke and ſnoute, they do call him Belus. They haue féete like vnto a Camell, and two very ſharp hornes, each of a cubit in length, the one ſtandeth in his forehead, and the other vpon his noſe. Alſo, there is another beaſt ſome what lõger thã a Hare, but in all proportions like, whiche they name ZEBET, and hath ſuch a ſtrong ſmell, that if at any time he rubbe himſelfe againſte any ſmall trée, he leaueth behinde hym ſuch a ſwéete ſauour, that thoſe that trauell and ſmell it, cutte off that part of the trée where the ſent is, and carrie it with them, and in ſmall péeces do ſell it déerer thã gold. Alſo they reported, that there is another kinde of Beaſt, of nine cubits in length, and ſixe foote in height, hauyng clouen féete like vnto an Oxe. Their body is a cubit in compaſſe, and much like in haire vnto the Libard, headed like vnto a Camell, and hathe a necke of four cubites in length. His tayle is very thicke, and muche eſtéemed, for the women do worke with it, embrodering it with precious ſtones, hanging them at their armes. They haue another wild beaſt, which they do take hunting, and he is to be eaten He is as bigge as an Aſſe, ſtriped with couloures redde and gréene, and hathe wreathed hornes vpward, of thrée cubites in length. Alſo, there is another, much like vnto a Hare, with little hornes, and of coulour redde, whiche giueth a greate leape. There is another muche like vnto a Goate, with his hornes vpon his buttockes more than two cubites pending, and for that the ſmoke of them healeth Feauers, they are ſolde for more than fortie Duckets a péece. There is another much like vnto this Beaſt, ſauing that he hathe no hornes His hayre is redde, hauing a necke of two cubites in length. There is another bodyed lyke vnto a Camell, and of the couloure of a Lybarde, hauing a necke of ſixe cubites in length. They ſayd he had a head like vnto a Déere Alſo they ſayd they had a bird of the height from the ground of ſixe cubites, ſmall legges, féete like a Gooſe, the necke and viſage like vnto a Henne. This bird flyeth little, but runneth faſter than a Horſe.

¶ Many other things they told me, whiche I leaue vnwritten, for that I finde my ſelfe wéerie. And they ſayd, that there were Serpents in the

Desertes without féete, of fiftie cubits in length, hauing a Scorpions tayle, and swallow a whole Caulfe at once. And in these things almost they did all agrée, and it séemed vnto me that they made no lie, séeing they had no cause why for to lye, and I thought good to write it, for y profit of those y list to rede.

FINIS

APPENDIX I

NOTES TO FRAMPTON'S TEXT OF MARCO POLO

NOTES

PROLOGUE

1. Page 15, line 1

The Prologue commences with a kind of invocation, which, however, reads quite differently in the Geographic Text (fr. 1116), and in all the more important MSS. It is, moreover, in the third person. I translate direct from fr. 1116:

"Governors, Emperors and Kings, Dukes and Marquises, Counts, Knights and citizens! and all those who would fain learn of the divers races of mankind and of the differences of the various regions of the World, take this book and read it."

2. Page 15, line 22 *Vſtacheo*

The Geographic Text of 1824 reads "Rustacians," which has now been corrected by Benedetto to "Rusticiaus." Yule used the form "Rusticiano," as being the "nearest probable representation in Italian form of the *Rusticien* of the Round-Table MSS...." He adds, however (Introd. p. 63), that it is highly probable that the Pisan's real name was Rustichello, which form he found in a long list of Pisan officials during the Middle Ages It is, therefore, satisfactory to see that Benedetto has come to the same conclusion (p. xiii) with a mass of fresh evidence.

3. Page 15, line 23 *Ianua*

Genoa. Written "Jene" in fr. 1116. Frampton follows the word with a comma, and continues immediately with, "raigning in Conſtantinople . ." This, however, is the beginning of the Travels, and should be quite distinct from what has preceded it. I have, accordingly, changed the comma to a full stop, and left a space of a few lines.

This starting-point in the four leading editions is: *Marsden*, Bk. I. Ch. I. Sect. 1; *Pauthier* and *Yule*, Prologue, Ch. 1; *Benedetto*, Ch. II.

In all future notes these four editions will be referred to simply as M., P., Y. and B. respectively.

4. Page 15, line 24

Raigning in Conſtantinople...

Both Ramusio (to be referred to in future as R.) and the Venetian MS called *V* by Benedetto (i.e Staatsbib Hamilton, 424[a]) give us the additional information that there also resided at Constantinople a magistrate representing the Doge of Venice—for the significance of this office of Podestà, see M p. 4; cf. also p. lx and note of his Introduction where he quotes a similar passage from the lost Soranzo MS (see No. 61 of the Yule-Cordier table, Y. Vol. II. p. 546).

5. Page 15, line 25

in the yeare of oure Lord .1250.

So read all the best texts, but, as Marsden and all subsequent editors have pointed out, this is clearly an error in copying. As we shall shortly see, the brothers were back at Acre in 1269, and on reaching Venice found Marco fifteen years old. Thus they must have left Venice in 1254 *at the very earliest.* Now both R. and F. (Frampton) state that Nicolo left his wife with child. This would date the departure at 1253–4. But neither fr. 1116, nor any of the MSS used by Pauthier or Yule mention this fact. Thus, if it is a later interpolation, Nicolo may have seen his son before leaving Venice. The only reason for mentioning this latter point is that as their start from Constantinople was in all probability 1260 we cannot account for how the years from 1254 to 1260 were passed, unless they were engaged all this time at their branch house which we know they had in Constantinople. In any case, we are quite justified in changing 1250 to 1260.

See further Notes 20 and 21.

6. Page 15, line 30 *Countrey of the Souldan*

The once prominent port of Soldaia, now Sudak, is meant. Rubruquis passed through it in 1253 on his great journey to Karakorum It lies on the south-east coast of the Crimea, between Uskyut and Otus. The elder Marco Polo had a house here.

7. Page 16, lines 1, 2

came to a Citie of the Lorde of the Tartarians, which is called Barcacan

Barka Khan was the third son of Juji, the firstborn of the famous Chinghiz (Genghis) Khan His two cities, mentioned in all the important texts, were Sara and Bolgara (Sarai and Bolghar), both on the Volga.

R. adds that Barka "had the reputation of being one of the most liberal and civilized princes hitherto known among the tribes of Tartary."

8. Page 16, line 10 *Alan*

This should read Ala*u*, i.e. Hulaku, brother of Kublai and Mangku Khan, and founder of the Mongol dynasty in Persia. The war between Barka and Hulaku is described in that portion of the Book which Pauthier and Yule have turned into Book IV (P. Bk. IV. Chs. CCXXI–CCXXVI; Y. Bk IV. Chs. XXV–XXVIII; B. Chs. CCXXIII–CCXXVIII). As Yule has reduced the chapters to a few lines, and they appear neither in R. nor F, I have added them in full in Appendix II. pp 337–39, taking them from Wright's edition of M.

9. Page 16, line 16 *Buccata*

Ucaca or Ukek, on the Volga halfway between Sarai and Bolghar. The name occurs in widely varying forms in the different MSS. It is "Ouchacca" and "Oucaca" in fr. 1116, "Euchatha" in Bib Nat 3195, "Oukaka" in R., and "Guthaca" in the Latin version published by Grynæus in the *Novus Orbis*, 1532.

10. Page 16, lines 17, 18

the Riuer... Tygris, whiche is one of the foure that commeth out of Paradise terrenall

By the "Tygris" is meant the Volga Yule (Vol I p 9) suggests that the connection in name arose out of some legend that the Tigris was a reappearance of the same river, and adds that the ecclesiastical historian, Nicephorus Callistus, appears to imply that the Tigris coming from Paradise flows under the Caspian to emerge in Kurdistan. Neither Y nor B mention the above passage which is found both in R. and the Berlin Latin text. Its inclusion in F. is very interesting, see P. p. 8. That portion of the text dealing with the Volga area corresponds to: M. Bk. I. Ch. I. Sect. I (cont.); P. and Y. Prol. Ch. II; B. Ch. III. See also Note 69.

11. Page 16, lines 22, 23

noble Citie called Bocora, and the same name hadde that Prouince, which the Kyng of that Countrey had, and the Citie was called Barache

This is obviously a mistake for ". . and the *Kyng* was called Barache," i.e. Borrak Khan, great-grandson of Chagatai. Bokhara was considered as a part of Persia.

12. Page 16, line 27

This Alan is otherwise called the greate Cane

This is, of course, a mistake Kublai was the great Khan, and the envoys were on their way back to him from Hulaku The French texts clearly say "qui aloit au grant sire de tous les Tartars." R. says the same, but adds: "qual stà ne' confini della terra fra Greco, & Leuante," "whose residence was at the extremity of the continent, in a direction between north-east and east." A similar passage is also found in *Z* and *V* (As already mentioned in Note 4, *V* = the Berlin Staatsbib. MS Hamilton, 424[a], of the fifteenth century. It was copied in 1793, and this copy forms Milan. Bib. Ambrosiana, Y. 162. P.S. This must not be muddled up with Bib. Amb. Y. 160. P.S. which = *Z*.)

13. Page 16, line 35 *friendſhippe of hym*

At this point end P. and Y. Prol. Ch. III and B. Ch. IV.

14. Page 16, line 40 and page 17, line 1

ſhall be declared in thys Booke

Here end M. Bk. I. Ch. I. Sect 1, P. and Y. Prol. Ch. IV and B. Ch. V

15. Page 17, lines 7, 8

Dukedomes in Chryſtendome, of theyr conditions

Here end P. and Y. Prol. Ch V and B. Ch. VI.

16. Page 17, lines 12, 13

Could ſpeake the Tartarie language

Here end P. and Y. Prol. Ch. VI and B. Ch VII

17. Page 17, line 26

Sepulchre of Ieſus Chriſte in Ieruſalem

Here end Y. Prol. Ch. VII. and B. Ch. VIII The French texts mention the Khan's ambassador, Cogatal, early in the chapter He appears as "Cocoball" in F. a few lines further on.

18. Page 17, line 37 *fell ſicke and dyed*

I can find no reference to his actual death in any of the other MSS. Fr 1116 merely says "chei amalaides," while R. reads, "s' ammalò grauemente."

19. Page 18, line 1 *a towne called Giaza*

I.e Ayas, once a famous port on the gulf of Scanderoon, thirty miles south-west of Adana.

Here end P. and Y. Prol Ch. VIII and B. Ch IX. They give us a little more information, which is also found in R

"...so great were the natural difficulties they had to encounter, from the extreme cold, the snow, the ice, and the flooding of the rivers, that their progress was unavoidably tedious, and three years elapsed before they were enabled to reach a sea-port town in the Lesser Armenia, named Giazza."

20. Page 18, lines 2, 3

in the yeare of our Lord God .1272.

This clearly shows a clumsy effort to make the date consistent with the wrong departure date, 1250 (see Note 5). F. should have gone further in his efforts, and when mentioning Marco's age a few lines lower down, should have changed it to twenty-two. R., giving the return as 1269 (which is quite correct), makes Marco's age nineteen, thereby justifying the previous 1250 date The French texts all have 1260 as the return date, which should be corrected to 1269.

Marco's age should be fifteen, as all the best texts have it. Roux, in his edition of fr. 1116 in 1824, mistook xv for xii, but this has been since corrected. See further next note.

21. Page 18, line 3 *the Pope Clement was dead*

None of the early texts give his name. It appears, however, in the Crusca Italian as Clement, while in Bib. Nat. 3195 it is "Clementem IV" and R. has: "che Clemente Papa Quarto nuouamente era morto." Now Clement IV died in November 1268, so that R.'s "recently dead" supports their return date as being 1269. Furthermore, the new Pope, Gregory X, was elected in 1271, in November of which year the three Polos made their second start from Acre

22. Page 18, lines 4, 5 *Miser Thebaldo*

Called "Teald de Plajence" in fr 1116, and "Tebaldo de' Vesconti di Piacenza" in R.

Fr. 1116 describes him as "legat por le yglise de Rome en tout le regne d'Egipte." F. is somewhat abbreviated at this point

23. Page 18, lines 9, 10 *Nigro Ponte*

Negroponte was the name given to the island and port of Eubœa in the thirteenth century; so-called, says Pauthier (p 16), because there was a bridge of five arches of which "l'arche du milieu était un *pont-levis* pour le passage des navires." After it had become the centre of Venetian influence in Romania, it formed a port of call on the Venice—Constantinople—Trebizond route.

24. Page 18, lines 15, 16

tarying the creation of a newe Pope

Here end M Bk I. Ch. I. Sect II (except that the sentence about the two years' wait begins Sect. III), P. and Y. Prol. Ch. IX and B Ch. X.

25. Page 18, lines 16, 17 *from Venice to Ierusalem*

I.e *via* Acre The full itinerary, including the double start, was Venice—Acre—Jerusalem—Acre—Ayas—Acre—Ayas and across Asia to K'ai-p'ing fu (the "Clemenfu" of R. and "Bemeniphe" of F.).

26. Page 18, line 26 *and so departed*

Here end P. and Y. Prol. Ch. x and B. Ch. xi.

27. Page 18, lines 33, 34

they sayled incontinente to the Pope

Here end P. and Y. Prol Ch xi and B. Ch. xii.

28. Page 18, lines 35, 36

two Friers, of the order of Sainct Dominike

Fr. 1116 simply says "deus freres precheors," so also P., Y. and R.

29. Page 18, lines 38, 39

disputations in the defense of the holy Catholike faith

Fr. 1116 mentions "brevileges et carte...," while R. is fuller:

"To them he gave license and authority to ordain priests, to consecrate bishops, and to grant absolution as fully as he could do in his own person. He also charged them with valuable presents, and among these, several handsome vases of crystal, to be delivered to the Grand Khan in his name and along with his benediction."

30. Page 19, line 1 *the Souldan of Babylon*

Most MSS mention his name: "Bondocdaire," "Bendocquedar," "Bundokdari," etc. "Babylon" means Cairo (*Bambellonia d' Egitto*).

For a good note on Bundūkdār's invasion of Cicilian Armenia see Y. Vol 1. pp 23, 24

31. Page 19, lines 4, 5 *they went not forwarde*

Here end P. and Y. Prol. Ch. xii and B. Ch. xiii.

32. Page 19, line 9 *a yeare and a halfe*

All the chief texts read three years and a half, or, to be exact, "bien trois anz et dimi." Thus if we take the second start from Acre as being in November 1271, F. lets them reach the Khan in May 1273 instead of the more correct May 1275.

33. Page 19, line 12 *fortie dayes iourney*

The best texts add that they were honourably entertained upon the road, and found at each place through which they passed every comfort provided for them.
See Marsden's interesting note (No. 43), pp. 23, 24.
Here end M. Bk. I. Ch. I. Sect. III, P. and Y. Prol. Ch. XIII and B. Ch. XIV.

34. Page 19, line 24 *Lordes and Gentlemen*

With these words F.'s "Prologue" ends, corresponding, at this point, to the end of P. and Y. Prol. Ch. XIV and B. Ch. XV.
From this point the corresponding chapter numbers of M., P., Y. and B. are not given in these notes, but will be found in square brackets immediately beneath those of F. in the text of this volume.

CHAPTER I

35. Page 22, line 2 (Chapter-headings, etc., are *not* counted in the numbering of the lines).

but alſo other thrée languages

Fr. 1116 reads "il soit de langaies et de quatre letres et scriture." B. supplies "quatre" before "langaies." The texts of both P. and Y. read "pluseurs languages" and "iiij. lettres de leur escriptures."
For a note on what the languages might have been see Y. Vol. I. pp. 28–30.

36. Page 22, lines 6, 7

in one of his Countreys, ſixe Monethes iourney

R., *V*, and *L* (which represents a Latin compendium as explained on p. xxix of this volume) give the name of the city as "Carazan," "Chiarenza," and "Caçaram" respectively.
Y. has shown (Vol. II p. 67) this to refer to the province of Yunnan.

37. Page 22, lines 20, 21 *Senior or Lorde*

This appears to be Santaella's attempt to convey the "mesere" of the text. But it is abbreviated—the full translation of the best texts being:

"From the time of this ambassage onwards the young man was called Messer Marco Polo, and so we shall call him in this our book And we have ample justification in so doing for he was both learned and of good breeding."

38. Page 22, line 23 *he was alwayes sente*

"And sometimes," adds R., "also he travelled on his own private account, but always with the consent, and sanctioned by the authority of the Grand Khan."

CHAPTER 2

39. Page 23, line 3

demaunded licence for to returne to Venice

We find much more detail in R :

"Our Venetians having now resided many years at the Imperial court, and in that time having realized considerable wealth, in jewels of value and in gold, felt a strong desire to revisit their native country, and, however honoured and caressed by the sovereign, this sentiment was ever predominant in their minds. It became the more decidedly their object, when they reflected on the very advanced age of the Grand Khan, whose death, if it should happen previously to their departure, might deprive them of that public assistance by which alone they could expect to surmount the innumerable difficulties of so long a journey, and reach their homes in safety; which on the contrary, in his lifetime, and through his favour, they might reasonably hope to accomplish. Nicolo Polo accordingly took an opportunity one day, when he observed him to be more than usually cheerful, of throwing himself at his feet, and soliciting on behalf of himself and his family, to be indulged with his Majesty's gracious permission for their departure ..."

The Khan refused, said he was hurt at the request and was willing to double all their possessions.

Now the point I would like to mention here is that in this longer account given only by R. (and not noted by Benedetto) is that he especially states that it was Nicolo, the head of the family, who made the request. The other texts say "*they* asked many times," yet a little further on *all* texts (except, of course, F. who abbreviates the whole account) agree that Marco had been away on a mission from which he suddenly returns at the psychological moment. Thus the reliability of R. seemed to be supported here.

40. Page 23, lines 5, 6

whose name was Balgonia

Read "Bolgana," i.e. Bulughan, wife first of Ābāka, and then of Arghun, the "Argon" of our text.

41. Page 23, line 12 *Onlora, Apusca, and Edilla*

Fr. 1116 reads: "le primer Oulatai, le segont Apusca, le tierces Coia."

42. Page 23, lines 15, 16

a Mayden, whiche was called Cozotine

Fr. 1116 reads "Cogacin," which Y. gives as "Cocachin." The correct form of the name would be "Kūkāchin"

43. Page 23, lines 27, 28

was content they should goe

Yet R. adds that he "...showed by his countenance that it was exceedingly displeasing to him, averse as he was to parting with the Venetians."

R also has a curious passage, which Yule says is no doubt genuine, in which he states that the party started on their way (without any of the Polos), but that owing to wars they were forced to return It was at this juncture that Marco chanced to return from his mission to India.

CHAPTER 3

44. Page 24, line 7

sent diuers Embassadors to the Pope..

So also fr. 1116 which reads "a l'apostoille .," but Y., following Pauthier's text, ignores the Pope and adds "the King of England"

45. Page 24, lines 9, 10

fouretéene great Shippes

So also in fr. 1116 and R., but Y. reads "thirteen." F. omits to say that the vessels can spread twelve sails. R. reads nine sails, and adds:

"Among these vessels there were at least four or five that had crews of two hundred and fifty or two hundred and sixty men."

I cannot trace the source of F.'s "sixe hundreth men," but see next note.

CHAPTER 4

46. Page 24, line 23

married that mayde to his sonne

According to fr. 1116 his name was "Casan," so also in P and Y. Here again R. gives us much more detail:

"Upon landing, they were informed that King Arghun had died some time before, and that the government of the country was then administered, on behalf of his son who was still a youth by a person of the name of Ki-akato [the "Archator" of F. and "Chia(ca)to" of fr. 1116]. From him they desired to receive instructions as to the manner in which they were to dispose of the princess, whom, by the orders of the late King they had conducted thither. His answer was, that they ought to present the lady to Kasan, the son of Arghun, who was then at a place on the borders of Persia, which has its denomination from the *Arbor secco*, where an army of sixty thousand men was assembled for the purpose of guarding certain passes against the eruption of the enemy. This they proceeded to carry into execution, and having effected it, they returned to the residence of Ki-akato, because the road they were afterwards to take, lay in that direction. Here, however, they reposed themselves for the space of nine months."

F. omits to mention the great loss of men suffered on the journey. Fr. 1116 says that of the six hundred that started, not counting sailors, only eighteen arrived. Y makes the number of survivors eight. R. has quite a different account, and says that of all the crews and other persons six hundred were lost, and of the three ambassadors, only one, named Goza, survived. Of the ladies and their female attendants only one died.

47. Page 25, line 3

foure Tables of gold

According to R., the inscription on the fourth Table began with invoking the blessing of the Almighty upon the Grand Khan, that his name might be held in reverence for many years, and denouncing the punishment of death and confiscation of goods to all who should refuse obedience of the mandate.

F. makes no further reference to the princess, neither does R. or most of the other MSS. Fr. 1116, however, has a most interesting passage which Y. has translated (Vol. I. p. 36) as follows.

"The Great Khan regarded them with such trust and affection, that he had confided to their charge the Queen Cocachin, as well as the daughter of the king of Manzi [also mentioned in Bib. Naz. Florence, II. iv. 88, the *Codex della Crusca*], to conduct to Argon the Lord of all the Levant. And those two great ladies who were thus entrusted to them they watched over and guarded as if they had been daughters of their own, until they had transferred them to the hands of their Lord; whilst the ladies, young and fair as they were, looked on each of those three as a father, and obeyed them accordingly.

Indeed, both Casan, who is now the reigning prince, and the Queen Cocachin his wife, have such a regard for the Envoys that there is nothing they would not do for them. And when the three Ambassadors took leave of that Lady to return to their own country, she wept for sorrow at the parting."

See Pauthier's note on the passage, p. 32

48. Page 25, line 9 *and ſo much they trauelled*

R. says that during their travelling they received news of the death of Kublai Khan. He had died in 1294.

This is probably a later addition, for, as P has pointed out, Polo always speaks of Kublai as if he believed him to be still alive.

CHAPTER 5

49. Page 25, lines 17, 18

Firſt and formoſt...as there is

These two lines seem to be unique to F Fr. 1116 begins directly with. "Il est voir qu'il sunt deus Harmenies." So also with P. and Y.

The Lesser Armenia roughly corresponds to the classical Cilicia.

50. Page 25, line 29 *a Citie called Gloza*

This is the "Laias" of fr. 1116, the modern Ayas. See Note 19, where F. spells it "Giaza" which is decidedly preferable to "Gloza" The evolution of the modern "Ayas" can be seen in the following forms

Gloza—Giaza—La Jazza—Lajazzo—Laias—Layas—Aiasso—Aïas—Ayas.

51. Page 25, lines 30–32

haue their Cellers and Warehouſes in that Citie, as well Venetians, and Ianoueys, and all other that do occupye into Leuant

This passage seems to be unique to F. All the chief MSS merely say that Venetians and Genoese come to trade at Ayas, which city is a starting-point for merchants travelling into the interior.

There is, however, a further passage found both in R. and *Z* (the newly found Latin MS, Bib. Amb. Y. 160. P.S), as well as in *V* (for which see p. xxix of the Introduction and Note 12) where it is somewhat abbreviated. It runs as follows:

"The boundaries of the lesser Armenia are, on the south, the land of Promise, now occupied by the Saracens; on the north, Karamania, inhabited by Turkomans, towards the north-east lie the cities of Kaisariah, Sevasta, and many others subject to the Tartars; and on the western side it is bounded by the sea which extends to the shores of Christendom."

Both *Z* and *V* add "Turchia" before "Cayssaria, & Sevasta."

CHAPTER 6

52. Page 26, line 2 *Torchomania*

This practically corresponds to Asia Minor, or perhaps better, to the Asiatic country of Rūm.

53. Page 26, lines 4, 5

and ſpeake the Perſian language

This is the first occurrence of a very curious mistake, which we shall find many times in Frampton's translation.

Santaella has translated "lingua per sì" as "lengua de persianos"; thus instead of speaking a language "peculiar to themselves," we shall find peoples all over the East speaking "the Persian language"!

54. Page 26, lines 8, 9

The other, or ſecond maner of people be...

This is a mistake for "The other two..." or some similar wording, as he has distinctly said there are three "maner of people," and no third is given.

55. Page 26, line 13 *Chemo, Iſiree, and Sebaſto*

It is not easy to see Konia, Kaisariya and Sivas in these corrupted forms.

"Chemo" must be a misprint for Chonio, the "Conio" of fr. 1116. "Isiree" is intended for Cassene, while "Sebafto," the least corrupted, is the "Savast" of Y. and "Sevasto" of fr 1116.

56. Page 26, line 14 *Saint Blase*

Blasius, not mentioned in fr. 1116 or Y., but occurring in R., became patron saint of wool-combers, owing, apparently to the fact that before being beheaded (in A.D. 316) his flesh was torn off him by wool-comber's irons.

57. Page 26, line 16 *he ſetteth gouernoures there*

All MSS end the chapter here, but some additional information is to be found in the *Imago Mundi* of Jacopo d' Acqui. This is reproduced by Benedetto as note (b) on p. 14 of his edition. See also pp. cxciii–cxcviii for details of this work, which he calls *I* The MS is in the Ambrosian Library at Milan (D 526)

CHAPTER 7

58. Page 26, lines 18, 19

a greate Citie called Armenia, where they doe make excellente Bochachims or Buckrams

A mistake for Arzinga or Arzingal, the modern Erzingan, ninety-seven miles west of Erzerum.

R. reads "bochassini di bambagio," but exactly what these buckrams are is uncertain. See Yule's interesting note, Vol. I. pp 47 *et seq.*

R and *Z* add that the city also produces "many other curious fabrics, which it would be tedious to enumerate."

59. Page 26, lines 22–24

Archinia... Archeten... Arzire

These appear in Y. as "Arzinga," "Arziron" and "Arzizi"; and in B. as "Arçingal," "Argiron" and "Darçiçi." They correspond to the modern Erzingan (which we have already had above), Erzerum, and Ardjish close to the north-eastern shores of Lake Van.

It was Erzingan which was the see of an archbishop, not Ardjish as stated by F

R. and *Z* mention a castle at Paipurth or Paperth (Baiburt), half-way between Trebizond and Erzerum, where is a silver mine. *V* also speaks of the mine, but does not mention Paipurth.

60. Page 27, line 2

ỳ Arke of Noe on a high Mountain

Further details are found in R., *Z* and *I*. The fullest account is that of *Z*, which is as follows:

"In the central part of Armenia stands an exceedingly large and high mountain, upon which, it is said, the Ark of Noah rested, and for this reason it is termed the mountain of the Ark. The circuit of its base cannot be compassed in less that two days. The ascent is impracticable on account of the snow towards the summit, which never melts, but goes on increasing by each successive fall.

In the lower region, however, near the plain, the melting of the snow fertilizes the ground, and occasions such an abundant vegetation, that all the cattle which collect there in summer from the neighbouring country, meet with a never failing supply."

61. Page 27, line 3

towardes the East called Mausill

In addition to Mosul, R. mentions "Maredin" (Mardin), while *Z* gives "Musul Mus et Meridin."

CHAPTER 8

62. Page 27, lines 13, 14

Nand Maliche...Dawnid

Here the text is muddled. The correct reading should be "Davit Melic... Davit roi."

63. Page 27, line 16 *with a token or signe*

F. omits to say what sign. It was that of an eagle, as all texts clearly state.

64. Page 27, line 17 *In this Countrey*

R., *Z*, *L* and *V* all give details of the country not found in other texts. See B p. 16

The passage in R. is as follows:

"One part of the country is subject to the Tartars, and the other part, in consequence of the strength of its fortresses, has remained in the possession of its native princes It is situated between two seas, of which that on the northern (western) side is called the Greater sea (Euxine), and the other, on the eastern side, is called the sea of Abakù (Caspian). This latter is in circuit two thousand eight hundred miles, and partakes of the nature of a lake, not communicating with any other sea It has several islands, with handsome towns and castles, some of which are inhabited by people who fled before the Grand Tartar, when he laid waste the kingdom or province of Persia, and took shelter in these islands or in the fastnesses of the mountains, where they hoped to find security. Some of the islands are uncultivated This sea produces abundance of fish, particularly sturgeon and salmon at the mouths of the rivers, as well as others of a large sort. The general wood of the country is the box-tree."

65. Page 27, line 28 *Tower and gate of yron*

The French texts add further details. Y. translates:

"This is the place that the Book of Alexander speaks of, when it tells us how he shut up the Tartars between two mountains; not that they were really Tartars, however, for there were no Tartars in those days, but they consisted of a race of people called Comanians and many besides."

66. Page 27, line 32 *Hawkes*

R. adds, "of a species named *avigi.*"

67. Page 28, lines 4, 5

a Monaſterie of Monckes of...Saint Bernarde

This seems to be a corruption from the "monasterio intitolato di San Lunardo di monachi" of R.

It was, however, a nunnery called St Leonard's.

The reading of fr. 1116 is: "un monester de nonain qui est apelé sant Lionard...." So also in P. and Y.

68. Page 28, lines 8, 9

Geluchelan...ſixe hundred Miles compaſſe

The Caspian; fr. 1116 and Y. read "seven hundred," the former definitely stating that figure is the circumference ("gire") of the lake. Benedetto, with the help of the *Z*, *L* and *V* MSS. has altered "VII^{c} to "II^{M}VII^{c}" which is undoubtedly correct. The Caspian is 760 miles long and its circumference has been given by Halbfass (*Peter. Mitt. Geog.* Erganz. No. 185, pp. 18–19) as 6000 kilos. Thus Yule's objection (Vol. I. p. 59) to 2700 miles as being too large is quite unfounded. It is just about 1000 miles too small

69. Page 28, line 11 *from Paradice terrenall*

Cf. with Note 10. Here again the passage seems to be unique to F.

For an article on the four rivers of Paradise, mentioned in Genesis ii. 10–14, see Hastings, *Dict. of the Bible*, under "Eden."

70. Page 28, line 13 *a ſilke called Gella*

Spelt "Gelle" in fr. 1116, written by B. as "G[h]elle"

The name given to the silk is undoubtedly derived from "Ghel," the Caspian.

CHAPTER 9

71. Page 28, line 20

a Patriarke, called Iacobia

Read "Jatolic" with B., P. and Y., standing for καθολικός. See Y. Vol. I. p. 61.

F. omits to add that he sends them out to India, Baudas, and Cathay, just as the Pope does in the Latin countries.

72. Page 28, line 24

and of other Merchandise

Apart from muslins and spices, Y. speaks of pearls and cloths of silk and gold.

At the end of the chapter R., *Z*, *L* and *V* mention both "Mus" (Mush) and "Merdin" (Mardin), and refer to the large quantities of cotton they produce. The "Cordos" are, of course, the Curds

CHAPTER 10

73. Page 29, line 3

chiefe gouernour & head

All the best texts add "as at Rome the Pope is of all the Christians."

74. Page 29, lines 10, 11

Betwéene Baldach and Chisi vppon the Riuer is a Citie called Barsera

The fact that in all MSS Polo makes the Tigris flow through Baghdad (Baldach of F., Baudac of fr. 1116), Basra *and* the island of Kais (Chisi), about 165 miles from the mouth of the gulf, is surely sufficient to show he is not speaking from personal knowledge. If he ever did visit Baghdad personally we would expect a much more definite proof of the fact See pp xxxiv *et seq* of the Introduction.

75. Page 29, line 13

cloth of Nasich, of Chrimson...

Fr. 1116 reads "nassit et nac et cremosi"—"stuffs of silk and gold." See Y. Vol. I. pp. 65, 66 The text then explains how the materials are wrought with figures of animals, but F., not realizing the connection, tells us that the country is well supplied with "foure footed Beastes" and "Fowles." See, further, Note 207.

76. Page 29, line 15

This Citie is one of the beft...

Before this statement R., *Z*, *L* and *V* have an additional passage. R. reads: "Almost all the pearls brought to Europe from India have undergone the process of boring, at this place. The Mahommedan law is here regularly studied, as are also magic, physics, astronomy, geomancy and physiogonomy."

77. Page 29, line 18

.1230....*Alan*...

Read with the best texts "1255...Alau. .."

78. Page 29, line 19

and toke it by force

R. contains considerably more detail about the capture of Baghdad and the death of the last of the Abbasides, Mosta'sim Billah As the passage in question is lengthy, it will be found in full in Appendix II No 1, pp. 263, 264.

CHAPTER II

79. Page 30, line 1

Totis is a greate Citie...of Baldach

Read "Toris [Tauris]...of Yrac ['Irak]."

F. has given this chapter before the account of the miracle of the mountain, which comes first in all the best MSS.

80. Page 30, lines 7, 8

and of Ofmafeilli, and of Cremes

I.e. of Mosul and Cremosor or Garmsir, the Hormuz district of the Persian Gulf. Y. omits "Mosul," which is curious as it is in both P. and fr 1116.

81. Page 30, line 13

robbers and killers

R. *Z* and *V* give additional information. I take the following from R :

"The Mahommedan inhabitants are treacherous and unprincipled. According to their doctrine, whatever is stolen or plundered from others of a different faith, is properly taken, and the theft is no crime; whilst those who suffer death or injury by the hands of Christians, are considered as martyrs If, therefore, they were not prohibited and restrained by the powers who now govern them, they would commit many outrages. These principles are common to all the Saracens.

When they are at the point of death their priest attends upon them and asks whether they believe that Mahommed was the true apostle of God. If their answer be that they do believe, their salvation is assured to them; and in consequence of this facility of absolution, which gives free scope to the perpetration of everything flagitious, they have succeeded in converting to their faith a great proportion of the Tartars, who consider it as relieving them from restraint in the commission of crimes. From Tauris to Persia is twelve days journey."

Ramusio then gives a short chapter: "Of the Monastery of Saint Barsamo, in the neighbourhood of Tauris." There is no need to give it here as Y. has included it in full (Vol. I. p. 77). It is curious, however, that he does not print it between square brackets, to show it is from R.

CHAPTER 12

82. Page 30, line 14

In Mofull, a Citie in the Prouince of Baldach

Here again our text is muddled. It should place the miracle as having taken place between the two cities.

Fr. 1116 gives the date, omitted in Y., as 1275. For a note on this see B. p. 20.

83. Page 32, lines 25, 26

he lyued and dyed like a true and faythfull Chriſtian

So ends F.'s abbreviated account of the legend. R. and Z have two unique passages Firstly, after the cobbler has offered up his prayer he cries in a loud voice: "In the name of the Father, Son, and Holy Ghost, I command thee, O mountain, to remove thyself!"

Secondly, the chapter in R. ends with the following:

"In commemoration of this singular grace bestowed upon them by God, all the Christians, Nestorians, and Jacobites, from that time forth have continued to celebrate in a solemn manner the return of the day on which the miracle took place, keeping fast also on the vigil."

CHAPTER 13

84. Page 32, line 30 *the thrée Kings*

Fr. 1116 and Y. distinctly read "three Magi." Their names are given as Beltasai, Gaspar and Melchior.

85. Page 33, line 7 *Calaſſa Tapeziſten*

Fr. 1116 has "Cala Ataperistan," so also P. and Y. Its locality has not yet been ascertained, but Y. would put it between Saveh and Abhār. See also Cordier, *Ser Marco Polo*, p. 18

Chapter 13 of F , representing an abbreviation of two corresponding chapters in P., Y. and B , is not found in R.

CHAPTER 14

86. Page 34, line 1 *eyght Kingdomes*

Viz. Kazvīn, Kurdistān, Lūristān, Shūlistān, Ispahān (the reading in Y. is "Istanit," an obvious corruption; B. has "Isfaan"), Shīrāz, Shabānkāra and Tūn-o-Kâin. Some of the readings in F. are sadly corrupted. He omits to mention that all the above kingdoms lie towards the south, except the last which lies in an easterly direction bordering on the country of the Arbre Sec, or Arbre Sol.

87. Page 34, line 7 *& courſers of great value*

F. omits to mention the actual prices; "...il vendent le un bien cc libre de tarnis" and the asses, "un trointe mars d'argent" says fr. 1116. Yule states that the *livre tournois* of Marco's time was equivalent to a little over 18 francs of modern (1903) French silver. Z and R. have a passage about the advantages of the ass over the horse. R. adds: "Camels also are employed here, and these in like manner carry great weights and are maintained at little cost, but they are not so swift as the asses."

88. Page 34, lines 9, 10 *Atriſo, & of Arcones*

These are very corrupted forms of Chisi and Curmosa.

89. Page 34, line 14 *robbed or taken priſoner*

R. adds: "A regulation is also established that in all roads, where danger is apprehended, the inhabitants shall be obliged, upon the requisition of the merchants to provide active and trusty conductors for their guidance and security, between one district and another; who are to be paid at the rate of two or three groats for each loaded beast, according to the distance "

Both R. and Z have an interesting passage about wine-drinking:

"Should anyone assert that the Saracens do not drink wine, being forbidden by their law, it may be answered that they quiet their consciences on this point by persuading themselves that if they take the precaution of boiling it over the fire, by which it is partly consumed and becomes sweet, they may drink it without infringing the commandment; for having changed its taste, they change its name, and no longer call it wine, although it is such in fact."

CHAPTER 15

90. Page 34, line 19 *Iasoy is a goodly Citie...*

I.e Yasdi, or Yezd.

91. Page 34, line 23 *another language*

This mention of language seems to be unique to F.

92. Page 34, line 24 *eyght dayes*

So also in R, but the French texts read "seven"

93. Page 35, line 1 *Crerina*

Read "Cherman," the "Kierman" of R, the "Creman" of P., the modern Kirmān.

94. Page 35, line 2

of a great and long inheritance

All the best texts give more details: " ..it was formerly governed by its own Princes in hereditary succession; but since the Tartars brought it under their dominion, the rule has ceased to be hereditary and they appoint as governors what lords they wish."

95. Page 35, line 5

plentie of Uayne, or Ore of Stéele, and of calamita

Read "plentie of veins of steel and ondanique"

E H Parker (*Journ. North China Br. Roy. As. Soc.* XXXVIII. 1907, p. 225) considers the "ondanique" to be the *pin t'ieh*, or "pig iron" of the Chinese. See Cordier, *op. cit* p 19

96. Page 35, line 12 *eyght dayes*

As before (Note 92) read "seven."

97. Page 35, line 19

haue ynough to do to liue

This corresponds to the end of M Bk. I Ch XIII, Y Bk I. Ch XVII and B. Ch. XXXV, but Z contributes an entirely fresh passage on Kirmān and its king. See B p 27 note a.

98. Page 35, lines 22–24

Camath...Reobarle

Fr. 1116 reads "Camandi...Reobar," i.e Camadi and Rudbar See the map facing p. xxxvi, of the Introduction.

99. Page 35, line 25

goodly frutes in great abundance

At this point R (also Z) reads: "Turtle doves are found here in vast numbers, occasioned by the plenty of small fruits which supply them with food, and their not being eaten by the Mahommedans, who hold them in abomination." All the best texts then continue:

"and on this plain there is a kind of bird which is called francolin, but different from other francolins of other countries, for their colour is a mixture of black and white, and their feet and beak are red."

100. Page 35, line 31

a greate tayle...that will weigh .32. pound

Fr. 1116 and all leading texts read "thirty" or "a good thirty." On the fat-tailed sheep of Persia see Cordier, *op. cit.* p. 19.

101. Page 35, line 33 *Caraones*

I.e. Ḳaraunaha, for which see Yule's note, Vol. I. pp. 101 *et seqq.*, and Cordier, *op. cit.* p 21

F omits to mention that they ride abreast to about the number of 10,000, but sometimes more or fewer.

102. Page 36, line 1

Their King is called Hegodar

I.e. Nogodar. F. omits to describe the most interesting inroad to Kashmir, as given in all the best MSS. See Introduction, p. xl, and accompanying map.

R gives a few extra details about the "Karaunas":

"...and these are the people who have since been in the practice of committing depredations, not only in the country of Reobarle, but in every other to which they have access"

After speaking of the magical darkness (dry fog and dust storm), he adds:

"Most frequently this district is the scene of their operations; because when the merchants from various parts assemble at Ormuz, and wait for those who are on their way from India, they send, in the winter season, their horses and mules [*sic* Marsden, but the text has "muli e camelli," mules and camels, very distinctly, "cavalli" is Ramusio's word for horses] which are out of condition from the length of their journies, to the plain of Reobarle, where they find abundance of pasture and become fat. The Karaunas, aware that this will take place, seize the opportunity of effecting a general pillage, and make slaves of the people who attend the cattle, if they have not the means of ransom."

103. Page 36, line 4 *a towne called Ganaſſalim*

Santaella's text gives "n" as the last letter. The 1503 edition has only one "s," i e. Canosalmi, as in fr. 1116; or Conosalmi as in Y. Its identification is not certain, but the suggestion of Houtum-Schindler, that it is the ruined town of Kamasal (Kahn-i-asal) near Kahn-i-panchar and Vokīlābād, seems to be much the best.

F. says Polo lost *many* of his companions Fr. 1116 has "sex" and P. has "seft."

104. Page 36, lines 5–7

and is of ſeauen dayes iourney, and at the end of them is a moũtayne, called Detuſtlyno, that is eighteene miles long...

All the best texts read "five" days journey, but there is nothing about a mountain called Detustlyno. Fr. 1116 reads

".. un autre clinee que convent que l'en aille pur au declin xx milles...."

Santaella's text clearly says "vn monte que llaman Detuſclino que dura en luengo ſeys leguas y media." Thus F. is true to his text, except that "t" has become "c," and "i" appears as "y." We can, however, see the "clinee" of the French texts in Sontaella's jumbled word.

105. Page 36, line 9 *the goodly playne*

I.e. Harmuza, the "Formosa" of the French texts.

106. Page 36, line 12 *Citie called Carmoe*

The "Cormos" of fr. 1116, and "Hormos" of Y. It is the Old Hormuz, on the mainland near the present Mīnāb, later transferred to the island. See A. T. Wilson, *The Persian Gulf*, 1928, pp. 100–109.

CHAPTER 16

107. Page 36, lines 17, 18

the king is called Minedanocomoyth

Written "Ruemedan Acomat" in fr. 1116 and "Ruomedan Ahomet" in Y. This is Rukn ud Din Muhammad (Wilson, *op. cit* p. 104) or possibly Rokn ed-Din Mahmud III (Cordier, *op. cit.* p. 24), but see further, Note 111.

108. Page 36, lines 19, 20

they doe make hauocke of all his goods

All the best texts state that the king confiscates the property of the deceased.

109. Page 36, line 27

And for the great heate in the Sommer...

F. has unfortunately omitted the important section about the boats being unseaworthy. This plays a part in our attempt to reconstruct the itinerary through Persia. See p. xxxvi of the Introduction.

The account of the summer heat is much abbreviated, and the terror of the hot wind ignored. Ramusio has an interesting story about the "ruler of Ormus" and how his body of troops was suffocated by the hot wind.

In view of the above, I have reprinted all these passages from R. in full. See Appendix II. No. 2, pp. 264, 265.

CHAPTER 17

110. Page 37, lines 8, 9

and not declaring any more, of the Indians

This is a mistake for "not go on to declare about India," or some such statement; the original corresponding passage in fr. 1116 is:

"Et ne vos contaron de Endie a cestiu point, car vos bien le conterai en notre livre avant, quant tens et leu sera."

111. Page 37, line 13 *Reu me cla vacomare*

This apparently represents a very corrupt form of Ruemedan Acomat, as found in fr 1116. There is considerable difficulty in determining which king of Hormuz is meant. It should be either Ruknuddin Masa'úd or Fakhruddin Ahmed (the spellings vary greatly), though it is possible that Polo has muddled the two and so produced his Ruemedan Acomat, or Ruomedam Ahomet (as in Yule's texts). Frampton has inadvertently connected this ruler with the Old Man of the Mountain whose cruelties are made the excuse for Polo's returning to Kirmān by a different route. There should be, of course, no connection whatever. Polo had experienced extreme cold on his journey to Hormuz, but apart from not wishing to endure this again unnecessarily, he would also have to encounter a "slope" that took over two days *to come down*. No wonder he preferred another route on the return journey. F. tells us nothing of this route It was in all probability *via* the Urzū district and Bāft See the Introduction, p xxxvii It is described as lying through fine plains with "abundance de viandes," including partridges, fruits (especially dates), wheaten bread, which is bitter owing to the water, and hot baths which cure the itch and other skin diseases.

112. Page 37, line 17

in the yeare of our Lord .1272. Alan...

Fr. 1116 reads "1262," and Y. "1252," which latter is the more correct date. See Y. Vol. I. p. 146. The name should, of course, read "Alau," for Hulaku. He took the fortress of Alamūt in 1265. By jumping to this part of the story of the Old Man of the Mountain, F. not only misses out all the first part of the tale, but omits the portion of the itinerary from Kirmān to Cobinan (Kuh-Banān) and Tunocain (Tun and Kain). See Introduction, pp. xxxvii–xxxviii.

I have, accordingly, restored all this in Appendix II. No 3, pp. 266–8.

CHAPTER 18

113. Page 38, lines 2, 3

with all things in it fitte for mans ſuſtenance

The French texts add that armies (les ost) gladly stay here on account of the great plenty that exists.

114. Page 38, lines 5, 6

ſpeake the Persian language

F. has this instead of "et les homes aorent Maomet" as found in fr. 1116 and other leading texts.

115. Page 38, line 7 *to go .40. miles*

Fr. 1116 has "Et alcune foies trouve l'en desert de LX milles, et de L, es quelz ne i se trove eive, .."

P. reads "...de soixonte milles ou de mains," while R. has "40 or 50." I rather suspect this latter is correct, and that the "LX" of fr. 1116 is a mistake for "XL."

Y. writes "50 or 60" without comment.

116. Page 38, line 10

a Citie called Sempergayme

This is the "Sapurgan" of all the best MSS, as well as R.

117. Page 38, line 11

There be excellente good Mellones

This is abbreviated. The French texts add that the fruit is cut in strips, and dried in the sun, when it becomes sweeter than honey. In this form it finds a large sale in all the country round.

CHAPTER 19

118. Page 38, line 14 *a Citie called Baldach*

In speaking of "Baldach," i e Balk, F. omits to mention its former greatness now reduced owing to the injury received at the hands of the Tartars "Many of its fine palaces and marble buildings are still visible," add the texts (including R.), "but only in a ruinous state"

119. Page 38, line 17

ſpeake the Perſian tong, and be all of the ſect of Mahomet

As before, in Note 114, the reference to the language is an addition of F The second part of the line, however, is found in fr. 1116: "Les gens aorent Maomet," but not in the Y. texts It must also be in the majority of the Pipino texts as it is in R as well as F

120. Page 38, lines 19, 20

And departing from this Citie...you ſhall goe two dayes...

Here, as in R., no fresh place is mentioned, and in two days we arrive at "Thaychan" (Talikan). In Yule and fr. 1116, however, we hear of "Dogana" (called

"Gana" in two of the P. texts), after leaving which a twelve days' ride brings the travellers to Talikan.

It is very difficult to determine where "Dogana" is and what is included by the term. I have discussed the question as best I can in the Introduction, p. xl.

121. Page 39, line 1 *a Towne called Thaychan*

The account of Talikan is abbreviated. The fullest account is found in R.: "...you reach a castle named Thaikan, where a great market for corn is held, it being situated in a fine and fruitful country. The hills that lie to the south of it are large and lofty. They all consist of white salt, extremely hard, with which the people, to the distance of thirty days' journey round, come to provide themselves, for it is esteemed the purest that is found in the world; but it is at the same time so hard that it cannot be detached otherwise than with iron instruments. The quantity is so great that all the countries of the earth might be supplied from thence. Other hills produce almonds and pistachio nuts, in which articles the natives carry on a considerable trade." [This last passage occurs only in R. and Z.]

CHAPTER 20

122. Page 39, line 9 *do ſpeake the Perſian language*

See Note 119, first paragraph.

123. Page 39, line 10 *ſingular good wines*

The French MSS add that the wine is boiled.

CHAPTER 21 (misprinted 12 by F.)

124. Page 39, line 13 *foure dayes iourney*

Read "three" days with all the best texts.

125. Page 39, line 14 *a Citie called Echaſen*

I.e. Casem, the modern Kishm. The French MSS add that it is subject to a count ("un cuens").

126. Page 39, lines 17, 18 *many wilde beaſtes*

F. omits to mention porcupines by name, which are described as rolling themselves into balls and shooting out their quills at the hunting-dogs.

127. Page 39, line 20 *doe ſpeake the Perſian tong*

Here is the old mistake for "lingua per si." See Note 53.

CHAPTER 23

128. Page 40, line 6

Ballaſia is a great prouince, & they do ſpeake the Perſian tong

This is Badashan, the modern Badakhshan, in north-east Afghanistan. For "Perſian tong" read "a language peculiar to themselves" See Note 53

R. adds that the kingdom is a full twelve days' journey in length.

129. Page 40, line 9

Darius king of Perſia... Culturi

F. has omitted "daughter of" before "Darius"

"Culturi" is a very corrupted form of "Zulcarnein," i.e. Zu-'lkarnain, the "two-horned," an Arabic epithet of Alexander which probably arose from the horned portraits on his coins. See Y. Vol I. pp. 160, 161.

130. Page 40, line 12 *ſtones, called Ballaſſes*

I.e. the balas-ruby, a rose-red spinel, deriving its name from Badakhshan and Balk. The mines are in the Ghāran country, which stretches along both sides of the Oxus. They have not been worked for a very long time. References are extremely scarce, but Romanowski, in his *Materialien zur Geologie von Turkestan*, 1880, p. 37, quotes A. Born, *Reise in der Bucharei, 1831–3*, III. Th. I. pp. 292–4, Moscow, 1849, where there is a brief description of the mines. See also Stein, *Innermost Asia*, Vol. II p. 877.

131. Page 40, line 18 *greate plenty of ſiluer*

I can find nothing concerning these mines at the Geological Society or elsewhere, save the notes by Wood, *Journey to the Source of the River Oxus*. They were known by the name of Lājwurd, and lay in the Korān valley of the Kolicha See Y Vol I p 162.

132. Page 40, line 18

there be very good courſers, or horſes

Here F. has abbreviated very much, while R., on the other hand, has several additions. The whole passage from R. is therefore given in full. See Appendix II. No. 4, pp. 269, 270.

CHAPTER 24

133. Page 41, lines 1, 2

eyght dayes iourney...a prouince called Abaffia

All texts read "ten." This is Pashai or Pasciai, which as we have seen (Introduction, p. xli) is an area in Kafiristan stretching as far east as the Kunar river. The old mistake about the "Perfian tong" follows. See Note 53

CHAPTER 25

134. Page 41, line 9

a Prouince called Thaffymur

Written "Thafsimur" in the chapter heading. This is, of course, Kashmir. See Note 53 re "Perfian tong," as before.

135. Page 41, lines 14, 15

The people of that Countrey be blacke and leane

Not "the people," but "the men"; and the best texts proceed: "but the women, although dark, are very beautiful"

136. Page 41, line 19 *There be alfo Hermites*

This is abbreviated. The fullest account is in R. In the following passage the part about the natives not killing animals is exclusive to *Z* and R.:

"They have amongst them a particular class of devotees, who live in communities, observe strict abstinence in regard to eating, drinking and the intercourse of the sexes, and refrain from every kind of sensual indulgence, in order that they may not give offence to the idols whom they worship. These persons live to a considerable age. They have several monasteries, in which certain superiors exercise the functions of our abbots, and by the mass of the people they are held in great reverence. The natives of this country do not deprive any creature of life, nor shed blood, and if they are inclined to eat flesh-meat, it is necessary that the Mohamedans who reside amongst them should slay the animal."

CHAPTER 27

137. Page 42, line 5 *threé dayes iourney*

This should read "twelve" with the best MSS.

138. Page 42, lines 9, 10

a Citie called Vochayn

The "Vocan" of fr. 1116 and "Vokhan" of Y, the modern Wakhān. Y. also tells us that the inhabitants are gallant soldiers, and have a chief whom they call None, which is as much as to say *Count*, and they are liegemen to the Prince of Badashan

CHAPTER 28

139. Page 42, lines 19–21

And of theſe hornes. . .mountaine called Plauor, you ſhall trauell tenne dayes iourney

This is the first time the Pāmirs are mentioned in any work. The form has been well preserved as both fr. 1116 and Yule read "Pamier." F 's corrupted form is merely due to the numerous languages and scribes through which it has passed.

Apart from the uses mentioned to which the horns of the Pāmir sheep (*Ovis Poli*) are put, Ramusio adds

"and with the same materials they construct fences for enclosing their cattle, and securing them against the wolves, with which, they say, the country is infested, and which likewise destroy many of these wild sheep and goats. Their horns and bones being found in large quantities, heaps are made of them at the sides of the road, for the purpose of guiding travellers at the season when it is covered with snow."

The passage about the wolves is also in *Z*. The latter portion attests to the veracity of Ramusio. The piles of heaped horns have been noted by several travellers; see Y. Vol I p. 176

For "tenne dayes iourney," read "twelve" with all the best MSS.

140. Page 43, lines 3, 4

the fire hath not the ſtrength to ſéethe their victuals, as in other Countries

In the first place, F omits to tell us that the country is so cold that you never see birds flying. (See Stein, *Innermost Asia*, Vol. II p. 860) In the second place, we note that R. introduces the remark about the fire by "and however extraordinary it may be thought, it was affirmed, that from the keenness of the air, fires when lighted do not give the same heat as in lower situations,. . . ."

This reads as if Polo had not seen the phenomenon for himself, but here we can surely see the hand of the editor who, while anxious to record what he reads, finds his own credulity a bit strained at times.

See the interesting notes on the subject collected by Y. Vol I p. 178

CHAPTER 29 (misprinted 39 by F.)

141. Page 43, lines 12, 13

And this Countrey is called Boſor

Written "Belor" in fr. 1116 and "Bolor" in Yule. Its exact modern equivalent is unknown, but the dreary route described by Polo must refer to the district east of Little Pamir, and his itinerary would have led him past Muztāgh-Ata and on towards the Gez defile.

It is interesting to note that this chapter is actually slightly fuller than the French texts.

CHAPTER 30 (misprinted 40 by F.)

142. Page 43, line 16

Cafchar

This is, of course, Kashgar. R. says the inhabitants produce flax and hemp as well as cotton, and that besides being covetous and sordid, they eat badly and drink worse!

F. makes his usual mistake about the "Perfian tong." See Note 53.

CHAPTER 31

143. Page 44, line 1

Sumarthan is a Citie great and faire

R. adds that Samarkand is adorned with beautiful gardens, and surrounded by a plain, in which are produced all the fruits that man can desire.

F. describes the city as being under the "great Cane," but fr. 1116 has "neveu dou grant can," while Y. adds his name as "Caidou." A little lower F. speaks of "a brother of the greate Cane" without giving his name. This is, however, given in most MSS as "Ciagatai" or "Sigatay." It should be pointed out that Chagatai was uncle, not brother, to the Great Khan.

F.'s account of the miracle is, as usual, somewhat abbreviated.

CHAPTER 32

144. Page 44, line 23

a prouince called Carcham

This is Yarkand, and appears practically the same, "Yarcan," in the French MSS.

There is the following interesting addition found in *Z*, *V*, *L* and R.:

"Provisions are here in abundance, especially cotton. The people are craftsmen. They are largely afflicted with swellings in the legs, and tumours in the throat, occasioned by the quality of the water they drink."

In support of the above, Sven Hedin notes that to-day three-fourths of the population of Yarkand are suffering from goitre.

CHAPTER 33

145. Page 45, lines 1, 2

Chota.. fiue dayes iourney

All the best texts describe Khotān as being eight days in length. It is very interesting to note that in common with R. alone F. speaks of wine, fruits, oil, wheat, and barley as well as cotton wool.

146. Page 45, line 7

good and valiaunt men of armes

Here F. has turned a negative into a positive. Fr. 1116 clearly reads "Il ne sunt pas homes d'armes."

CHAPTER 34

147. Page 45, line 8 *Poym is a small prouince*

Also written "Pem" (fr. 1116) and "Pein" (Y.). It is the "Pimo" of Hiuen-Tsiang, and is probably to be identified with the modern Keriya. See Introduction, p. xlii.

148. Page 45, line 15 *fiftéene or thirtie dayes*

All the best MSS distinctly say "twenty"

CHAPTER 35

149. Page 45, line 20

a great Citie called Ciarchan

This place has preserved its name unaltered—Charchan or Chachan. The form given by F. closely resembles the "Ciarcian" of fr. 1116.

150. Page 46, line 6 *fering the ill people*

There is some mistake here. The texts clearly read "when an army passes through the land...."

CHAPTER 36

151. Page 46, lines 14, 15

a gret Citie called Iob

I.e. Lop, the modern Charkhlik, on the edge of the desert. See Stein, *Innermost Asia*, Vol. I. pp. 163 *et seq.*, and Map 30 in Vol IV; also *Serindia*, Vol. I. pp. 311 *et seq.*

152. Page 46, line 23 *the ſound of Tabers...*

The fullest account of these spirits of the desert is to be found in R.:

"...In the night-time they are persuaded they hear the march of a large cavalcade on one side or the other of the road, and concluding the noise to be that of the footsteps of their party, they direct theirs to the quarter from whence it seems to proceed; but upon the breaking of day, find they have been misled and drawn into a situation of danger.... It is said also that some persons, in their course across the desert, have seen what appeared to them to be a body of armed men advancing towards them, and apprehensive of being attacked and plundered have taken to flight.... They find it necessary also to take the precaution before they repose for the night, to fix an advanced signal, pointing out the course they are afterwards to hold, as well as to attach a bell to each of the beasts of burthen for the purpose of their being more easily kept from straggling." Stein (*Innermost Asia*, Vol. I. p. 306) says this fear of danger from evil spirits is as lively to-day as ever.

CHAPTER 37

153. Page 47, lines 2, 3

Sangechian...Tanguith

I.e. Sachiu or Saciou, the modern Sha-chau or Tun-huang on the Tang-ho. Tanguith is a corruption for Tangut, the modern Kansu.

154. Page 47, lines 12–14

the which they can not doe, for they haue neyther mouth nor ſenſe, and ſeeing their Idols do not eate it...

This has been added in deference to the Christian readers who could never believe that the idol really ate the food! but fr. 1116 makes it clear, whereas Yule misses the point by saying, "And, if you will believe them, the idol feeds on the meat that is set before it!" Fr. 1116 reads: "et dient que le ydre menuient *la sostance* de la cars." This we can all understand. R., *Z*, *L* and *V* add "The priests of the idol have for their portion the head, the feet, the intestines, and the skin, together with some parts of the flesh."

155. Page 47, line 20

all clothed in cloth of golde and ſilke

It was not the mourners who were so dressed, but a small house was draped with cloths of silk and gold, before which the body was set down and offered food and drink.

156. Page 48, line 2 *and put it in a coffin*

R. adds. "the joints or seams they smear with a mixture of pitch and lime, and the whole is then covered with silk"

157. Page 48, line 8

remoue him to ſome other ſide of the houſe

Here F. misunderstands the text. A hole was to be made in the wall through which the body was to be taken. R adds a passage saying that any mishaps which occur later are attributed to some breach of the etiquette

CHAPTER 38

158. Page 48, line 10 *Chamul is a prouince*

I e. Hami, lying off the main route to the north-west. This may have been the route followed by the elder Polos See Introduction, p xliii, where I have discussed this digression and the additional mention of "Carachoco" as found in the Z text.

159. Page 48, line 26

the greate Cane that is paſte

The best texts give his name, Mongu or Mangu Kaan

CHAPTER 39

160. Page 49, line 6 *Hingnitala is a prouince*

Fr. 1116 has "Ghinghintalas" Its exact locality has not been determined, but following Polo's description closely I have suggested (Introduction, p xliv) that it should in all probability be looked for in the neighbourhood of Barkul.

161. Page 49, lines 13, 14

that whiche is called Salamandra

F. omits to mention that steel and ondanique are also found. In most texts Polo explains that it was a Turkish friend of his, named Zurficar, who told him all about the Salamanders, or asbestos See Laufer, *T'oung Pao*, Vol. xvi. 1915, pp. 299–373.

162. Page 49, line 27 *a prouince called Sachur*

The ten days must be taken from Tun-Huang (or Sachiu) to which Polo now returns. Sachur is the "Succiu" of fr. 1116 and the "Sukchur" and "Sukchu" of Y. It is to be identified with the modern Suhchau, or Su-chow.

163. Page 49, lines 31, 32

and carry it to all places to ſel

R. and Z add an interesting passage. R. reads as follows:

"It is a fact that when they take that road, they cannot venture amongst the mountains with any beasts of burthen excepting those accustomed to the country, on account of a poisonous plant growing there, which, if eaten by them, has the effect of causing the hoofs of the animal to drop off; but those of the country, being aware of its dangerous quality, take care to avoid it.... The district is perfectly healthy, and the complexion of the natives is brown."

CHAPTER 40 (misprinted 44 by F.)

164. Page 50, line 1 *Campion is a greate Citie*

This is the "Canpicion" of fr. 1116, the modern Kan-chau, chief city of Kansu.

165. Page 50, line 5

the Idolators haue also Monaſteries

F. omits the interesting account of the idols. R. gives the best account: "and in these [monasteries] a multitude of idols, some of which are of wood, some of stone, and some of clay, are covered with gilding. They are carved in a masterly style. Among these are some of very large size, and others are small. The former are full ten paces in length, and lie in a recumbent posture; the small figures stand behind them, and have the appearance of disciples in the act of reverential salutation. Both great and small are held in extreme veneration."

166. Page 50, line 6

more chaſte and comly than the other

F., or rather Santaella, omits details about intercourse. The latter part is unique to R. and Z:

"The unlicensed intercourse of the sexes is not in general considered by these people as a serious offence; and their maxim is, that if the advances are made by the female, the connection does not constitute an offence, but it is held to be such when the proposal comes from the man." The French texts, but not R., say that if a man take pleasure with a woman against nature he is condemned to death.

167. Page 50, line 13 *ſeauen yeres*

All texts agree that the period was *one* year, not seven. Fr. 1116 and R. state that Nicolo was there also. P. makes them "en legation," but fr. 1116 merely says "por lor fait que ne fa a mentovoir."

CHAPTER 41

168. Page 50, lines 15, 16

a Citie called Eufina. .in a fielde of the Defert called Sabon

I.e. Etzina on the edge of the sandy desert F. reads "Sabon" for "sablon" and takes it as a proper name. It should be identified with Kara-Khoto on the Etsin-gol, and should be clearly distinguished from Kara-Khoja, the old capital of Turfān. See Introduction, p. xliii

169. Page 50, line 19 *other cattell withall*

No mention is made of the falcons. "faucons lanier et sacri assez et sunt mout bones."

170. Page 51, line 1 *a Citie called Catlogoria*

I.e "Caracoron" of the texts, the famous Karakorum F becomes muddled and repeats the name as that of the "firft Prince or Lorde among the Tartars," whereas all the text means to convey is that Karakorum was the first city possessed by the Tartars R and *Z* add a few further details:

"It is surrounded with a strong rampart of earth, there not being any good supply of stone in that part of the country On the outside of the rampart, but near to it, stands a castle of great size, in which is a handsome palace occupied by the governor of the place."

B and Y name Ciorcia, or Chorchia (the Manchu country) as the place where the Tartars first dwelt. R speaks of "Giorza and Bargu," while F. writes simply "the North."

171. Page 51, lines 6, 7

and do pay tribute to Prester Iohn

The best texts have "Unc Can," which is given as the native form of the name. See Yule's long note, Vol. 1 pp 231–7, also the article on "Prester John" by T. Barns in Hastings, *Ency. Rel. Eth* Vol x. pp 272 *et seqq.*

172. Page 51, line 12

fhould not be of fo greate a power

R. makes a further addition:

"With this view also, whenever the occasion presented itself, such as a rebellion in any of the provinces subject to him, he drafted three or four in the hundred of these people, to be employed on the service of quelling it; and thus their power was gradually diminished He in like manner despatched them upon other expeditions, and sent among them some of his principal officers to see that his intentions were carried into effect."

173. Page 51, lines 19, 20

Chenchis....1187.

So also B. and Y., but R. gives 1162 as the date of the election of Chinghiz (Genghis) Khan to the throne This agrees with the Chinese Annals. R. alone mentions his eloquence in the catalogue of his virtues.

174. Page 51, lines 26, 27

conquered eighte Kingdomes

So also B and Y., but R. has "about nine provinces," and adds:

"Nor is his success surprising, when we consider that at this period each town and district was either governed by the people themselves, or had its petty king or lord; and as there existed amongst them no general confederacy, it was impossible for them to resist, separately, so formidable a power "

CHAPTER 42

175 Page 52, line 4

in the yeare of oure Lord God .1190.

Both B. and Y. read 1200. R. gives no date. F. has altered the subsequent passages to *oratio obliqua*

176. Page 52, line 15

a great plaine called Tanguth

Read "Tanduc" with all the best MSS. The identification of this place is uncertain, and the accounts have become muddled.

The final defeat of Prester John (Aung Khan) was at Chacher Ondur, near the modern Urga (Orgo, or Hurae as the Mongols call it), on a tributary of the Tola river, about 700 miles north-west of Peking

177. Page 52, line 17 *but in the end...*

F has here omitted the account of the consultation of the astrologers by Chinghiz Khan as to the issue of the battle. See Marsden, Ch. XLIV (Murray, Ch. XLV); Y. Ch. XLIX, and B. Ch LXVII. It tells how the Saracens were unable to foretell the result of the battle, but that the Christians split a cane lengthways and let one piece represent Chinghız and the other Prester John. The piece representing the conqueror would of itself move on to the top of the other one. As Chinghiz watched he saw the cane bearing his own name move on to the other piece, and so was greatly delighted

178. Page 52, lines 21, 22

laying ſiege to a Caſtell, was hurte in the Knée. and...dyed

Fr. 1116 gives the name of the Castle as "Caagiu," the "Caaju" of Y. The historical accounts of his wounding and death vary. Chinese Annals, however, agree that he was wounded by a stray arrow at the siege of Ta-t'ung fu in 1212, and died at the travelling palace of Ha-la T'u on the Sa-li stream in 1227. See Cordier, *Ser Marco Polo*, p. 57.

179. Page 52, line 23 *one called Cane*

F. omits his name—Cuy or Cui. He repeats Chenchis and omits the "Oktai" and "Mongu" of fr. 1116 and the "Alacon" of Y As Y has pointed out neither Batu nor Hulaku have any right in the list, as the former was Khan of Kipchak and the latter Khan of Persia. The real succession ran: Chinghiz—Okkodai—Kuyuk—Mangku—Kublai.

180. Page 52, line 31

a mountaine called Alchay

I.e. Altay or Altai, a name apparently applied to the Kenter-Khan, north-east of Urga. *Khan* here means "mountain." See Yule's long note Vol. I. pp. 247–50.

181. Page 53, line 7 .300000. *men*

The figure has increased! Read "more than 20,000" with the best texts.

182. Page 53, lines 12, 13

where it is freſhe and pleaſaunt aire

At this point R adds

"and their cattle are free from the annoyance of horse-flies and other biting insects. During two or three months they progressively ascend higher ground, and seek fresh pasture, the grass not being adequate in any one place to feed the multitudes of which their herds and flocks consist"

183. Page 53, line 14

and theſe houſes they carry with them...

The French texts also tell us that the frames of the tents are made very strong but at the same time light for carrying. R. adds that they are carried on a sort of cart with four wheels. Yule gives a picture of one facing p 254 of Vol I.

184. Page 53, lines 22, 23

They do eate all manner of fleſhe

F. omits to mention "des rat de faraon" which is found in large quantities on the plains. Possibly it is a variety of marmot or dormouse, but "Pharaoh's Rat" was the name given by foreigners to the Egyptian ichneumon (*Herpestes ichneumon*), the mongoose of India

185. Page 53, line 24

as manye wiues as they will

"They may take 100 wives if they want to," say the French texts. R adds that the women are most faithful, and although ten or twenty of them are all living together peace and quietness reign supreme.

CHAPTER 43

186. Page 54, line 2 *an Idoll called Nochygay*

Written "Nacygai" in fr. 1116, and by Y. as "Natigay." The idols are apparently identical with the Ongons of the Buriats. See the article on them by Demetrius Klementz in Hastings, *Ency. Rel Eth* Vol. III. p. 12

R. starts his corresponding chapter with the words: "They believe in a deity whose nature is sublime and heavenly. To him they burn incense in censers, and offer up prayers for the enjoyment of intellectual and bodily health. They worship another likewise, named Natigay,. . ."

According to Banzaroff all the relatives and forefathers of Chinghiz Khan have become Ongons, and usually consist of pieces of material sometimes decorated with owl-feathers or various furs.

187. Page 54, line 15

and is called with them Cheminis

This is, of course, Kimiz, Kumiz, or Koumiss made of mare's milk to which a little sour cow's milk is added It is continually churned in a vessel made of horse-skin. See further Y. Vol. I. pp. 259 *et seq.*

188. Page 54, line 17 *Their harneſſe...*

F. omits to mention the weapons—bows and arrows, the sword and mace.

189. Page 54, line 31 *Milke made like dry paſte*

R. tells us how it is made: "They boil the milk and skimming off the rich or creamy part as it rises to the top, put it into a separate vessel as butter, for so long as that remains in the milk, it will not become hard The latter is then exposed to the sun until it dries. Upon going on service they carry with them about ten pounds for each man, and of this, half a pound is put, every morning, into a leathern bottle or small *outre*, with as much water as is thought necessary. By their motion in riding the contents are violently shaken, and a thin porridge is produced, upon which they make their dinner."

190. Page 55, line 3
When the Tartares wyll ſkyrmiſhe wyth theyr enimies...

The account given by F. is much abbreviated He makes no mention of the method by which the Tartar princes convey orders to their armies, or how justice is administered by blows with a stick. The description of the marriage after death, however, is a full account.

CHAPTER 44

191. Page 55, lines 27, 28
Cuthogora... Acay

As we have already seen, F. previously spelt these names Catlogoria, and Alchay.

192 Page 55, lines 30–32
Barga... Mecrith

I.e. Bargu.. Mecrit, or Mescrift. The country in question lay near Lake Baikal, while the tribe, the Merkit, inhabited the district to the south-east of the lake.

193. Page 56, line 3
They haue neyther breade nor wine

R. adds: "They feed likewise upon the birds that frequent their numerous lakes and marshes, as well as upon fish It is at the moulting season, or during summer, that the birds seek these waters, and being then, from want of their feathers, incapable of flight, they are taken by the natives without difficulty."

CHAPTER 45

194. Page 56, line 8 *called Peregrinos*

After mentioning the Peregrine falcons, F. omits to say that it is so cold that you find neither man nor women, but only a bird called Bargherlac, or Barguerlac on which the falcons feed. ' They are about the size of a partridge," says R., "with tails like the swallow, claws like those of the parrot kind, and are swift of flight When the Grand Khan is desirous of having a brood of peregrine falcons, he sends to procure them at this place."

CHAPTER 46

195. Page 56, line 19 *a kingdom called Erguil*

Having returned to Kan-chau, Polo continues to Erginul or Erguiul which has been identified with Liang-chau fu. F. spells it with a "y" a few lines lower.

196. Page 56, line 24 *a great Citie called Syrygay*

Called "Sinju" in Y, which B. would write "Silingiu," the modern Si-ning fu.

197. Page 57, line 5 *called Del Efpinazo*

This is, of course, the yak, *Bos grunniens*. R. adds. "Their hair, or rather wool, is white, and more soft and delicate than silk. Marco Polo carried some of it to Venice, as a singular curiosity, and such it was esteemed by all who saw it."

198. Page 57, line 13 *and that is the Mufke*

R. tells us that the animal is taken during full moon, "when they cut off the membrane, and afterwards dry it, with its contents, in the sun.... Marco Polo brought with him to Venice the head and the feet of one of them dried." See further, Note 281.

199. Page 57, line 16 *of .25. days iourney*

This agrees both with fr. 1116 and R, but Y. has "26"

200. Page 57, line 17 *Feyfants, and very greate*

As Yule says, this is probably the China pheasant known as Reeve's Pheasant. F says naively "I think thefe be Peacocks"

CHAPTER 47

201. Page 57, lines 25, 27

Egregia . Chalacia

Written in fr. 1116 as "Egrigaia" and "Calacian." As stated in the Introduction, p. xlvi, I take these to represent Ning-sia fu and Ting-yuan-ying respectively.

CHAPTER 48

202. Page 58, lines 1, 3

Arguill, . . . Tanguthe

F. has got muddled and brings in Liang-chau fu again, which he now spells "Arguill." See Note 195.

As before (see Note 176) we should read "Tanduc" for "Tanguthe" The province must have included the district lying at the great northern bend of the Hwang ho.

203. Page 58, line 4

called George by his proper name

Both R. and Z tell us he was a Christian and a priest, and that many of the inhabitants were also Christians.

See Pelliot, *T'oung Pao*, 1914, pp 632 *et seqq.* and Yule and Cordier, *Cathay and the Way Thither*, Vol. III p. 15 n.

204. Page 58, lines 9, 10 *Lapis laguli*

The French texts merely say: "En ceste provence se trouve les pieres dont l'azur se fait"

205. Page 58, line 14 *Argarones, or Galmulos*

I.e. Argons, which the French call Guasmul. The texts differ and the exact meaning of the passage is not clear. See Y. Vol I pp 289–92.

The word Guasmul was used by Franks in the Levant as a name for half-breeds sprung from their unions with Greek women

206. Page 58, line 21 *Gog and Magog*

F omits to mention that the natives call it Ung and Mungul, the former being the name of the inhabitants of the country, and the latter a name applied to the Tartars.

CHAPTER 49

207. Page 58, line 27 *cloth of gold*

The names have been omitted—in fr. 1116 we read "...que l'en apelle nascisi fin et nac, et dras de soie de maintes maineres. Ausint com nos avon les dras de laine de maintes maineres, ausint il ont dras d'ores et de soie de maintes maineres."

Y. writes them "Nasich" and "Naques," and says he thinks they correspond to the mediaeval "Tartary cloth."

In Ch. 10 F. speaks of "cloth of Nafich, of Crimfon, and of diuers other coloures and fafhions." See Note 75.

208. Page 59, lines 1, 4 *Sindathoy...Idica*

Neither of these places has been identified, although Yule would see Siuen-hwa fu in the former.

I have already (Introduction, p xlvi) given my reasons for not accepting it. Fr. 1116 gives the locality of the mines as "Ydifu."

F. omits to add that the country is well stocked with game.

CHAPTER 50

209. Page 59, line 6 *a Citie called Gianorum*

L and R. tell us that it means "White Pool" The form found in F is corrupted from Ciagannor or Chagan-Nōr. It probably lay to the east of Anguli-Nōr.

210. Page 59, line 18

when hée goeth into that Countrey

F.'s account is abbreviated, but there is an interesting unique passage in R.: "Nigh to this city is a valley frequented by great numbers of partridges and quails, for whose food the Grand Khan causes millet, panicum, and other grains suitable to such birds, to be sown along the sides of it every season, and gives strict command that no person shall dare to reap the seed; in order that they may not be in want of nourishment. Many keepers, likewise, are stationed there for the preservation of the game, that it may not be taken or destroyed, as well as for the purpose of throwing the millet to the birds during the winter. So accustomed are they to be thus fed, that upon the grain being scattered and the man's whistling, they immediately assemble from every quarter. The Grand Khan also directs that a number of small buildings be prepared for their shelter during the night; and in consequence of these attentions, he always finds abundant sport when he visits this country, and even in the winter, at which season, on account of the severity of the cold, he does not reside there, he has camel-loads of the birds sent to him, wherever his court may happen to be at the time."

CHAPTER 51

211. Page 59, line 21 *Liandei*

This is Chandu, or K'ai-p'ing fu, Kublai's summer residence.

212. Page 59, line 24 *in compaſſe fiftéene miles*

All the best texts read "sixteen." F. is abbreviated here.

213. Page 60, line 3 .40000.

This is unique to F. The leading MSS read "more than 200." The "graye Lyon" mentioned two lines lower down should be "leopard."

214. Page 60, line 10 *And this houſe...*

F.'s description of the pavilion is abbreviated. R. gives us most details. "In the centre of these grounds, where there is a beautiful grove of trees, he has built a royal pavilion, supported upon a colonnade of handsome pillars, gilt and varnished. Round each pillar a dragon, likewise gilt, entwines its tail, whilst its head sustains the projection of the roof, and its talons or claws are extended to the right and left along the entablature "

He also speaks of the precautions made against high winds, etc.

215. Page 60, line 24 *and ſome others*

"appellés Horiat" says fr 1116. Y. writes "Horiad" for "Uirad" or "Oirad," a tribe from the head waters of the Kem or Upper Yenisei.

CHAPTER 52

216. Page 60, lines 29, 30

that milke poured out, is the holye Ghoſte

Apparently F. has taken "espirt" in the sense of Holy Ghost, and has made it identical with the milk. The texts, however, make it clear that a libation of milk is being offered to the spirits and idols as a propitiation for a continuance of future blessings.

217. Page 61, line 1 *ỷ .29. day of Auguſt*

A mistake for "28" as found in all the best texts.

218. Page 61, line 7 *by vertue of his Idols*

F. omits to tell us that the men who perform these wonders come from Tibet and Kashmir, and that they persuade people to believe that they obtain their power because of the holiness of their lives.

R. adds that they exhibit themselves in an indecent state, and are filthy, and squalid in appearance Most texts give these men the name of "Bacsi," probably a corruption of the Sanskrit *Bhikshu*, a title used for wandering Buddhist ascetics.

219. Page 61, line 13

they demaunde of the greate Cane...

F.'s account, much abbreviated, has altered the text into *oratio obliqua*, which in all the best texts reads as a speech made by the "Bacsi" to the Khan.

R. tells us that the Khan occasionally invites people to witness the performance of the moving cups.

220. Page 61, line 25 *.400. Monkes*

The French texts read "more than 2,000," which is no exaggeration. See Y. Vol I. p 319. F omits to mention the fact that the name given to the ascetics who "do eate the branne and the meale kneaded togither" (probably the Tibetan parched barley) is Sensin, a corruption, or rather transcription, of Sien-seng, the name given by the Mongols to the Tao-sze. F. tells us they "do lye in Almadraques, ſharpe and harde beds" This appears to be unique to Santaella who seems to have in mind the spike beds of Hindu ascetics (see *Ocean of Story*, Vol I p 79 n. 1). "Almadraque" is an obsolete Spanish word for "bed" or "matress."

F. omits to mention the fact that the idols of the Tao-sze have female names.

CHAPTER 53

221. Page 62, line 15

and was fiue and forty yeares old when he was made Emperor

So far the dates in this chapter have agreed with the best texts Here slight differences occur Fr. 1116 says by 1298 he had reigned forty-two years, and that now (1298) he was eighty-five years old, while Y. adds that he must have been about forty-three years of age when he first came to the throne.

222 Page 62, line 18

but alwayes ſent his ſonnes,...

So also in R , but not in Y. or B at this particular place In these versions the statement will be found in Bk. II. Ch VI. and Ch LXXXI respectively.

223. Page 62, line 20 *a nephew of his*

F. omits to give his name, Nayan or Naian In all texts the relationship is mixed. Nayan was the great-great-grandson of Chinghiz's brother Uchegin Fr. 1116 and R. say he could put 400,000 men in the field. Y. makes this 300,000.

224 Page 62, line 24

whyche was called Cardin

This is, of course, Kaidu, Kublai's cousin and enemy.

225. Page 62, line 32 *two and twenty days*

This agrees with fr. 1116, but some texts have read "xxii" as "x.xii," and translated, as Yule did, "ten or twelve"

F.'s 300,000 fighting men should be 360,000 cavalry and 100,000 footmen, as in all the best texts

R. adds a passage explaining that the Khan found it necessary to maintain garrisons throughout his dominions to preserve order, and that if he decided to summon only half the men thus employed, the number would be incredible.

226. Page 63, line 13

ſet forwarde on his way with his people...

Both R. and F. omit to say that he marched for twenty days

The French texts give Nayan's host as consisting of some 400,000 horse.

227. Page 63, line 23

thought he might well take his reſt that nighte

F omits to mention that Nayan was in the arms of his favourite wife. As this is found in R. and the Pipino texts, it has probably been purposely left out by the Catholic Santaella.

228. Page 63, line 28

a great frame vpon an Elephant

All the best texts read "four elephants" Fr 1116 says he was mounted "sor une bertresche ordree sor quatre leofans." Yule correctly translates "bertresche" or "bretesche" by "bartizan," the Old English derivate applied to any boarded structure of defence or attack.

229. Page 63, line 30 .25000. *in a battell*

The French texts have 30,000. The fullest account is found in R.:

"His army, which consisted of thirty battalions of horse, each battalion containing ten thousand men, armed with bows, he disposed in three grand divisions; and to those which formed the left and right wings he extended in such a manner as to out-flank the army of Nayan. In front of each battalion of horse were placed five hundred infantry, armed with short lances and swords, who, whenever the cavalry made a show of fight, were practised to mount behind the riders and accompany them, alighting again when they returned to the charge, and killing with their lances the horses of the enemy."

230. Page 63, line 32

and caused his trumpets to blowe

F. abbreviates here, and tells us nothing of the music and singing which preceded the battle, or of the beating of the Nakkaras, the great kettledrums of war. His description of the battle is reduced to a minimum, but, as Yule suggests, the style is very reminiscent of similar battle scenes found both in Eastern and Western histories There are descriptions even in the *Thousand and One Nights*, as well as in the *Kathā-sarit-sāgara*, which contains practically identical sentences.

231. Page 64, line 6

Furciorcia, Guli, Baston, Scincinguy

Written in the French texts as "Ciorcia," or "Charcha"; "Cauli"; "Barscol"; and "Sichintingin," or "Sikintinju" The two first have been identified with Churchin, the Manchu country, and Kao-li, Korea respectively. The other two are still uncertain. See Y. Vol. I. p. 345.

F. makes no mention of the fact that Nayan was a Christian, and ignores the chapter on the rewards given to Jews, Christians, Mahommedans and his nobles. This is given in full in Appendix II. No 5, pp. 270–72.

CHAPTER 54

232. Page 64, lines 14, 15

and his eldest sonne, that he hath by his first wife,
doth kepe Court by himselfe

Here the point is missed. The French texts clearly tell us that the eldest of his sons by any of his four wives is the heir to the throne.

The number of the queens' retainers is given as 10,000, but F. follows the Crusca edition in reducing it to 4000.

233. Page 64, line 21 *Origiathe*

The "Ungrac" of fr. 1116, and "Ungrat" of Y.—the Mongol tribe of Ḳungurat. F.'s account of the concubines is sadly abbreviated. The fullest description is found in R., and is of very great interest to the student of sociology. It is reprinted in full in Appendix II. No. 6, pp. 272, 273

CHAPTER 55

234. Page 65, lines 4, 5

noble Citie called Cambalu

Into this chapter F. has crowded Polo's highly interesting description of Kublai's capital, together with the manners and customs of its inhabitants. The fullest account is that given by R. to which reference should be made, as it is impossible here to reprint every addition that the text affords. I have, however, reprinted his description of the New City of Tai-du in Appendix II. No. 7, pp. 273–5, which is entirely omitted by F.

235. Page 65, lines 26, 27

Without this Citie be twelue ſuburbes...

This portion of the text about the suburbs and large number of prostitutes is a very brief résumé of part of B Ch. XCVI and Y. Bk. II. Ch. XXII. The rest of these chapters corresponds to the last six lines of F. Ch. 58, p. 71.

236. Page 66, lines 4, 5

They be called Chiſitanos

Written in fr. 1116 as "Quesican," and in Y as "Keshican." It is a Mongol term to designate the Khan's lifeguard. See the long note in Y. Vol. I pp 379–81.

237. Page 66, lines 15, 16

At the ſaide Tables commonly do ſitte foure thouſand perſons, or very néere

The sense is wrong here. The French texts clearly explain that the tables are arranged in such a way that the Khan can see them all at a glance R. then adds that most of the soldiers and officers sit on carpets, and that on the outside stand a great multitude of persons who come from different countries, bringing with them many rare and curious articles. The French texts, however, state that the crowd is *outside* the hall, and numbers more than forty (not four) thousand.

238. Page 66, lines 28, 29

hathe a cuppe of golde before hym to drinke in

F is not clear here. Fr. 1116 reads " .Et se metent deus homes que sieent a table un. Et chascun de cesti deus homes hont une coppe d'or a maneque; et con celle cope prennent dou vin de cel grant vernique d'or. Et ausint en ont entre deus dames unde celz grant [verniques] et deus coupes comant ont les hómes " The etymology of "vernique" is uncertain (see Y Vol. I p. 384), but it was probably a large lacquered bowl from which the smaller cups would be filled—corresponding in some degree to our punch bowls and sets of cups

All the best texts mention the amazing quantity and value of the Khan's plate R. adds an interesting passage in which he explains how strangers are informed of the etiquette of the court, and how two enormous officers stand at each door in order to see that no one touches the threshold with their feet on entering the hall. It is considered a bad omen if this happens, and the offence is punished by blows or else by the person's garment being taken to be redeemed by payment. The rule does not hold *after* a banquet, as the guests would not have control over their feet!

239. Page 68, line 19

to ſay thirtie dayes iourney

Fr. 1116 has "sixty days' journey," and Y. "some forty days' journey." R. tells us how the animals are killed: "all persons possessed of land in the province repair to the places where these animals are to be found, and proceed to enclose them within a circle, when they are killed, partly by dogs, but chiefly by shooting them with arrows."

F. omits the chapter telling of the lions, leopards and wolves used by the Khan for the chase. See M. Bk. II. Ch XIV; Y. Bk II. Ch. XVIII and B. Ch. XCII.

CHAPTER 56

240. Page 68, lines 27, 28

Baian, and the other Mytigan, and they be called Cinitil

The first name is correctly written, the second is "Mingan" in the French MSS. Their title appears as "Ciunci" in fr. 1116; and "Chinuchi (or Cunichi) in Y. Laufer considers the word to be derived from the Tibetan *čang-k'i*, "wolf-dog," while Pelliot would read the word *Cuiuci*, and connect it with the verb *guyu* or *guyi*, "to run" See Cordier, *Ser Marco Polo*, p. 70.

241. Page 68, lines 31, 32

all apparelled in one liuerye of whyte and redde

This is a mistake The whole point was that the two companies should be distinguished in the field. The French texts clearly state that one lot was clad in vermilion and the other in blue.

Some few lines lates, R. has the following addition:

"The two brothers are under an engagement to furnish the court daily, from the commencement of October to the end of March, with a thousand pieces of game, quails being excepted; and also with fish, of which as large a quantity as possible is to be supplied, estimating the fish that three men can eat at a meal as equivalent to one piece of game"

CHAPTER 57

242. Page 69, line 13 *fiue thousand*

Read "500" with all the best texts

243. Page 69, line 18 *they bée called Tustores*

Read "Toscaor" or "Toscaol," the Turki meaning "guardian" or "watcher."

244. Page 69, line 25

one of those barrõs his brethren

F. omits to tell us that his title was "Bularguci," keeper of lost property, and that all articles found must be delivered up to him. He took his stand in a prominent part of the camp, and displayed his banner to attract attention, and proclaim his whereabouts.

CHAPTER 58

245. Page 70, line 11

which they do call Caziamon

The "Cacciar Modun" of fr. 1116, and "Cachar Modun" of Y

Yule considered it must be in the region north of the eastern extremity of the Great Wall. Pelliot says it must be the "Ha-ch'a-mu-touen" of the *Yuan Shih*, Ch 100, fo. 22.

246. Page 70, line 15 2000. *knights*

Read "1000" with all the best texts.

247. Page 71, between Chs. 58 and 59

At this point Yule has given the very interesting account of the oppressions of Achmath (Achmac, or Ahmad) the Bailo, and the plot that was formed against him. It is only found in the Ramusio text and is accordingly reproduced in full in Appendix II. No. 8, pp. 275–8. The revolt against the oppression of Kublai's tyrannical minister is fully substantiated in the Chinese records, as well as in the contemporary Persian version by Rashīd-ud-dīn (the *Jámi'u't-Tawáríkh*), whose account tells us of two separate attempts to murder Ahmad, and, curiously enough, connects them with the siege of Saianfu (Siang-yang), for which see Note 353, p. 228, and Appendix II. No. 23, pp. 301, 302.

After carefully weighing over all the evidence, the Rev. A C Moule has come to the conclusion (*Journ. North China Br. Roy. As. Soc.* Vol. LVIII. 1927, pp. 1–35) that if the story does not come direct from Polo, it cannot in any case be much later in date than his lifetime He considers, however, that Murray's arguments against Polo's authorship perhaps deserve more attention than they have hitherto received. Murray's argument (*Travels of Marco Polo*, pp. 32 and 124) is that the Achmath chapter contains a statement that the Cathayans detested the Khan's rule, while in a passage of undoubted Polian authorship they are said to "worship him as he were God." Such contradictory statements, he suggests, would never have been countenanced by Polo. But I doubt whether they really are contradictory statements The Mongol conqueror, Kublai, would doubtless inspire fear, and possibly hate, in a vanquished foe, as well as "worship," when he extended munificence to the poor. Unfortunately the newly discovered *Z* text is silent, so R. still remains the sole authority for the story.

CHAPTER 59

248. Page 71, line 20

The greate Cane causeth his money to be made in this manner...

F.'s account of the paper-currency is unfortunately abbreviated. See M. Bk. II. Ch. XVIII; Y. Bk. II. Ch. XXIV; and B. Ch. XCVII. To Yule's notes, we must add those given by Cordier, *op. cit.* pp. 70–72.

CHAPTER 60

249. Page 72, lines 10, 11

tenne Barons...to gouerne .64. prouinces and countries...

Here F. has got his figures wrong. It should be *twelve* Barons and *thirty-four* provinces

R. gives us some extra details concerning the duties of these men with respect to the army, the conferring of benefits, etc.

CHAPTER 61

250. Page 73, line 1 *are called Senich*

The "scieng" of fr. 1116, and "shieng" of Y. Two similar words, "Sing" and "Sheng," were apparently applied both to the High Council of State as well as to the provincial governments. We meet with "Sing" as the denomination of Yang-chou. See Note 350.

CHAPTER 62

251. Page 73, line 8 *The Citie of Cambalu...*

This short disjointed passage about the gates of the city, and the subsequent one on the "blacke ftones," is all that remains in F. of seven chapters. As R.'s account is the fullest, it is reprinted in Appendix II. No 9, pp. 278–86.

CHAPTER 63

252. Page 73, line 18 *fourtéene moneths*

Read "four" with all the best texts.

253. Page 74, lines 2, 3

tenne dayes iorney, I founde a very great riuer which is called Poluifanguis

Read "miles" instead of "dayes." The river referred to is the Hun-ho, called "Pulis-anghin(z)" in the French texts.

254. Page 74, line 13 *two hundred pillers*

This seems to be unique to F. R. gives us a few further details about the bridge, but says nothing about the number of the pillars.

CHAPTER 64

255. Page 74, lines 15–17

tenne miles...Goygu

Read "thirty miles." Fr. 1116 gives the name as "Giongiu," and Y. as "Juju." It corresponds to the modern Cho-chau.

CHAPTER 65

256. Page 75, line 1 *Countrey of the Magos*

Read "countrey of Manzi," i.e. China south of the Yellow and Huai rivers.

CHAPTER 66

257. Page 75, line 5

a Citie named Tarafu

This is T'ai-yuan fu, called "Taianfu" in the French texts. For a note on the vines referred to by Polo, see Cordier, *op cit.* pp. 75 *et seq.*

R tells us of Acbaluc (Ch'êng-ting fu) which, he says, is reached in five days from Cho-chau, and to which the limits of the Khan's hunting-grounds extend, "and within which no persons dare to sport, excepting the princes of his own family, and those whose names are inscribed on the grand falconer's list; but beyond these limits, all persons qualified bŷ their rank are at liberty to pursue game...."

CHAPTER 67

258. Page 75, lines 10–13

eighte dayes...Paymphu

Read "seven days" The name of the city should read "Pianfu," the modern P'ing-yang fu

259. Page 75, lines 14, 15

a fayre Towne named Caychin

This should read "Castle of Caiciu," or Caichu, the exact situation of which is so puzzling See Introduction, p. xlviii.

F omits any sort of description of the king of "Caychin" to whom he gives the name of Bur (for Dor). The fullest account is given by R After mentioning the collection of paintings of the princes who have ruled at the castle, he continues:

"A remarkable circumstance in the history of this King Dor shall now be related He was a powerful prince, assumed much state, and was always waited upon by young women of extraordinary beauty, a vast number of whom he entertained at his court. When, for recreation, he went about the fortress, he was drawn in his carriage by these damsels, which they could do with facility, as it was of a small size They were devoted to his service, and performed every office that administered to his convenience or amusement In his government he was not wanting in vigour, and he ruled with dignity and justice. The works of his castle, according to the report of the people of the country, were beyond example strong He was, however, a vassal of Un-khan, who, as we have already stated, was known by the appellation of Prester John, but, influenced by pride, he rebelled against him."

CHAPTER 68

260. Page 75, line 16

This Bur

As mentioned in the last note, this is a corruption of Dor, i.e Roi D'Or, a literal translation of the Mongul Altun Khān, the Emperor of the Kin or Golden Dynasty. There appears to be no historical foundation to the legend briefly related in this chapter. The number of the "yong Gentlemen" agrees with that given in fr. 1116, but Y. makes it seventeen.

None of the conversation between King Dor and Prester John, as found in the French MSS, is retained in F.

CHAPTER 69

261. Page 76, line 12

a great Citie named Casiomphur

I e. Cachanfu, the modern P'u-chau fu F has muddled his distances He omits to mention the Caramoran (Hwang ho) which Polo crosses twenty miles after leaving the castle, and then after two days he reaches P'u-chau fu. See the Introduction, p xlviii and map opposite.

CHAPTER 70

262. Page 76, line 15

eight dayes

So also in all the best MSS, but R. has "seven."

263. Page 76, lines 17, 18

with goodlie and faire Gardens

F. omits to mention the abundance of mulberry trees. R and Z add: "The inhabitants in general worship idols, but there are also found here Nestorian Christians, Turkomans, and Saracens"

264. Page 76, lines 20, 21

a faire Citie whiche is called Bengomphu

I.e. Si-ngan fu, famous in Chinese history

265. Page 76, lines 22, 23

one of the great Canes sonnes, who is called Magala

I.e. Mangalai, third son of Kublai Khan.

266. Page 76, line 25

the which with the Wal of the Citie is tenne myle compaſſe

Fr. 1116 simply says the palace wall was five miles in compass. R. tells us that the palace stood in a plain five miles from the city and that the wall of the park was five miles in circumference, enclosing all kinds of wild animals, both beasts and birds, which were kept for sport. He adds that the halls and chambers of the palace were ornamented with paintings in gold and the finest azure, as well as with great profusion of marble.

CHAPTER 71

267. Page 77, line 8 *the prouince of Chinchy*

I.e. the southern portion of Shen-si in the neighbourhood of Han-chung fu.

268. Page 77, lines 11, 12

Lions, and plentie of other wilde beaſtes

I.e., of course, *tigers*, etc. Most texts give more details of the "other wilde beaſtes," viz. bears, lynxes, fallow deer, antelopes and stags.

CHAPTER 72

269. Page 77, lines 14, 15

a Citie named Cyneleth

In the chapter heading F spells it Cineleth It is a corruption of Acbalec. The district referred to is doubtless the river valley of the Han kiang. R. speaks of "Ach-baluch Manji, which signifies the White City on the confines of Manji...."

270. Page 77, line 24

Lions and beares, beſides other wilde beaſtes

Yule's texts do not mention these animals, but fr 1116 reads "lionz et ors et leus cerver, dain, cavriolz et cerf," i e "lions [tigers], and bears and lynxes, fallow deer, roebuck and stags."

CHAPTER 73

271. Page 78, line 2 *the countrey named Cindarifa*

For Cindarif*u*, i.e. Sindafu, the modern Ch'êng-tu fu. F spells it "Sindarifa" a few lines lower down.

272. Page 78, line 4 *diuided it into thrée partes*

F. omits details about the division of the city. All the best MSS relate how when the old king was dying he divided his city into three parts, one of which he gave to each of his three sons Each son walled off his part of the city and became a powerful king, but the Great Khan conquered them all.

For a possible explanation of this see Cordier, *Ser Marco Polo*, p. 79.

273. Page 78, line 18

tenne thousande Bisancios of God

I.e. "bezants of gold." R. reads "100," and fr. 1116 "1000"

The description of the watering of the city and of the bridge is given much fuller in R. and *Z*, where certain unique passages also occur. In the following extract those portions found only in R. are given in italics.

"The city is watered by many considerable streams, which descending from the distant mountains, surround and pass through it in a variety of directions. Some of these rivers are half a mile in width, others are two hundred paces, and very deep; over which are built several large and handsome stone bridges, eight paces in breadth, their length being greater or less according to the size of the stream. From one extremity to the other there is a row of marble pillars on each side, which support the roof, for here the bridges have very handsome roofs, constructed of wood, ornamented with paintings of a red colour, and covered with tiles. Throughout the whole length also there are neat apartments and shops, where all sorts of trade are carried on One of the buildings, larger than the rest, is occupied by the officers who collect the duties upon provisions and merchandise, and a toll from persons who pass the bridge. In this way, it is said, his majesty receives daily the sum of an hundred besants of gold. *These rivers uniting their streams below the city, contribute to form the mighty river called Kian, whose course, before it discharges itself into the ocean, is equal to an hundred days' journey; but of its properties occasion will be taken to speak in a subsequent part of this Book*" [i.e. F. Ch. 94; M. Bk. II. Ch. LXIII; P Bk. II. Ch. CXLVI; Y. Bk. II. Ch. LXXI; B. Ch. CXLVIII. See Note 355].

CHAPTER 74

274. Page 78, lines 21, 22

it indureth fiue dayes iourney

At this point most texts add a few lines stating that the inhabitants live by agriculture, that many savage beasts, such as lions and bears, are found, and that the people of "Sindu" live by manufactures, particularly of fine cloths, silks and gauzes.

275. Page 78, lines 23, 24

Cheleth, which was deſtroyed by the great Cane

I.e. Tibet, yet in the next chapter it is called "Thebet." Evidence of the "destruction" of Tibet appears to be wanting. It seems to have been a case of peaceful occupation and surrender without fighting. See Y. Vol II. p 46, but cf De Guignes, *Histoire Générale des Huns*, Book xv. p 123. R. (also *L*) adds "To the distance of twenty days journey you see numberless towns and castles in a state of ruin, and in consequence of the want of inhabitants, wild beasts, and especially tigers, have multiplied to such a degree, that merchants and other travellers are exposed there to great danger during the night." The *whole* passage is not unique to R and *L*, as the French MSS refer to the "multiplication of wild beasts" somewhat later.

276. Page 78, lines, 24, 25

Canes which are called Berganegas

Polo is referring here to the bamboo, but F. is alone in using the word "Berganegas." Prof. Dr Sten Konow considers it to be Iranian, apparently for "Bargānaga," where *bargāna* would mean "leafy" from the Persian "barg," *ga* being the otiose suffix. But this seems to have no connection with bamboo Likewise the Hindustani, *bargā*, "rafter," can surely have nothing to do with it A more probable derivation is through the Spanish "caña de Bengala," a cane. The feminine ending *-ega* would make "Bengalega," whence the corruption to "Berganega" is not forced. See Diez, *Etymologisches Worterbuch der romanischen Sprachen.*

277. Page 79, line 1

the ſound is harde many miles off

Fr. 1116 and Y's texts give a definite distance, ten miles.

R reads, however, "duoi miglia," which seems more reasonable, especially as Yule (Vol II. p 46) considers Polo's account somewhat exaggerated.

F.'s description of the burning bamboos is abbreviated The French texts explain how that the horses became so alarmed by the noise that the men tie all four legs and peg them down firmly with ropes, as well as wrapping up the heads and eyes of the animals. R. alone adds that the merchants provide themselves with iron shackles.

278 Page 79, line 10

do carrie prouiſion for thoſe twenty daies iourney

Both R. and *Z* add "unless perhaps once in three or four days, when you take the opportunity of replenishing your stock of necessaries."

CHAPTER 75

279. Page 79, lines 12, 13

a Prouince or Countrey, that is full of Cities and Townes

R and Z say that at the end of the twenty days "you begin to discover a few castles and strong towns, built upon rocky heights or upon the summits of mountains, and gradually enter an inhabited and cultivated district where there is no longer any danger from beasts of prey."

280. Page 79, lines 13, 14

And the custome in this Countrey is...

So also fr. 1116 merely says, "Et hi a un tiel costumes . ," but R. (following Pipino here I imagine) is evidently shocked, for he says: "A scandalous custom, which could only proceed from the blindness of idolatry, ..." Again, somewhat later, he (as also F.) omits the frivolous remark found in the French texts about it being a fine place for young bachelors to visit Details of the custom in question vary in the different MSS. For some unexplained reason Yule never mentions the interesting variant readings in R., while Benedetto omits to record R. as saying, "che questo piace alli loro Idoli" He gives, however, an interesting passage from Z which appears to be the origin of R. It reads (B p. 111, note e)·

"Nam mulier sive donucella que non fuerit ab aliquo viro cognita dicitur apud eos diis fore ingrata quare propter hoc homines aborrent eas et de ipsis non curant, quare si eorum ydolis essent grate eas homines concupiscerent et affectarent"

The next point of interest (unrecorded by Y. or B.) is that R speaks of "a caravan of merchants" who are visited "after they have set up their tents for the night." This mention of a "Carouana di mercanti" should have found a place in *Hobson-Jobson* in view of its etymological interest. Fr. 1116 speaks of "lor tendes" to which the old women take their daughters to numbers varying between twenty and forty. P. and Y. read (somewhat lower down) "twenty or thirty." R. gives no figures at all In our text, we see that F has changed matters round, as he says "and sometimes there lyeth with hir ten, and with some other twenty"

It has long since been pointed out that customs similar to that described by Polo exist in many parts of the world. The most recent, and largest number of references I have seen is given by Briffault, *The Mothers*, Vol. III. pp. 313 *et seq.* He points out that several reasons for the dislike of marrying virgins are found among different tribes. In primitive society marriage is an economic measure rather than an avenue to sexual life. The necessity of fertility in the woman is, therefore, a *sine qua non* in the bride. Prostitution with strangers is among many peoples considered a means of getting into touch with the powers whence her fertility is truly derived (see further, *op. cit.* p. 317).

To return to our text, F reads, "And ſhe that hathe vſed hir ſelfe with moſte ſtrangers, it ſhall be knowen by the moſt quantitie of jewels that ſhe weareth aboute her necke, and ſhe moſt ſooneſt ſhall finde a mariage, and ſhall be moſt prayſed and loued of hir huſband." R. is fuller than other texts:

"When, afterwards, they are designed for marriage, they wear all these ornaments about the neck or other part of the body, and she who exhibits the greatest number of them is considered to have been the most attractive in her person, and is on that account in the higher estimation with the young men who are looking out for wives; nor can she bring to her husband a more acceptable portion than a quantity of such gifts. At the solemnisation of her nuptials she accordingly makes a display of them to the assembly, and he regards them as a proof that their idols have rendered her lovely in the eyes of men...."

Part of the above probably came from *Z* (see B. p. 111 note g). P. and Y. tell us that a girl must have at least twenty tokens before she can get married

In concluding this note, I might refer to the fact, well known to most readers, that the lady with the lovers' tokens appears in the frame story of the *Thousand and One Nights*, whence she came from India, in Somadeva's great collection, the *Kathā-sarit-sāgara*. See my *Ocean of Story*, Vol. v. p. 122. For further examples to those I give in the note on that page, see Wesselski, *Marchen des Mittelalters*, Berlin, 1925, pp. 185–7.

281. Page 79, line 29

and ſpecially of Muſkettes

R (and *Z*) give us the fullest account of the methods of obtaining musk from the musk deer:

"Here are found the animals that produce the musk, and such is the quantity, that the scent of it is diffused over the whole country. Once in every month the secretion takes place, and it forms itself as has already been said, into a sort of imposthune or boil full of blood, near the navel; and the blood thus issuing, in consequence of excessive repletion, becomes the musk Throughout every part of this region the animal abounds, and the odour generally prevails. They are called *gudderi* in the language of the natives, and are taken with dogs."

Cf. the interesting account given by Chardin (see p. 151 of the Argonaut Press edition, 1927). One of the earliest accounts is that by Cosmos, the Egyptian monk (c. A.D. 545), who says it is "called in the native tongue *Kastouri*. Those who hunt it pierce it with arrows, and having tied up the blood collected at the navel they cut it away. For this is the part which has the pleasant fragrance known to us by the name of musk The men then cast away the rest of the carcase."

The gland producing the musk is only found in the male, and in a sac about three inches in diameter situated beneath the skin of the abdomen, the orifice being immediately in front of the preputial aperture. See the articles "musk" and "musk-deer" in the *Ency Brit.* 11th edition, Vol. XIX p 90

The word musk appears to be derived from the Sanskrit *mushka,* meaning "scrotum."

282. Page 79, lines 30, 31

Canuas, and Cowhydes, and the ſkinnes of wilde beaſtes

The French texts also mention "bocoran," buckram, for a note on which see Y. Vol. I. pp 47, 48.

F. omits to mention that no coined money is used by the Tibetans, nor even the Khan's paper money, but that salt (or, according to R., coral) is in circulation instead.

CHAPTER 76

283. Page 80, line 1

Maugi is a great prouince and Countrey

I.e. Manzi or Mangi, Southern China F has got muddled. He is still speaking of Tibet in this chapter, but was misled by the mention of "Maugy" at the very end of Ch. 75 He describes it later in its proper place (Ch. 88), where he correctly calls it Mangi.

284. Page 80, line 3

gold of Payulſa

I.e. gold dust, the "or de paliolle" of fr 1116, and "oro di paiola" of R. It corresponds to the modern French "paillettes d'or."

The French texts also mention that "canele," cinnamon, grows in great abundance.

285. Page 80, line 7

and of Chamlet great plenty

The camlet here mentioned is the "giambelot" of fr. 1116, and the "Zambellotti" of R. It was a stuff of camel's hair and originally, as in our text, only referred to such a product. In time, however, the term was applied to stuffs containing both wool and silk.

286. Page 80, lines 8, 9

Inchanters, and euill diſpoſed men

R. (and Z) give us more detail:

"These people are necromancers, and by their infernal art, perform the most extraordinary and delusive enchantments that were ever seen or heard of. They cause tempests to arise, accompanied with flashes of lightning and thunderbolts, and produce many other miraculous effects."

287. Page 80, line 9 *Masties as bigge as Asses*

R. tells us that these Tibetan mastiffs are "strong enough to hunt all sorts of wild beasts, particularly the wild oxen, which are called *beyamini*, and are extremely large and fierce."

For a possible explanation of *beyamini* see Y. Vol. II. p. 52 and Cordier, *Ser Marco Polo*, p. 83.

All the best texts now mention the breeding of lanner falcons and the good sport they have with them. R. also refers to "sakers, very swift of flight."

The French texts follow on with a paragraph explaining that all these provinces now being described are subject to the Great Khan, and that this fact must be understood even if it is not mentioned.

CHAPTER 77

288. Page 80, line 11 *Candew*

F. calls it "Candon" in the chapter heading. It is "Gaindu" in fr. 1116; and "Caindù" in R. (written by M. as Kain-du). The exact etymology is uncertain (see Y. Vol. II, p. 70), but its identification has been determined. The name was applied both to a district and to its chief town. This is definitely stated both in R. and *L*. The district is the Kien-ch'ang valley watered by the Ngan-ning (or An-ning) which meets the Yalung Kiang just before the latter joins the Kin-sha kiang about 120 miles N.N.W. of Yunnan fu. The town of Caindu is the modern Ning-yuen fu on the Ngan-ning, roughly half-way between Ya-chow, and Yunnan fu (I see it is called Ling-yuen in map 62 of *The Times Atlas*).

289. Page 80, line 11

that lyeth towards the Occident

R. is much more explicit: "Kain-du is a western province which was formerly subject to its own princes, but since it has been brought under the dominion of the Grand Khan, it is ruled by the governors whom he appoints. We are not to understand, however, that it is situated in the western part (of Asia), but only that it lies westward with respect to our course from the north-eastern quarter. Its inhabitants are idolaters. It contains many cities and castles, and the capital city, standing at the commencement of the province, is likewise called Kain-du. Near to it there is a large lake of salt water, in which are found abundance of pearls, of a white colour, but not round."

290. Page 80, lines 19, 20

the custome of the people in this Countrey is...

Cf. with Note 280. *Z* adds a few interesting details, see B. p. 114 note e.

291. Page 81, lines 5, 6

The people of this Coũtrey do vſe money made of gold...

The last lines of this chapter are sadly abbreviated. The fullest version is to be found in R. but as the passage is lengthy it will be given *in toto* in Appendix II No. 10, pp. 286, 287.

CHAPTER 78

292. Page 81, lines 12–14

And at the tenne dayes iourneys end, you come vnto a greate Riuer, whiche is named Brus, at the which endeth the Countrey and prouince named Candew

R. reads "fifteen days" Here the valley of Kien-ch'ang is referred to The "Brus," corrupted from "Brius" of fr. 1116 and all the best texts, is the Kin-sha kiang already mentioned in Note 288. It is the Tibetan portion of the Yang-tze kiang which "falleth into the Occean Sea" in lat. 32° N

CHAPTER 79

293. Page 81, line 17

a Prouince named Caraia

I.e. the "Caragian" of fr 1116; "Carajan" of Y.; "Caraian" of P and "Karaian" of R. This is the province of Yunnan

294. Page 81, lines 19, 20

one of the greate Canes ſonnes, named Esentemur

Isentimur was the grandson, not son, of Kublai. See further, Note 303.

295. Page 81, line 24 *great plentie of Horſes*

Here all the texts add that the province had a language peculiar to itself and very difficult to understand. Santaella probably omitted it on purpose, for, as we have seen, he always translated "lingua per sì" as "lengua persiana." Apparently he hesitated in describing the Chinese as talking Persian, but when dealing with the islands of the archipelago his qualms disappeared!

CHAPTER 80

296. Page 82, lines 1, 2

a Citie which is named Ioci

I.e. Yunnan fu, called "Iaci" in fr. 1116, "Yachi" by Y., and "Jacin" by P. In the chapter heading F. wrongly calls it a "Prouince "

297. Page 82, line 2

and full of people Idolaters...

R. adds that the Idolaters are the most numerous; while all the best texts have a passage about their wheaten bread being unwholesome, that rice takes its place, and that they make a spiced drink which has the effect of alcohol.

298. Page 82, lines 4-6

fine ſhelles white...and foureſcore of them are worth a Sazo of gold...

A mistake for "a Sazo of *siluer*," as is obvious when F. immediately afterwards tells us the next unit of value is 8 Sazos of silver = 1 ounce = 1 Sazo of gold. It is hard to say what a "Sazo" is. R. calls it "Saggio," and fr. 1116 writes "Saje," and says that "un saje d'arjent" corresponds to "deus venesians gros." Yule's text adds that two Venetian groats = 24 "piccoli" ("livres" in P.).

For a very important note on the cowrie shell (*Cypræa moneta*) see Stein, *Kalhaṇa's Rājataraṅgiṇī*, Vol. II. pp. 323, 324. The ornamental and sexual side of the use of the cowrie is dealt with by Briffault, *The Mothers*, Vol. III. pp. 275-8. See my *Ocean of Story*, Vol. IX. p. 17 n. 2.

299. Page 82, line 8

There they do make Sault of the water of Welles great plẽty

All the best texts mention the fact that the king derived a large revenue from the duty on salt.

300. Page 82, lines 12, 13

They cut it in ſmall péeces, and ſauce it with Garlike and ſpices...

It is interesting here to compare the passage in several of the texts, for it would appear that R. shows a closer relationship to fr. 1116, while F., Y. and P. show signs of inferior readings.

From fr. 1116 it is obvious that Polo is trying to explain that the rich and poor each dress their meat a distinctly different way. R. also clearly appreciates this. The other texts, however, have mixed up the two accounts as one. Let us look at fr. 1116 first:

"Encore vos di que il menuent la char crue de galine et de mouton [B. reads mon*t*on] et de buef et de bufal: car les povres homes se [corrected by B. to s'en] vont a la becarie, et prenent le feie crue tant tost con [B. reads co*m*] se trai hors de la bestes, et le trence menu, puis le met en la sause de l'aille et le menuie mantenant. Et ausi font de toutes les autres chars. Et les gentilz homes menuient encore la cars crue, mes il la font menussier menuemant, puis la metent en la sause de l'aille meslee con bone espece, puis la menuient ausi bien con nos faison la coite."

The words from "Et les gentilz" to "con bone espece" are omitted in P. and Y. Hence the comparison intended in the passage is lost. In R., however, we find no reference to the "becarie," shambles, but the differentiation clearly made:

"The people are accustomed to eat the undressed flesh of fowls, sheep, oxen, and buffaloes; but cured in the following manner. They cut the meat into very small particles, and then put it into a pickle of salt, with the addition of several of their spices. It is thus prepared for persons of the higher class, but the poorer sort only steep it, after mincing, in a sauce of garlic, and then eat it as if it were dressed."

I consider that a comparison of the above passages clearly shows, if further proof be needed, the high importance of R. For although the MSS used by P. and Y. resemble fr. 1116 closely at first, they not only add nothing further, but have grave *lacunae*.

But if R. omits a passage found in fr. 1116 it often adds a sentence that gives further information and shows a grasp of the original which we do not find in the other texts. This is by no means a solitary example. Many could be cited.

CHAPTER 81

301. Page 82, lines 14, 15
tenne dayes iourney

Fr. 1116 and all the best texts have "por ponent," or the equivalent, "in a westerly direction."

302. Page 82, line 15 *Prouince named Chariar*

F. omits to mention that the city also bore the same name as the province. Chariar, the second city of Yunnan, is undoubtedly Ta-li fu. Full evidence will be found in Yule's notes on this and the previous chapters. Fr. 1116 calls it "Caragian"; M. writes "Karazan" and Y. "Carajan"

303 Page 82, lines 17, 18 *named Chocayo*

I.e. a corruption of Cogacin or Cogachin. Obviously Polo has got muddled here as he told us in the last chapter that "Esentemur" was ruler of "Caraia," i.e. Yunnan.

Hukāji (Chocayo) was succeeded by Isentimur.

304. Page 82, line 19 *great plenty of gold*

F. omits to mention gold dust, the "gold of Payulsa" as he calls it in Ch. 76. See Note 284.

305 Page 82, line 22 *there be certayne Serpents*

F.'s account of the crocodiles is sadly abbreviated.

The fullest version is found in R which also contains additional passages towards the end of the chapter. The rest of R. Ch XL will, therefore, be found in full in Appendix II No 11, pp. 287–9. Some of the passages hitherto considered unique to R are also found in *Z*. See B. pp. 116, 117.

306. Page 83, lines 6, 7 *But .95. yeares hitherto*

Read "35 yeares hitherto."

CHAPTER 82

307. Page 83, lines 11, 12

another Prouince named Nocteam, and also the Citie named Nociam

Both names are quite unrecognizable in such corrupted forms. "Nocteam" is the "Çardandan" of fr. 1116; the "Cardandan" of R. which M., owing to the absence of the cedilla, gives as "Kardandan" instead of the more correct "Zardandan" of P and Y. On p. 85, line 1, read "Nocteam" for "Chanan."

The name means "Gold-teeth" being the Persian *zăr-dandān*, equivalent to the Chinese *chin-ch'ih*. The exact locality of this Province cannot be stated for certain, but roughly speaking it embraced a district in Western Yunnan having the Salween (Nu-wu or Lu Kiang) from about 24° to 27° N. lat. as a longitudinal centre. The Kachins inhabited the western part, while the Northern Shans were due south. See *Cathay and the Way Thither*, Vol III p 131 n. 1. In "Nociam" we see the corrupted "Vocian" of fr. 1116, the "Vochan" of Y. and "Vochang" of R All forms are attempts at the Chinese Yung-ch'ang, which town lies nearly half-way between the Mekong and the Salween

308. Page 83, lines 14, 15

téeth couered with golde

R. alone adds a passage about tattooing;

"The men also form dark stripes or bands round their arms and legs, by puncturing them in the following manner. They have five needles joined together, which they press into the flesh until blood is drawn, and then they rub the punctures with a black colouring matter, which leaves an indelible mark. To bear these dark stripes is considered as an ornamental and honourable distinction."

Cf. the short account of Conti, p. 131 of this volume, and see Temple's article "Burma" in Hastings' *Ency. Rel. Eth.* Vol. III. p. 31, while for a comprehensive article on the subject see "Tatuing" by Grant Showerman, Hastings' *Ency. Rel. Eth.* Vol. XII. pp. 208–15.

309. Page 83, lines 19, 20

the women of this Countrey haue this custome...

F.'s account is slightly longer than that found in most texts. It closely resembles that in R.

To the notes on the custom known as *couvade* given by Yule, add those by Cordier, *Ser Marco Polo*, pp. 85–7, and W. Crooke, *Religion and Folklore of Northern India*, 1926, p. 214. In his article on "Burma" mentioned in the last note, Sir Richard Temple, in speaking of the Karens, says:

"Among Red and White Karens there are curious traces of the *couvade*. Among the Red Karens only the father may act as midwife, and he may not speak to any one after the birth of his child. Among the White Karens (Mēpū) no one may leave the village after a birth until the umbilical cord is cut, this event being announced by bursting a bamboo by heating [see Note 277]. This custom is said to be extended to the birth of domestic animals. No stranger may enter the house of a woman during her confinement. No customs seem to exist connected with the umbilical cord, except that the Red Karens hang up all the cords of the village in sealed bamboo receptacles (*Kyedauk*) on a selected tree."

A curious example of the *couvade* entering into local legendary history is afforded by the story of the invasion of Ulster by the Fír Bolg and how the male inhabitants were unable to defend the kingdom of Conchobar, being *en couvade*. The situation was saved by the help of Cúc'hulainn, the sun-hero It has been suggested that the custom is here used to explain the annual birth of the sun-god just within the Arctic circle. See further T. Barns, Hastings' *Ency. Rel. Eth.* Vol. IV. p 749.

310. Page 84, lines 1, 2

for that they dwell among the moyst mountaynes,
corrupted with euill ayres

After describing the *couvade*, F abbreviates considerably. In the first place the chief texts tell us that the inhabitants eat rice with their meat, and manufacture a wine from rice to which a mixture of spices is added. Their ignorance of writing, etc., is thus described in R.:

"They have no knowledge of any kind of writing, nor is this to be wondered at, considering the rude nature of the country, which is a mountainous tract, covered with the thickest forests. During the summer season the atmosphere is so gloomy and unwholesome, that merchants and other strangers are obliged to leave the district in order to escape from death. When the natives have transactions of business with each other, which require them to execute any obligation for the amount of a debt or credit, their chief takes a square piece of wood and divides it in two. Notches are then cut on it, denoting the sum in question, and each party receives one of the corresponding pieces; as is practised in respect to our tallies. Upon the expiration of the term, and payment made by the debtor, the creditor delivers up his counterpart, and both remain satisfied."

311. Page 84, line 33 *If the paciente heale...*

From this point to the end of the chapter we have a passage not found in fr. 1116, P. or Y.

An even fuller account, however, appears in R.:

"and if through God's providence the patient recovers, they attribute his cure to the idol for whom the sacrifice was performed; but if he happens to die, they then declare that the rites had been rendered ineffective, by those who dressed the victuals having presumed to taste them before the deity's portion had been presented to him. It must be understood that ceremonies of this kind are not practised upon the illness of every individual, but only perhaps once or twice in the course of a month, for noble or wealthy personages. They are common, however, to all the idolatrous inhabitants of the whole provinces of Kataia and Manji, amongst whom a physician is a rare character. And thus do the demons sport with the blindness of these deluded and wretched people."

312. Pages 84, 85 *Chaps.* 82 *and* 83

Between these two chapters F. has omitted the account of the battle called by the Burmese the Battle of Ngas-aunggyan, 1277 [not 1272 as in the MSS] which was fought on the Taping river about seventy miles above Bhamo.

The fullest account is given by R., which will be found in full in Appendix II. No. 12, pp. 289–92.

In order to appreciate the errors made by Polo, readers should see Cordier, *Ser Marco Polo*, pp. 87, 88, and especially G. E. Harvey, *History of Burma*, 1925, pp. 64–70, 333 and 336.

CHAPTER 83

313. Page 85, lines 1, 2

a greate penet or hill

The word "penet" troubled me for a while, as I had hoped to discover in it a clue to Frampton's county. All dialect dictionaries, however, yielded negative results. It seems, therefore, to be a misprint for "pente," the modern French for a descent, slope, or declivity. This use in English must be very rare. Murray has no note of it. "Pent" is given as short for "Penthouse."

314. Page 85, line 2

two dayes iourney

All the best texts read two-and-a-half days.

315. Page 85, line 3

fauing one towne

Fr. 1116 reads "une grant place," while P has "une moult grant place." In R., however, we have "una pianura ampla, & spatiosa," which Marsden translates as "a spacious plain."

316. Page 85, line 5

a Sazo of golde for fyue of filuer

F. omits details here. R. continues: "The inhabitants are not allowed to be exporters of their own gold, but must dispose of it to the merchants who furnish them with such articles as they require, and as none but the natives themselves can gain access to the places of their residence, so high and strong are the situations, and so difficult of approach, it is on this account that the transactions of business are conducted in the plain."

"Nor will they allow," says Y., "anybody to accompany them so as to gain a knowledge of their abodes."

317. Page 85, line 6

you doe come vnto the prouince named Machay

Fr. 1116 reads "Mien est apelés." Y. (and also P.) have "Amien." R. has "la città di Mien," i.e Burma.

Unfortunately at the end of this chapter F. omits six short chapters (corresponding to Y. Bk. II. Chs. LIV–LIX; B Chs. CXXVI–CXXXI; and M. Bk. II Chs. XLIV–XLIX) dealing with the city of Mien (probably Old Pagan, i e Tagaung) Bengal, Laos, Tonking, and the route back to Sindafu (Ch'êng-tu fu) and so to the point where the two roads meet near Juju, i.e. the "Cinguy" of F.'s next chapter.

All these missing chapters are given in full in Appendix II. Nos. 13–18, pp. 292–296 from the R. texts, with notes between square brackets showing varying readings or additional matter from other texts.

318. Page 85, line 11

When they wil take any Elephant...

For some unexplained reason the account given here really refers to tigers, and not elephants, and has been taken from the middle of one of the omitted chapters (Y. Bk. II. Ch. LIX; B. Ch. CXXXI; and M. Bk. II. Ch. XLIX).

CHAPTER 84

319. Page 85, lines 19, 20

another prouince named Cinguy

Here we are back at the road bifurcation near Cho-chau, the "Giongiu" of fr. 1116; the "Juju" of Y.; and "Gingui" of M.; corrupted to "Cinguy" by Santaella and F.

320. Page 85, line 22

a greate Citie named Cancafu

This is Ho-kien fu ("Cacianfu" in fr. 1116; "Cacanfu" in Y. and P.; and "Pazan-fu" in M.) in Chih-li, one hundred miles nearly due south of Peking and seventy-five miles from the coast at Chi-kow on the gulf of Pe-chih-li. Here the new itinerary starts through the eastern provinces of Cathay and Manzi to the city and harbour of Zayton (Chüan-chau fu, Tsiuan-chau fu).

R. gives more details of the city than any other text; he says it "belongs to Kataia [Cathay] and lies towards the south in returning by the other side of the province. The inhabitants worship idols, and burn the bodies of their dead. There are here also certain Christians, who have a church [only in R. and Z]. They are subjects of the Grand Khan, and his paper money is current amongst them. They gain their living by trade and manufacture, having silk in abundance, of which they weave tissues mixed with gold, and also very fine scarfs. The city has towns and castles under its jurisdiction.

A great river flows beside it, by means of which large quantities of merchandise are conveyed to the city of Kanbalu; for by the digging of many canals it is made to communicate with the capital [only in R. and, with differences, in Z]. But we shall take our leave of this, and proceeding three days journey, speak of another city named Chan-glu."

CHAPTER 85

321. Page 86, line 3

fiue dayes iourney

As we have just seen, R. makes it "three days journey," which agrees with all the best texts.

322. Page 86, line 4 *Citie named Cianglu*

At last we have a form which exactly corresponds to that of fr. 1116, as well as P. Yule, following M., reads "Changlu." It would appear to be the modern Tsang-chau, about thirty miles E.S.E of Ho-kien fu F 's account is abbreviated, and the best is that found in R. (also in part in *Z*):

"Chan-glu is a large province, situated towards the south, and is in the province of Kataia. It is under the dominion of the Grand Khan. The inhabitants worship idols, and burn the bodies of their dead. The stamped paper of the emperor is current among them. In this city and the district surrounding it they make great quantites of salt, by the following process In the country is found a salsuginous earth. Upon this, when laid in large heaps, they pour water, which in its passage through the mass, imbibes the particles of salt, and is then collected in channels, from whence it is conveyed to very wide pans, not more than four inches in depth. In these it is well boiled, and then left to crystallize. The salt thus made is white and good, and is exported to various parts. Great profits are made by those who manufacture it, and his majesty derives from it a considerable revenue.

This district produces abundance of well-flavoured peaches, of such a size that one of them will weigh two pounds troy-weight. We shall now speak of another city named Chan-gli."

CHAPTER 86

323. Page 86, line 9

Sixe dayes iourney beyonde the Citie named Cianglu

But from Cianglu Polo went to Ciangli which took five days. It seems that Santaella's copyist thought the two places were identical

There is, however, some excuse for the mistake, especially as "Ciangli" and the next place mentioned, "Candrafra," appear to have been mixed up. See next note.

324. Page 86, lines 10, 11

a Citie named Candrafra

This has been identified with Yen-chau (35° 37′ N. lat, 116° 50′ E long). It appears in fr 1116 as "Tandinfu." Other readings are: Candinfu, Condinfu and Cundinfu R. has "Tudin-fu" As Yule has pointed out (Vol 1. p 137) Yen-chau was of only second importance, and the description and history applied to it really belong to "Ciangli" which, as we have seen, is omitted by F. "Ciangli" of fr 1116 is the "Chinangli" of Y. and must be identified with Tsi-nan fu (36° 43′ N lat, 116° 57′ E. long.).

See Appendix II. No. 19, pp. 296–7.

325. Page 86, lines 11, 12

had vnder it...twelue Cities

All the best texts read "eleven."

326. Page 86, line 15

a fayre Citie named Singuymata

Here the name is well preserved. Fr. 1116 has "Singiumatu," P. "Singuy matu," and Y. "Sinjumatu." It is the "Sunzumatu" of Friar Odoric (see *Cathay and the Way Thither*, Vol. II. p. 214 n. 2). It is to be identified with the modern Tsi-ning-chau. See A. C. Moule, *T'oung Pao*, July 1912, pp. 431–3.

Frampton has badly abbreviated Polo's account of "Candrafra" and "Singuymata," and has entirely omitted the story of Liytan Sangon (R.'s Lucansor). After "Singuymata" the texts of Y. and B. continue the itinerary to "Linju," "Piju" and "Siju," the two latter of which we can probably identify with Pei-chau and Su-t'sien respectively. "Linju" is a difficulty and I can see no way of accepting Yule's Lin-ch'ing as its modern equivalent. I would suggest Sü-chau fu (34° 12 N., 117° 20′ E.) as fitting the conditions the best. See p. lii of the present volume.

For a full account of "Singui-matu" and the passages omitted by F. see Appendix II. Nos. 20, 21, pp. 297–9.

CHAPTER 87

327. Page 87, line 1

Going from Singuymata ſeuentéen dayes

This represents an attempt to bridge the gap caused by the omitting of "Linju," "Piju" and "Siju." It might be a misprint, I think, for "sixteen" as found in R. This would be correct as the distances to be added are: 8 + 3 + 2 + 3. In the next few lines F. gives the usual formula about the people being subject to the Khan, etc., which, however, includes: "Their language is Persian." This is, of course, the mistake already referred to (Note 53), but none of the best texts mention the language here at all.

328. Page 87, lines 9, 10

Vpon this riuer the great Cane hath fiftéene great ſhips

F. omits to tell us that quantities of large fish are found in the river. Santaella apparently dislikes possible exaggerations, and reduced the "15,000" of all the texts to "15." He does the same with the "twentye poore ſtriplings" of Ch. 88 (see Note 335).

329. Page 87, line 12 *and fiftéene mariners*

Read "twenty" here with all the best texts.

330. Page 87, lines 14, 15

The biggest of them is named Choyganguy, and the other Caycu and they be both a dayes iourney from the sea

"Choyganguy," (printed "Choygamum" by F. in the chapter heading), is the "Koi-gan-zu" of R.; "Coguiganguy" of P ; "Coiganju" of Y.; and "Coigangiu" of fr. 1116.

It is to be identified with Hwai-ngan-chau, now -fu (usually spelt Hwaianfu in modern maps) in c. lat. 33° 30′ N., long. 119° 10′ E. Its recent official name is Huaian-hien, though commonly called Huai-ch'êng.

In a certain semi-confidential Admiralty publication (from which I have permission from the Controller of H.M. Stationery Office to quote, but not to mention by name) there is a Gazetteer of all the chief places in Kiangsu. In this, Hwai-ngan-chau is described as being very poor in population, while the whole city lies below flood-water level, so that sometimes in summer the gates have to be kept continually closed and backed up with earth ramparts for protection against floods. All the surrounding country is low-lying, swampy and liable to floods.

This information fully prepares us for our inability to discover any trace of "Caycu," the small town opposite. Perhaps it has been swept away in course of time. Hwai-ngan-chau actually consists of three walled cities, and we can imagine that its preservation has been only due to the constant combating of the floods by closing the gates as mentioned above

"Caycu" is the "Caigiu" of fr. 1116 and the "Caiju" of Y It should be clearly distinguished from Yule's "Cayu," which we shall come to soon. (See Note 349)

F. tells us that both places are a day's journey from the sea This information is apparently an addition. Hwai-ngan-chau is about seventy-five miles from the Yellow Sea (Hwang-Hai).

CHAPTER 88 [Misprinted 80 by F.]

331. Page 87, line 16

Passing the saide riuer, you enter into ỹ prouince of Mangi . . .

R. and Z alone give us the following further information about Cathay before Polo goes on to describe Manzi or Southern China:

"Upon crossing this river, you enter the noble province of Manji· but it must not be understood that a complete account has been given of the province of

Kataia. Not the twentieth part have I described. Marco Polo in travelling through the province has only noted such cities as lay in his route, omitting those situated on the one side and the other, as well as many intermediate places, because a relation of them all would be a work of too great length, and prove fatiguing to the reader.

Leaving these parts we shall, therefore, proceed to speak, in the first instance, of the manner in which the province of Manji was acquired, and then of its cities, the magnificence and riches of which shall be set forth in the subsequent part of our discourse."

332. Page 87, line 17

where raigneth a king named Fucufur

I e. Facfur, which, however, was only a title; being the Persian equivalent of the Chinese *Tien-tzŭ*, "Son of Heaven."

333. Page 87, lines 22, 23

being brode and full of water

R. says "a bow-shot wide"; Y. translates, "more than an arblast-shot in width."

334. Page 88, lines 1, 2 *was very leacherous*

R. says "He maintained at his court and kept near his person about a thousand beautiful women in whose society he took delight," while Y. has "...all their delight was in women, and nought but women; and so it was above all with the King himself, for he took thought of nothing else but women, unless it were of charity to the poor." Fr 1116 simply has: "mes sun delit estoit de fenmes et fasoit bien a povres jens"

335. Page 88, lines 6, 7

twentye poore ſtriplings

Fr. 116 has "XXm," so also Y. and MS C. of P.

336. Page 88, line 9

in the yeare of our Lord .1267.

The date differs in the various MSS. All are wrong. It should read "1276." See Y Vol. II. pp 148 *et seq*

337. Page 88, lines 12, 13 *named Gaiſſay*

I.e. Kinsay, the Chinese *King-sze* meaning "Capital." Its proper name was Lin-ngan, and is now Hang-chau. See further Note 367.

338. Page 88, line 14

Baylayncon Can

Fr. 1116 writes "Baian Cincsan," and Y. has "Bayan Chincsan." F. seems to have arrived at his corrupted form by imagining that "csan" was meant for "Can."

"Bayan" means noble and the Chinese form *Pe-yen* could punningly mean "100 eyes." The second part of the name is merely the title *Chinsiang,* "minister of State."

339. Page 88, lines 17, 18

ſauing one named Sinphu...

This city, which in Ch. 92 F calls "Saimphu," is not mentioned until its proper place in the French texts or R. Unfortunately F. misses the whole point of the surrender of Kinsay as he omits the incident about the horoscope, and the Queen's consequent superstition. The account given by R is much better arranged than that of either Y. or B. and for this reason, and also for the sake of comparison, is given in full in Appendix II. No. 22, pp 299–301.

CHAPTER 89. [Misprinted 91 by F.]

340. Page 88, lines 23, 24

Citie...named Coygangui

F. apparently fails to realize that this is the "Choyganguy" to which he has already referred in Ch 87. See Note 330

341. Page 88, lines 26, 27

and haue the Perſian tongue

Here again the best texts have no mention of the language at all. See Note 327.

CHAPTER 90. [Misprinted 92 by F.]

342. Page 89, lines 4, 5

and there be very déepe waters on ech ſide of the cawſey

R. (and Z) gives more detail R says: "On both sides of the causeway there are very extensive marshy lakes, the waters of which are deep, and may be navigated; nor is there besides this, any other road by which the province can be entered. It is, however, accessible by means of shipping, and in this manner it was that the officer who commanded his majesty's armies invaded it, by effecting a landing with his whole force."

343. Page 89, line 6

a citie named Pangui

Fr. 1116 reads "Pauchin"; Y. "Paukin"; and R. "Pau-ghin." It corresponds to the modern Pao-ying, a *hien* city dependent on Yang-chau.

344. Page 89, lines 8, 9

and here is greate ſcarcitie of corne...

Here F. is wrong. The text reads "a great plantee."

345. Page 89, line 11

named Cayn

Fr. reads "Caiu," Y. "Cayu" and R. "Kain." It is the modern Kao-yu-chau, recently officially named Kao-yu-hien, having an estimated population of 15,000.

346. Page 89, line 13

for the value of ſixe pence

The texts have "a Venetian groat," which was equal to 5*d.*, or 4·99*d.* to be exact. See Y. Vol. II. p. 591.

CHAPTER 91. [Misprinted 93 by F.]

347. Page 89, line 15

the grounde of Tinguy

A misprint for *city* of Tinguy (?). Fr. 1116 writes "Tigiu" to which B. supplies an n: Ti[n]giu; Y. has "Tiju" and R. "Tin-gui." It is apparently Tai-chau an important city of about 70,000 inhabitants. It is a great centre of the salt industry, and is entirely surrounded by a moat. F. omits to mention "Tinju" of which Y.'s texts says:

"And there is a rich and noble city called Tinju, at which there is produced salt enough to supply the whole province, and I can tell you it brings the Great Kaan an incredible revenue. The people are Idolaters and subject to the Kaan." Yule would identify it with Tung-chau, but this is over ninety-eight miles from Tai-chau and is not a salt centre at all, being famous for cotton and silk. The suggestion made by J. C. Ferguson (*Journ. North China Br. Roy. As. Soc.* Vol. XXXVII. 1906, p. 190) that it is Hsien-nü-miao (Siennümiao) seems much more probable. It is only twenty-three miles from Tai-chau, is an important salt centre, and fits in better with the itinerary.

CHAPTER 92. [Misprinted 94 by F.]

348. Page 90, line 3 *a Citie named Manguy*

Spelt "Mangui" in the chapter heading. It is the "Yangiu" of fr. 1116, the "Yanju" of Y., and "Yan-giu" of R., and is to be identified with Yang-chau on the west bank of the Grand Canal, eleven miles north of the Yangtze. It is the centre of the salt administration of the Liang Huai district. The principal industries are lacquerware and silverwork, a possible echo of the harness-making mentioned by Polo [see Note 350 below].

349. Page 90, line 4 *and ſpeake the Perſian tong*

Here again F. is alone in mentioning the language. See Note 327.

350. Page 90, lines 4, 5
This Citie hathe vnder it ſeauentéene Cities...

Read "twenty-seven" with all the best texts. F. suppresses all details. R. is as follows:

"...which, having twenty-seven towns under its jurisdiction, must be considered as a place of great consequence. It belongs to the dominion of the Grand Khan. The people are idolaters, and subsist by trade and manuel arts. They manufacture arms and all sorts of warlike accoutrements, in consequence of which many troops are stationed in this part of the country. [Y. translates: "a great amount of harness for knights and men-at-arms."] The city is the place of residence of one of the twelve nobles, before spoken of, who are appointed by his Majesty to the government of the provinces; [R. omits "car elle est esleue por un des XII sajes," which Y. translates, "for it has been chosen to be one of the twelve *Sings*"] and in room of one of these [unique to R.], Marco Polo, by special order of His Majesty, acted as governor of this city during the space of three years." See next note.

351. Page 90, lines 5, 6
and I Marcus Paulus did gouerne this vnder the great Cane thrée yeares

We must not take this too literally, even if we accept the reading "Seigneurie" of fr. 1116 and fr. 5631 (Pauthier's "A") instead of the "Sejourna" of fr 5649 (P.'s "C"). R. reads: "di commissione del gran Can, n'hebbe il gouerno tre anni...," and it is due chiefly to this that subsequent editors have made him "Governor-General." At most he held the post of governor of the Lu, or circuit, of Yang-chau. Y. suggests the three years in question must have been between 1282 and 1287–8.

In order to appreciate the whole argument it is necessary to study Pauthier, pp. 467, 492; Y. Vol. II. p. 157; and Pelliot, *T'oung-Pao*, 1927, pp. 164–8, in his review of Charignon's edition of *Le livre de Marco Polo.*

352. Page 90, line 7

a prouince or Citie named Manguy

This must be a misprint as we have just finished with "Manguy." It is probably meant to be "Nanguy," a corrupted form of the "Nan-ghin" of Y. and R. and the "Nanchin" of fr. 1116. Here Polo leaves his itinerary to describe "two great provinces of Manzi which lie towards the west." The first of these is "Nanchin," that is the Ngan-king or Anking fu of modern maps; the second is "Saimphu," for "Saianfu," the modern Siang-yang fu. F. omits to mention that the Khan derives a large revenue from the city.

CHAPTER 93

353. *Howe this prouince was wonne by the great Cane*

Here we have a distorted *précis* of the surrender of Siang-yang. The much fuller accounts found in R. (for which see Appendix II. No. 23, pp. 301, 302 and the best French MSS are, however, equally difficult to explain.

They tell us that Siang-yang held out three years after the rest of Manzi had surrendered. This is in exact contradiction to fact. The siege of Siang-yang was the prologue, not the epilogue, to the conquest of Southern China. But this is not all, for not only does the claim made by Polo of being personally responsible for its surrender seem exaggerated, but the Chinese records clearly prove that Polo could not have been at the siege at all. In the annals of the *Sung shih*, the siege is continually mentioned. It started in the winter of 1268–9 and ended on March 17th, 1273. Now the three Polos did not reach Kublai Khan till 1275, or late in 1274 at the very earliest. It will be noticed in R.'s account (Appendix II. No. 23) that Marco is not mentioned. It has therefore been suggested that Nicolo and Maffeo were at the siege before their first return home. But, as we already know, the brothers had reached Acre by April, 1269, and were in Venice during the next two years.

Thus none of the Polos could possibly have been at the siege. It is quite contrary to the whole character of Marco Polo to imagine that he is purposely lying in order to get credit for himself, his father and his uncle.

We can only suspect the romantic pen of Rustichello. In order to appreciate how easy it would be to substitute the Polos for the people who *did* make "mangonels or trebuchets" reference must be made to the excellent paper by A. C. Moule, "The Siege of Saianfu and the Murder of Achmach Bailo," *Journ. North*

China Br. Roy. As. Soc. Vol. LVIII, 1927. We have already referred to the murder of the Bailo in Note 247, and mentioned how in the *Jámi'u't-Tawárikh*, Rashīd-ud-dīn gives a curious version connecting the two events. Among other evidence from Chinese sources, Moule gives extracts from the biographies of two of the Moslems mentioned as the catapult-makers: A-lao-wa-ting (Alau'd Din) of Mu-fa-li (Mosul?) and I-ssŭ-ma-yin of Hsu-lieh or Shih-la (Shiraz?). In the biography of the latter, we learn that after his death his son Pu-pai held his office assisted by his brother I-pu-la-chin and his colleague Ma-ha-ma-sha. In these three men, Moule would recognize the three brothers mentioned by Rashíd-ud-dín as (A)bu bak(r), Ibrahim and Muhammad.

Taking all the evidence given by Moule from Chinese sources as a whole, it is impossible to doubt the accuracy of the stories told both by Polo and Rashíd as far as the main events are concerned, but there is no thread of evidence that the Polos had anything, or could possibly have had anything to do with the siege. We can only imagine that Rustichello, the editor and translator of Romances, was thoroughly determined that the heroes of such an entertaining tale should not be three men with unknown names, and Moslems to boot! What could be easier than substituting the three heroes of the whole book?

CHAPTER 94

354. Page 91, lines 12–14

Goyng from Siamphu [*sic*], *and trauelling fiftéene dayes iourney towardes Syroco, or to the Eaſte ſoutheaſt, you come vnto the Citie named Singuy*

The break in the itinerary in order to speak of Nanchin and Saianfu has caused trouble in the different MSS

F. and R. talk as if the itinerary had not been broken, and make "Singuy" fifteen days from Saianfu, which is reasonable.

Fr. 1116 and Y. make him travel fifteen *miles* from Yanju to "Singuy" on the Yangtze which is also reasonable.

Several editors, however, have muddled the two up and made Polo reach the Yangtze after a journey of fifteen miles from Saianfu!

Y is troubled about Polo's direction. The text says he went "sceloc" or "yseloc," south-east; whereas if we identify "Singuy" with Icheng or I-ching-hien (which seems correct) the direction was south-west. However, he dismisses the point on the grounds that Polo's style of orientation must not be taken too literally But may not the explanation be that Polo is thinking of the direction from Saianfu? In order to get back to his route the name of the place might be altered as well as "jornee" becoming "milles," but the direction remain unaltered.

Fr. 1116 reads "angiu" as the point of the renewal of the itinerary. This has been written by B. as [Y]angiu, and taken to be the "Yangiu" of his Ch. CXLV. This seems entirely justifiable. Icheng is a walled town connected by a creek with the Yangtze, one and a half miles to the south. Another creek, the San-ch'a ho (I can find no trace of Y.'s "*two* branch canals"), connects the town with the Grand Canal.

355. Page 91, lines 16, 17

riuer. . . named Tuognrou. . . . 17. miles in breadth, and one hundred dayes iorney in length

"Tuognrou" is a misprint for the "Tnoguron" as printed by F. in the margin. I cannot suggest how the corruption was arrived at, unless it is meant for "Ta-kiang," "great river," one of the best known names of the Yangtze.

In Ch. 73 F. called it "Champhu." As regards its breadth, it is, of course, exaggerated. Most MSS give varying distances of ten, eight and six miles. In point of fact, the Yangtze averages from three-quarters to two miles in width during its course through the province of Kiangsu. Below Tungchau it is ten miles wide, and even exceeds this at Woosung on the coast opposite Shang-hai. The length is by no means exaggerated. The latest estimates put it somewhere between 3200 and 3500 miles.

356. Page 91, line 20

fiue thousande shippes or barkes

So also in R. The number is missing in fr. 1116, and B. has supplied "V^M^." Pauthier's text says that Polo heard from the Khan's revenue officer that 200,000 ships passed up-stream in a year, without counting those going down. Y. has included this in his translation in addition to the statement that "Messer Marco Polo said that he once beheld at that city 15,000 vessels at one time." The smaller number, as in our text, seems much more likely to be the correct one.

R. contains several lines not in the French MSS and is valuable for comparison. It is therefore given in full in Appendix II. No. 24, pp. 302, 303.

Z also contains an important addition; see B. p. 140 note c.

CHAPTER 95

357. Page 92, line 1

Cianguy is a small Citie

"Cuguy" of P.; "Caiju" of Y.; and "Caygiu" of B. This is, without any doubt, Kwa-chau on the north bank of the Yangtze, thirteen miles E.S.E. of Icheng.

358. Page 92, lines 7, 8

for it is better prouided with barkes than with cartes, or horſes

So ends the chapter, omitting, however, the very interesting passage on the Grand Canal and Golden Island.

Y. translates as follows:

"You must understand that the Emperor hath caused a water-communication to be made from this city to Cambaluc, in the shape of a wide and deep channel dug between stream and stream, between lake and lake, forming as it were a great river on which large vessels can ply. And thus there is a communication all the way from this city of Caiju to Cambaluc; so that great vessels with their loads can go the whole way. A land road also exists, for the earth dug from those channels has been thrown up so as to form an embanked road on either side.

"Just opposite to the city of Caiju, in the middle of the river, there stands a rocky island on which there is an idol-monastery containing some 200 idolatrous friars, and a vast number of idols And this Abbey holds supremacy over a number of other idol-monasteries, just like an Archbishop's see among Christians."

Both accounts, the Grand Canal and Golden Island, are accurate.

In the semi-confidential Admiralty publication we read:

"The embankments of the Grand Canal consist of earth actually thrown up when the bed of the canal was cut, further reinforced by soil taken from the adjacent plain. The eastern embankment measures about 100 ft. at the base and 30 ft. at the top. The western embankment is somewhat narrower (about 80 ft. at the base, 10 ft. at the top).... The top of the embankment provides a convenient towpath, but the available room is greatly reduced by the numerous houses which line the water-way."

The changing nature of the river-bed at this point has continually altered the position of Chin Shan, or Golden Island. In 1823 it was described as being on the left bank; in 1842 it was an island in the middle of the river; in 1862 it was joined to the right bank by a spit; in 1907 it was nearly 700 yds. inside the low river edge. To-day it can be described as a precipitous rocky hill on the right bank of the river. It is covered with temples and crowned by a pagoda 213 ft. high.

CHAPTER 96. [Misprinted 98 by F.]

359. Page 92, line 9

Pingramphu is a Citie...

A much corrupted form of Chinghianfu (Chin-kiang fu), a walled city of the usual type on the south bank of the Yangtze, three and a quarter miles S.E. of Kwa-chau.

360. Page 92, lines 10–12

edified by Marsar Conoſtor...in the yeare of oure Lord .1288.

The best texts read "1278." Fr. 1116 repeats the date twice, and reads the name "Marsarchis." Y. has "Mar Sarghis" (or Dominus Sergius) which he says appears to have been a common name among Armenian and other Oriental Christians. Our text omits the usual details: that the city consists of idolaters, that they are subject to the Khan, use paper money, live by trade, have abundance of victuals, and make stuffs of silk and gold.

361. Page 92, lines 16, 17

the citie of Tigningui

"Chinginguy" of P.; "Chinginju" of Y.; "Cangiu" of B.; while R. preserves a form somewhat similar to that of our text: "Tin-gui-gui." It is the modern Chang-chau, forty-eight miles south of Chinkiang, on the Grand Canal. R. explains that the walls of the city were surrounded by a double wall. If so, no trace now remains. The present walls, however, date from the Ming period and are four and a quarter miles in perimeter, 25 ft. high, surrounded by a moat 5 to 15 yds. wide and 3 to 8 ft. deep.

362. Page 92, line 18

men named Alanos

The Alans, the remnants of whom were settled on the northern skirts of the Caucasus. See Y. Vol. II. pp. 179, 180; Cordier, *Ser Marco Polo*, pp. 95, 96 and the references given in those pages.

363. Page 92, line 21

Barayn

Here we recognize the "Baylayncon" of Ch. 88, the "Baian" of fr. 1116 and "Bayan" of Y.

CHAPTER 97. [Misprinted 99 by F]

364. Page 93, line 1

Singuy is a...noble citie

The "Sugiu" of fr. 1116; "Suju" of Y.; and "Sin-gui" of R. This is the modern Su-chau (Soochow), the capital of Kiangsu, with an estimated population of 280,000. Polo's description applies to-day as it did in the thirteenth century. The city is celebrated for its silk-weaving, and is an important educational centre.

365. Page 93, lines 1–7

.40. miles in compasse... .7000. bridges of ſtone

Y. reads "60 miles," but F. agrees with fr. 1116. Our "7000" is a misprint for "6000" as found in the best texts.

366. Page 93, lines 10, 11

for fixe pence they doe giue more than fiue pound of Ginger

Here again, as in Ch. 90 (see Note 346), "sixe pence" is given for "un venesian gros." All the best texts read "40 pounds" instead of "fiue" as in our present version.

367. Page 93, line 11

there be .17. Cities greate and fayre

All the best texts read "16," and add that Su-Chau and Kinsay mean "City of Earth" and "City of Heaven" respectively.

This false etymology is probably due to a local "vulgar error." Before proceeding to describe Kinsay, most texts mention two intermediate cities. Fr. 1116 mentions three—Vugiu, Vughin and Ciangan. These I take to be P'ing-wang, Hu-chau, and Ka-shing. See the Introduction p. liv and map facing it.

CHAPTERS 98–100. [Misprinted 97 by F.]

368. Page 93, line 16

a noble and famous Citie named Quinsay

These three chapters constitute Frampton's entire description of Kinsay, which is given in such great and most interesting detail by Ramusio. No edition of Polo could possibly afford to ignore it, and it will be found in full as passage No. 25 in Appendix II. pp. 303–14 The "Ganfu" of F. is probably to be identified with Ning-po, but see Pelliot's suggestion in Cordier, *op. cit.* p. 98.

CHAPTER 101

369. Page 96, lines 2, 3

what rent the greate Cane hath yearely

So abridged is this chapter that it is necessary to give it almost complete from Ramusio: "In the first place, upon salt, the most productive article, he levies a yearly duty of eighty tomans of gold, each toman being eighty thousand saggi, and each saggio fully equal to a gold florin, and consequently amounting to six millions four hundred thousand ducats This vast produce is occasioned by the vicinity of the province to the sea, and the number of salt lakes or marshes, in which, during the heat of summer, the water becomes crystallized, and from whence a quantity of salt is taken, sufficient for the supply of five of the other divisions of the province. There is here cultivated and manufactured a large quantity of sugar, which pays, as do all other groceries, three and one-third per cent. The same is also levied upon the wine, or fermented liquor, made of rice. The twelve classes of artisans, of whom we have already spoken, as having each a thousand shops, and also the merchants, as well as those who import the goods into the city, in the first instance, as those who carry them from thence to the interior, or who export them by sea, pay, in like manner, a duty of three and one-third per cent.; but goods coming by sea from distant countries and regions, such as from India, pay ten per cent.

So likewise all native articles of the country, as cattle, vegetable produce of the soil, and silk, pay a tithe to the King. The account being made up in the presence of Marco Polo [Pauthier's text says that the Khan sent Polo to inspect the amount of the revenues], he had an opportunity of seeing that the revenue of His Majesty, exclusively of that arising from salt, already stated, amounted in the year to the sum of two hundred and ten tomans (each toman being eighty thousand saggi of gold), or sixteen million eight hundred thousand ducats."

CHAPTER 102

370. Page 96, lines 11, 14, 19

Thampinguy... Vguy... Greguy

I.e. "Tanpi[n]giu... Vugui... Ghiugiu" of fr. 1116, and "Tanpiju... Vuju... Ghiuju" of Y.

I would identify them with Fu-yang, Tung-lu and Yeng-chau respectively (according to Phillips) rather than with Shao-hsing, Kin-hwa and Kiu-chau as suggested by Y. See further in Introduction of this volume, p. liv.

371. Page 96, lines 22 *et seqq.*

and many Lyons

This account of catching lions [tigers] appears to be unique to F. Anyway, I can find no trace of it in any of the leading texts. It may, perhaps occur in one of the innumerable Pipino versions.

CHAPTER 103. [Misprinted 101 by F.]

372. Page 97, lines 12, 15, 16

Cinaugnary... Signy... the Realme of Fuguy

I.e. "Cianscian... Cugiu... the kingdom of Fugiu" of fr. 1116, and "Chanshan... Cuju... the kingdom of Fuju" of Y. Here again the true identification of the places is difficult, but, as already stated in the Introduction (p. lv), I much prefer Phillips' "Lan-ki... Kiu-chau" to Y.'s "Sui-chang... Chu-chau" for the first two places.

F. gives the distance between them as four days. This should be corrected to "three" with all the best texts.

His "Realme of Fuguy" is Fu-chau, to which we return very soon.

373. Page 97, line 20 *and ſuger ſo plenty*

This is a mistake for "ginger."

374. Page 97, lines 29, 30

the Citie named Belimpha, which hath foure bridges of marble...

I.e. Kien-ning fu, the "Quenli[n]fu" of fr. 1116 and "Kelinfu" of Y. All texts have "three bridges" instead of "four."

375. Page 98, line 6

At theſe ſixe dayes iourneys ende, ſtandeth the Citie named Vguca...

Here F. has got muddled in his distances. We are dealing with the second half of the six days' journey, and fr. 1116 reads: "Et au drean de ceste trois jornee a xv milles...." Thus it is clear that after travelling from Kiu-chau (Cugiu) to Kien-ning fu in three days, Polo goes on for another three days. At the 15th mile on the third day he reaches "Vguca," the "Unquen" of fr. 1116, and "Unken" of Y. Continuing a further fifteen miles he gets to "the noble city of Fugiu...chief of the kingdom of Choncha" So at this point Polo was travelling thirty miles a day. Thus "Unquen" should be seventy-five miles from Kien-ning fu and fifteen miles from Fugiu. This fact, added to the agreement of the description and direction, has made me (Introduction, p. lv) suggest Yüyüan as its modern equivalent, rather than Min-tsing (Y.) or Yung-chun (Phillips).

376. Page 98, line 7 *great plentye of ſuger*

Both Z and R. have additional information. Marsden translates: "Previously to its being brought under the dominion of the Grand Khan, the natives were unacquainted with the art of manufacturing sugar of a fine quality, and boiled it in such an imperfect manner, that when left to cool it remained in the state of a dark brown paste. But at the time when this city became subject to His Majesty's government, there happened to be at the court some persons from Babylon [i.e. Cairo] who were skilled in the process, and who, being sent thither, instructed the inhabitants in the mode of refining the sugar by means of the ashes of certain woods."

CHAPTER 104

377. Page 98, lines 10, 11

the Citie named Friguy, which is the head of ẏ Realme of Tonca, which is one of the nyne Kingdomes of Mangi

As we have seen, the "Realme" was mentioned by F. in the last chapter as "Fuguy," and is to be identified with Fu-chau in the province of Fu-kien. "Tonca" is the "Choncha" of fr. 1116 and "Chonka" of Y. Its etymology has not been satisfactorily explained. See Y. Vol. II. p. 232.

F. now correctly speaks of "nyne kingdomes," but in Ch. 100, line 1, he said, "Mangi was diuided into .8. kingdomes, ..."

378. Page 98, lines 12, 13

a Riuer of seauen miles in breadth

Read "one mile" with the best texts. The rest of the text is much abbreviated. Apart from the passage from R. quoted below, *Z* has 70 unique lines (see Benedetto, pp. 157–8) of considerable interest. They deal with lion-hunting with the help of dogs, "animalia vocata papiones," and the religious views of the people of Fugiu as described in a conversation with Marco and Maffeo.

The passage from R. (also with slight differences in Y., etc.) is as follows:

"In this place is stationed a large army for the protection of the country, and to be always in readiness to act, in the event of any city manifesting a disposition to rebel. Through the midst of it passes a river, a mile in breadth, upon the banks of which, on either side, are extensive and handsome buildings. In front of these, great numbers of ships are seen lying, having merchandise on board, and especially sugar, of which large quantities are manufactured here also. Many vessels arrive at this port from India, freighted by merchants who bring with them rich assortments of jewels and pearls, upon the sale of which they obtain a considerable profit. This river discharged itself into the sea, at no great distance from the port named Zai-tun. The ships coming from India ascend the river as high up as the city, which abounds with every sort of provision, and has delightful gardens, producing exquisite fruits."

CHAPTER 105

379. Page 98, lines 20, 21

hauing abundance of all victuals

F. omits to mention that many of the trees supply camphor, and that all the people are traders and craftsmen, subjects of the Khan, and under the government of Fugiu.

380 Page 99, line 2

for one Shippe that commeth vnto Alexandria

Read, "one ship-load of pepper that commeth... "

After mentioning the various percentages, Polo speaks of tatooing, the manufacture of porcelain at Tiungiu [? Jau-chau fu], the language and writing of the province of Manzi, etc. All this is omitted by F.

The process of the porcelain manufacture is the most interesting, and is found both in *Z* and R It is thus translated by Marsden:

"They collect a certain kind of earth, as it were, from a mine, and laying it in a great heap, suffer it to be exposed to the wind, the rain, and the sun, for thirty or forty years, during which time it is never disturbed. By this it becomes refined and fit for being wrought into the vessels above mentioned. Such colours as may be thought proper are then laid on, and the ware is afterwards baked in ovens or furnaces. Those persons, therefore, who cause the earth to be dug, collect it for their children and grandchildren. Great quantities of the manufacture are sold in the city, and for a Venetian groat you may purchase eight porcelain cups."

Before passing on to speak of Japan, Polo gives us a chapter on the merchant ships of Manzi This is omitted by F., but is to be found in full from R. in Appendix II. No. 26, pp. 314–15.

CHAPTER 106

381. Page 99, line 15

the Iland named Ciampagu

Written variously Cipangu, Chipangu, Cipingu, Zipingu, etc., representing the Chinese Jih-pên-kwé, Japan.

382. Page 99, line 20

ſpeake the Perſian tong

See Note 53. It will be unnecessary to refer again to this oft-recurring mistake.

383. Page 99, line 25

a piece of two Ryals of plate

This I take to mean "of the diameter of a two-real piece of silver." Mr G. F. Hill, of the Dept. of Coins and Medals at the British Museum, tells me that the two-real piece was a common denomination of the Spanish coinage in Santaella's time. Under Ferdinand and Isabella it measured 1·15 in. in diameter, which is the exact measurement of "two fingers thick" of the average man's hand, as found in the best texts.

384. Page 99, line 26

greate plenty of precious ſtones

F. omits to mention the pink pearls (perles...rojes), which are described as being equal in value to the white ones.

R. and Z add that some of the dead in the Island are buried, and others are burnt, and that when a body is burnt one of the pearls is placed in the mouth.

385. Page 99, line 31

Abatan, and the other Vonsaucin

Written almost identically in the French texts. The "u" in the latter name should be "n." See note to line 22 of B. pp. 163, 164; also Cordier, *op. cit.* p. 103.

386. Page 100, line 13

and in that inſtant...

This certainly makes a better story, but in none of the MSS I have seen is the killing of the eight men in any way connected with the storm. In fact, in the French texts the incident about the pieces of iron comes later on. In R. we find the text rearranged as in F. Conti also speaks of the use of iron inserted under the skin for the same purpose. See p. 144 (last few lines) of this volume.

387. Page 100, line 16

tenne miles

Read "four miles" with the best texts.

388. Page 100, lines 23, 24

retired backe by the Ilande

R. (and also Z) has a more detailed account:

"The Tartars, on their part, acted with prudent circumspection, and, being concealed from view by some high land in the centre of the island, whilst the enemy were hurrying in pursuit of them by one road, made a circuit of the coast by another, which brought them to the place where the fleet of boats was at anchor."

389. Page 100, line 30

and ranſacked it

F. omits "except the pretty women, whom they kept for their own use."

390. Page 100, line 36

in the yeare of our Lorde .1248.

The date varies greatly in the different MSS. Fr. 1116 has 1268, Y. 1279, R. 1264. Kublai made many unsuccessful attempts to conquer Japan from 1266 to 1274, but the final disaster (only briefly related by Polo, but with additional facts which apparently have no historical basis) came in 1280–1. See Y. Vol. II. pp. 260, 261. F. omits to tell us the fate of the two commanders.

I quote from R. which alone with *Z* gives the additional information about the mode of punishment on the island of Zorza.

"The Grand Khan having learned some years after that the unfortunate issue of the expedition was to be attributed to the dissension between the two commanders, caused the head of one of them to be cut off; the other he sent to the savage island of Zorza, where it is the custom to execute criminals in the following manner. They are wrapped round both arms, in the hide of a buffalo fresh taken from the beast, which is sewed tight. As this dries, it compresses the body to such a degree that the sufferer is incapable of moving or in any manner helping himself, and thus miserably perishes."

391. Page 101, line 8 .7448. *Ilandes*

This number agrees with that given in fr. 1116. Y. has 7459, and R. 7440. In the Catalan map, where the information was almost certainly derived from Polo, the number is given as 7548 See Yule and Cordier, *Cathay*, Vol. I. p. 302

F. omits to tell us that the name of the sea containing the islands is "The Sea of Chin," which is the same as saying "The Sea over against Manzi"

CHAPTER 107

392. Page 101, line 17

a Countrey named Cyaban

This is the "Cianba" of fr. 1116, and "Chamba" of Y., the mediaeval "Champa," corresponding to the southern half of Annam. See M. G. Maspero, *Le Royaume de Champa*, Paris, 1928

It is interesting to note that in R., and also *Z*, we find an interpolation immediately before the mention of Champa. As Y says, Marsden's translation is forced so as to describe the China sea His only rendering is as follows (ii. 266):

"Leaving the port of Zayton you sail westward and something south-westward for 1500 miles, passing a gulf called Cheinan [? 'An-nan, i.e. Tong-king], having a length of two months' sail towards the north. Along the whole of its south-east side it borders on the province of Manzi, and on the other side with Anin and Coloman, and many other provinces formerly spoken of. Within this gulf there are innumerable Islands, almost all well-peopled, and in these is found a great quantity of gold-dust, which is collected from the sea where the rivers discharge. There is copper also, and other things, and the people drive a trade with each other in the things that are peculiar to their respective Islands. They have also a traffic with the people of the mainland, selling them gold and copper and other things; and purchasing in turn what they stand in need of. In the greater part of these Islands plenty of corn grows. This gulf is so great, and inhabited by so many people, that it seems like a world in itself."

393. Page 101, line 20 .1248.

Read 1278 with the French MSS. The name of the "great Baron" is given in fr. 1116 as "Sogatu," and in Y. as "Sagatu." In the Chinese history he appears as "Sotu."

394. Page 101, line 24 *his tribute*

F. omits to tell us what the tribute was. The French texts give it as twenty large elephants, to which R. and Z add, "and a very large quantity of lignum-aloes.' R. gives the king's name as Accambale.

395. Page 101, line 25 .1275.

The best texts read 1285 Maspero thinks the actual date of Polo's visit to Champa was 1288. See Cordier, *Ser Marco Polo*, p. 104

396. Page 101, lines 26–28

he had .325. *Among his ſonnes he hadde* .25. . .
men of armes

All leading texts read "326" and "150" capable of bearing arms.

397. Page 101, line 29

and great Mountaynes of blacke Ebbanie

Fr. 1116 reads "Il ont maint bosches dou leigne que est apellés bonus, que est mout noir, dou quel se font les escace e les calamans [les échecs et les écritoires]."

The "bonus" is "Abenuz" in Spanish, from the Persian "Abnús," hence our "ebony" F. makes no mention of the chessmen or pen-cases.

CHAPTER 108

398. Page 102, lines 2, 3

.1400. miles...in compaſſe thrée thouſand miles

Read "1500" with the best texts. P. gives the circumference of Java as an even more exaggerated figure—5000 miles.

F.'s "ſeauen" Kings should read "a great King."

CHAPTER 109

399. Page 102, lines 9, 12, 13

Sayling ſeauentéene myles. ....two hundreth miles ...Iocathe

F. is very far out in his distances. For "17" read "700," and for "200" read "500." After passing the Condor group, Polo touched at some point on the N.E. coast of the Malay Peninsula which the best texts call "Locac," F.'s "Iocathe."

400. Page 102, line 20

ſtandeth out of the way

F. omits to add that the king discourages visits to the island, on account of the treasures and other resources it contains

CHAPTER 110

401. Page 103, lines 1, 2

fiue miles .Penthera

Read "five hundred" miles. "Penthera" should read "Pentain" or "Pentam," in which we must recognise Bentan. F becomes very hard to follow here, as he obviously has no idea what Santaella means. His mileages are all wrong, and even in the best texts are not easy to understand I need only refer here to Dr Blagden's remarks on pp. lvi *et seq* of this work

402. Page 103, lines 13, 14

Beyonde it ſtandeth the Realme of Ferlech

The French texts tell us that Ferlec, or Perlec (in Malay Pĕriak), was so overrun by Saracen merchants that the natives were converted to Mahommedanism.

CHAPTER 111

403. Page 103, line 19 *Baffyna*

Written also by F. as Baffina and Baxina. It is the "Basman" of fr. 1116 and R.; and the "Basma" of Y. Blagden considers this to be undoubtedly Pasai, though the etymology is hard to explain. See p. lix.

404. Page 103, line 24 *and Unicornes*

This, of course, is the rhinoceros. F. omits a short passage well worth quoting. Y. translates:

"'Tis a passing ugly beast to look upon, and is not in the least like that which our stories tell us of as being caught in the lap of a virgin; in fact, 'tis altogether different from what we fancied."

This mediaeval legend is said to have arisen from Aelian, xvi 20, where mention is made of the gentleness of the unicorn to its mate at mating time. Personally I am inclined to attribute the legend to the well-known folk-lore belief of the power of virginity. For a good general article on the unicorn see *Ency. Brit.* 11th Edit. Vol XXVII. p. 581, where many useful references are given.

405. Page 104, line 2 *Cieno or miery puddel*

It would appear that F. has left the Spanish *cieno*, mire, untranslated by an oversight.

CHAPTER 112

406. Page 104, line 12 *Samara*

For Samatra, which probably gave its name to the whole island. R. gives us much fuller details of the precautions taken by Polo against the natives whom he thought were cannibals:

". . Marco Polo established himself on shore, with a party of about 2,000 men; and in order to guard against mischief from the savage natives, who seek for opportunities of seizing stragglers, putting them to death and eating them, he caused a large and deep ditch to be dug around him on the land side, in such a manner that each of its extremities terminated in the port, where the shipping lay. This ditch he strengthened by erecting several blockhouses or redoubts of wood, the country affording an abundant supply of that material; and being defended by this kind of fortification, he kept the party in complete security during the five months of their residence. Such was the confidence inspired amongst the natives, that they furnished supplies of victuals and other necessary articles according to an agreement made with them."

407. Page 104, lines 19, 20

and from them commeth water, as it commeth from the vyne...

Once again we must turn to R. who gives the fullest description of the tree, the *Areng Saccharifera*, which supplies the toddy

"So wholesome are the qualities of this liquor, that it affords relief in dropsical complaints, as well as in those of the lungs and of the spleen. When these shoots that have been cut are perceived not to yield any more juice, they contrive to water the trees, by bringing from the river, in pipes or channels, so much water as is sufficient for the purpose; and upon this being done, the juice runs again as it did at first. Some trees naturally yield it of a reddish, and others of a pale colour. The Indian nuts also grow here, of the size of a man's head, containing an edible substance that is sweet and pleasant to the taste, and white as milk. The cavity of this pulp is filled with a liquor clear as water, cool and better flavoured and more delicate than wine or any other kind of drink whatever. The inhabitants feed upon flesh of every sort, good or bad, without distinction."

The passage about the cocoa-nuts appears in F as "In this Iland there groweth great plentie of the *Indian* nuts," but even so has become misplaced as it was not Dagroian but Samara that is described as producing nuts

VB also contains the passage about the "noxe de India grosso quanto el chapo de l'omo..." etc.

408. Page 104, line 22

which is named Deragoya

Written "Dagroian" in the French texts. It is still unidentified, but must have been near Samara on the same line of coast.

CHAPTER 113

409. Page 105, lines 6–8

Lambry...great plentie of ſpices...men that haue feathers about their priuities...

"Lambri," cf. fr. 1116, was somewhere near Kota Raja at the N.W. end of Sumatra. The description of the region as given by Polo has proved too incredible for Santaella.

He omits to mention the brazil of which Polo brought some seed to Venice, and tried in vain to grow.

The tailed men of Lambri have become "men that haue feathers about their priuities...," while the unicorns and other beasts are ignored.

410. Page 105, line 10 *Samphur*

The "Fansur" of the French MSS is to be identified with Baros, famous for its camphor.

Z and R. (also *VB* to a lesser extent) have a much fuller account of the sago tree than is found in other texts.

See Appendix II. No. 27, pp. 315, 316, where R.'s version is given in full.

CHAPTER 114

411. Page 105, lines 16–18

Going from Lambry ſayling .140. myles...the one is named Necumea, and the other Nangania

F. makes no mention of the "ysle molt pitete que est apellé Gauenispola," lying very close to Lambri. Although Polo says he will tell us about the island we hear nothing more of it. This has caused confusion in some of the texts, for a few lines later he speaks of "two islands, of which one is called Necuveran."

Some editors have made Gauenispola the "other" island.

All is clear, however, in fr. 1116.

All the best texts read "about 150 miles" as the distance from Lanbri to the Nicobars.

F.'s "Necumea" and "Nangania" (which he spells variously "Nangama" and "Nangana") are corruptions of Necuveran and Angaman, in which it is not difficult to see the Nicobars and Andamans.

412. Page 105, line 18

The people of Necumea, liue like beaſtes

Here F.'s account of the immorality of the inhabitants appears to be unique. He omits, however, to mention that they are idolaters.

The *Z* text has an interesting passage describing how the natives buy most beautiful kerchiefs or face-napkins of silk ("taveleas sive facitergia de Syrico") from passing traders. They make no use of them except to keep them in their houses hung over poles. They value them as if they were pearls or precious stones, and those who possess the most and finest are held to be nobler and greater than the rest.

Mr E. H. Man, C.I.E., the well-known expert on the Nicobars and Andamans, informs me, through Dr Blagden, that Polo's description is perfectly correct, and that to this day the natives will eagerly store up every gaudy silk handkerchief or piece of cloth which they can obtain from the traders. Plated goods, German silver spoons, cruet stands, chains and other similar objects have been added since Polo's day All these are found hanging up inside the huts. No other use is made of them except to excite the admiration and envy of the less fortunate neighbour. As it is incumbent on mourners to destroy the personal property of their deceased relatives at their death, one sees valuable wooden chests and such objects as have been mentioned above covering the graves of the recently deceased. They are, moreover, specially damaged in some way or other as to render them useless in the future.

413. Page 106, lines 1, 2

mountaines of Sandolos or Saūders, and of nuts of India, and of Gardamonia, and many other ſpyces

Fr. 1116 has "il sunt sandal vermoil [*VB* and R. say both red and white varieties are found] e noces d'Inde e garofal et berci e maintes autres bonnes arbres."

The *Z* text also adds apples of Paradise (? plantains, see Y Vol. I p. 99), to which F. refers in Chs. 15 (p 35) and 118 (p. 111).

It is hard to say where F gets his "mountaines" from, we must read "woods containing...."

For a note on sandalwood, see my *Ocean of Story*, Vol. VII pp. 105–7. I imagine that "Gardamonia" is some corrupted form of the "Garofal" of the best texts. This latter word needs a little explanation Other forms are *Garophul* and *Karpophul*, it apparently became Hellenised as *Caryophyllum*, whence the modern French *girofle*. The English *clove* was derived from *clou*, nail, which name was given by the French in 1770 when they introduced the clove-tree into Mauritius See further, *Ocean of Story*, Vol VIII. p 96 n 2. It is of interest to note that Polo mistook the ports whence cloves were shipped for the home of the plant, whereas Nicolò de' Conti was the first traveller to describe it correctly as coming from the Moluccas (or rather Banda) to Java and Sumatra, see p 133 of this volume, where "Clauas" is a mistake for "Iauas," the Greater and Lesser Java.

414. Page 106, line 6 *great plentie of ſpices*

R. mentions "Indian nuts, apples of Paradise, and many other fruits different from those which grow in our country "

Z also adds an interesting passage on the strength of the currents, and how ships find it impossible to anchor, and become entangled with the large amount of trees and roots which are washed into the gulf

CHAPTER 115

415. Page 106, lines 10, 11

being in compaſſe thirtie thouſand myles

This is a compromise of F. The French texts say 2400 miles, and in ancient times 3600 miles, but that part of the island has become submerged by the strong winds. Y.'s text adds: "For you must know that, on the side where the north wind strikes, the Island is very low and flat, insomuch that in approaching on board ship from the high seas you do not see the land till you are right upon it."

Although thirteenth-century writers have greatly reduced the exaggerated estimates of the circumference of Ceylon, they still made it nearly four times too much!

416. Page 106, line 11 *a very rich king*

F. omits to mention his name, Sendemain. It is not clear to whom Polo refers here. The native king from 1267 to 1301 was Pandita Prakama Bahu II. See further Cordier, *Ser Marco Polo*, p. 111.

417. Page 106, line 14 *and of the Wyne of trées*

F. omits to mention brazil, sappan-wood, which is described as being very abundant and the best in the world.

418. Page 106, line 17 *and of diuerſe other kindes*

The French texts also name sapphires, while *Z* and R. add garnets. F. omits to tell us that the Khan tried to procure the great ruby from the king of Ceylon, but was unable to obtain it at any price.

CHAPTER 116

419. Page 107, line 2 *fortie myles*

All the best texts read "sixty miles."

420. Page 107, lines 9, 10

named Sendarba...king of Nor

Fr. 1116 has "Sender Bandi Devar," and Y. "Sonder Bandi Davar." F.'s rendering must have been due to an error in Santaella's MS, such as we find in the Latin text (Bib. Nat. lat. 3195) where the Tuscan is corrupted by Pipino's version (see p. xxiv of this volume). Here we read "Senderba, rex de Var," which at once enables us to see how F. has arrived at his corruption.

As to the possible identification of the king, see Y. Vol. II. pp 333 *et seq.*

421. Page 107, line 11 *The fiſhermen do fiſh*

F. abbreviates here sadly. The French texts give a fairly detailed account of the methods employed by the pearl-fishers. R. adds:

"These [oysters] they bring up in bags made of netting that are fastened about their bodies, and then repeat the operation, rising to the surface when they can no longer keep their breath, and after a short interval diving again. In this operation they persevere during the whole of the day, and by their exertions accumulate (in the course of the season) a quantity of oysters sufficient to supply the demands of all countries. The greater portion of the pearls obtained from the fisheries in this gulf are round, and of a good lustre. The spot where the oysters are taken in the greatest number is called Betala, on the shore of the mainland; and from thence the fishery extends sixty miles to the southward."

422. Page 107, line 17 *a hundereth & foure*

Apparently a mistake for "108," the mystical number among Brahmans and Buddhists. See Y. Vol. II. p. 347; and *Ocean of Story*, Vol. IX. p. 145. Here I mention a suggested interpretation offered by M. Pelliot, viz. that 108 represents a multiplication of the 12 months by the 9 planets. R. adds that their prayer consists of the words "pacauca, pacauca, pacauca" [? Pagavâ, "Lord"].

423. Page 107, line 20 *tenne riche Cities*

Read with fr. 1116, "une bone cité." F. omits to tell us of the restrictions enforced by the king against taking pearls out of the kingdom, and of the big prices he gives for those brought to him.

424. Page 107, lines 31–33

countrey named Cormos, at the price of fiue ounces of gold euery horſe,... The merchaunts of Quinſay, of Suffer, and of Beden,...

F. has muddled the text. Fr. 1116 reads: "...les mercant de Curmos e de Quisci et de Dufar et d'Escer e de Aden.... Il vendent le un bien V[c] saje d'or que vaillent plus de c mars d'arjent." These places we now recognize as Hormuz, Kais [which F. has taken to mean "Quinfay"], Dhofar, Sohār, and Aden.

The 500 "saje" or "saggi" is probably intended for dinars. See Y. Vol. II. p. 349; and Stein, *Rājataraṅgiṇī*, Vol. II. pp. 308–28.

425. Page 108, line 24 *Thefe people*

F. omits to mention the name of the caste, which is given in the French texts as "govi" or "gavi." It almost certainly corresponds to the modern Paraiyan caste of the Tamil country. See Thurston, *Castes and Tribes of Southern India*, Vol. IV. pp. 77–139.

426. Page 108, line 33 *In thys prouince...*

The account of the manners and customs of the people of "Maabar" [the Coromandel coastal regions] is so greatly abbreviated by F. that I have given the full account from R. in Appendix II. No. 28, pp. 316–19.

CHAPTER 117

427. Page 109, line 19 *Mvfuly is a Region*

Read "Mutfili" with all the best MSS. It is to be identified with Motupalli, a port in the Guntur district of Madras, 170 miles north of the capital.

F., whose accounts of the Indian Provinces are all abbreviated, omits to tell us that the country was formerly under the rule of a king, but that since his death his queen [Rudrama Devi] had ruled with great justice for forty years.

The chief food of the inhabitants is flesh, rice and milk; to which R. and *Z* add fish and fruits.

428. Page 109, line 25 *they do find the Adamants*

F. has muddled the account about the methods of obtaining the diamonds, and has entirely omitted any mention of the famous legend, so well known from Sindbad the Sailor's second voyage, of the eagles and the flesh to which the diamonds stick. Full reference to this incident will be found in V. Chauvin, *Bibliographie des Ouvrages Arabes*, VII. pp. 10, 11. We shall meet Sindbad's huge bird, the rukh, or roc, when we come to Madagascar.

The *Z* text tells us there are many other methods as well as those mentioned by which the diamonds are obtained.

F. also omits to mention the fine buckrams which are described as looking like the tissue of a spider's web. See Cordier, *Ser Marco Polo*, p. 118.

429. Page 110, line 6 *Thomas Dauana*

We can see from this latter word how the "Ananias" which we find in R has been created. We should read "Avarian" with the French texts, which has been explained as a corruption of the Arabic *Hawāriy*, "An apostle of the Lord Jesus Christ." For traditions relating to St Thomas see Cordier, *op. cit* pp. 116, 117; and M. Longworth Dames, *Duarte Barbosa*, Hakluyt Society, 1921, Vol II. pp. 98 (and note)–101, 126–9.

F omits to mention the legend of St Thomas' death by a chance arrow intended for a peacock.

The French texts, as well as R, tell us that the colour of the earth where the Saint was martyred is red, and *Z* adds that Marco Polo took some of it to Venice with him.

The date of the miracle should read "1288" with all the best texts

R. adds a passage about the "Indian Nuts." It is much more detailed in *Z*. See B. p. 187 note a.

430. Page 110, line 23 *oyle of Aiomolly*

Read "oleo de sosiman" with fr. 1116 etc., "oil of sesame."

CHAPTER 118

431. Page 111, lines 2, 3

a Prouince named Lahe, and there dwell the men named Bragmanos

For "Lar," or more correctly "Lāṭ-desa," an early name for Guzerat and North Konkan

"Bragmanos" is a corrupted form of "Abraiaman" as found in the French texts, apparently an Arabized form of Brahman

432. Page 111, line 4

They will not lye for all the worlde

Both *Z* and R. have an additional passage. R. reads:

"When any foreign merchant, unacquainted with the usages of the country, introduces himself to one of these, and commits to his hands the care of his adventure, this Brahman undertakes the management of it, disposes of the goods, and renders a faithful account of the proceeds, attending scrupulously to the interests of the stranger, and not demanding any recompense for his trouble, should the owner uncourteously omit to make him any gratuitous offer."

433. Page 111, line 9 *They do honour the Idols*

Before speaking of this, the best texts mention the sacred thread (see *Ocean of Story*, Vol. VII. pp. 26–8), and the king who sends to Soli (Chola) for pearls and precious stones.

434. Page 111, line 14

doe eate and drinke temperately

F. omits various other superstitions; e.g. observing if a tarantula advances from a lucky quarter, sneezing, and the flight of a swallow; and also the mention of betel chewing. See further, p. 321.

435. Page 111, line 16 *Cingnos*

Written "Ciugui" in fr. 1116, and "Chughi" in Y. The sect of *yogis* is, of course, meant. F. omits the passage about the novices and the test they have to undergo at the hands of dancing-girls.

Both R. and Z have a curious passage about the care taken by the *yogis* to scatter their ordure. See Wright, p. 404, and B. p. 192 note d.

This chapter should be followed by a short one on Ceylon and another on the city of Cail [Kail, a port on the Tinnevelly coastal region]. Both are given in full from R. in Appendix II. Nos. 29, 30, pp. 319–321.

CHAPTER 119

436. Page 112, lines 1, 2

Orbay is a Kingdome...beyond Marbar fiue miles

This should be written "Coilum," our "Quilon." For "fiue" read "fiue hundred."

F.'s account is much abbreviated. He makes no mention of brazil or ginger, nor does he name the animals: lions, parrots, peacocks, cocks and hens.

CHAPTER 120

437. Page 112, last line *Apes that are like men*

In speaking of this country around Cape Comorin, Polo finishes the chapter with the following sentence (omitted by F.):

"Il hi a gat paul si devisés que ce estoit mervoille. Lions, liopars, lonces, ont en abondance." It is hard to say exactly what is meant by "gat paul," but as Y. has pointed out (Vol. II. p. 385) it must refer to some variety of monkey. The P. MSS read "granz paluz et moult grans pautains," swamps and marshes; being entirely ignorant of the word. See P. Vol. II. p. 646.

CHAPTER 121

438. Page 113, lines 1–3

thirtie miles,.. and come to the Region of Hely, where they are all Idolaters

Read "three hundred miles." Hely, or Ely, lay about sixteen miles north of Cananore on the Malabar coast. Monte d'Ely is famous as being the first Indian land sighted by Vasco da Gama in 1498. The meaning of the word is not easy to discover, and many suggestions have been made (see M. L. Dames, *Duarte Barbosa*, Vol. II. pp 1, 2). Later forms of the word have substituted "D" for the "H," due, it would appear, to confusion with the *tali* in Rāmantali, a name given to-day to the country around Monte d'Ely. After saying that the people are idolaters, F. omits to tell us that there is no proper harbour in the country, but that the rivers have good estuaries. Pepper, ginger, and other spices are also mentioned. The ginger is described by Conti (see p. 127 of this volume) as being called "Bellyedy, Gebelly and Belly," known as Belledi or Baladi to the Italians of the fourteenth century. Marco Polo appears to have been the first traveller to have seen the plant alive. It was first described by John of Montecorvino in 1292, and was exported to Europe as a sweet in the Middle Ages.

439. Page 113, lines 8, 9

that it is no finne to robbe them

F. omits to add that ships arriving from Manzi and other places lay in their cargoes in six or eight days owing to the lack of a port and danger of sandbanks. The ships from Manzi, however, have large wooden anchors which hold in the worst weather.

CHAPTER 122

440. Page 113, lines 16, 17

a hundred Shippes togither,.. they roue into the Countrey a hundred miles

F. has muddled the sense here. The account found in R. is reliable

"In order that no ships may escape them, they anchor their vessels at the distance of five miles from each other, twenty ships thereby occupying a space of a hundred miles. Upon a trader's appearing in sight of one of them, a signal is made by fire or by smoke, when they all draw close together, and capture the vessel as she attempts to pass. No injury is done to the persons of the crew; but as soon as they have made prize of the ship, they turn them to provide themselves with another cargo, which, in case of their passing that way again, may be the means of enriching their captors a second time."[1]

Mention is also made of copper, gold brocades, silks, gauze, drugs, etc. brought by the ships from Manzi.

CHAPTER 123

441. Page 114, line 7

they giue them ſo great tormentes...

The French texts tell us they are made to drink "tamarendi et eive de mer," apparently some fruit mixed with the salt water, which causes them to void any pearls or precious stones they may have swallowed.

F. makes no mention of the Guzerat (Gieſurath) pepper, ginger, indigo or cotton which is found in all the best texts.

CHAPTER 124

442. Page 114, line 10

Thoma & vnto Sembelech

I.e. Tana and Semenat, the modern Thāna and Somnath. F. has greatly abbreviated the two original chapters into one very short one, and omitted Cambaet, Cambay.

He makes his usual error about the "language and fayth of Perſia." Considerably more detail is found in the leading texts. Tana is described as being an important shipping centre, with a large export of leather, buckram and cotton. The king has an arrangement with the corsairs whereby he obtains a supply of horses. The exports of Cambaet are given as indigo, buckram, cotton and hides, while the chief imports are gold, silver and copper to which R. adds *tutia*, for making kohl for the eyes. Semenat is merely described as a great kingdom where the people are honest and enjoy good trade by their industry.

CHAPTER 125

443. Page 114

In this chapter F. merely tells us he will not weary his readers by describing the places inland. He omits, however, to mention Kesmacoran, or Mekran, although he refers to it in Ch. 126 as Beſmaceian. R. says of it:

"Some of the inhabitants are idolaters, but the greater part are Saracens. They subsist by trade and manufactures. Their food is rice and wheat, together with flesh and milk, which they have in abundance. Many merchants resort thither, both by sea and land."

Thus the brief description of the seaports of India ends. As I have mentioned

[1] To the list of spices, which includes turbit (*Radex Turpetti*), we should add cinnamon and nuts of India. R. reads "cubebs" instead of "turbit."

in the Introduction, except at one or two points we are on no fixed itinerary and the muddled order of the places given must not surprise us. See Y. Vol. II. p. 403.

CHAPTER 126

444. Page 115, line 2 .25 *miles*

Read "500 miles."

445. Page 115, line 8

Augu∫t, September, and October

Read "March, April, and May."

446. Page 115, line 11 *vntill they be ∫eauen*

The best texts state that if the children be girls they stay with their mothers, but if boys they go to their fathers on arriving at the age of fourteen

For Chau Ju-kwa's remarks on the Male and Female islands see Cordier, *Ser Marco Polo*, p. 120

CHAPTER 127

447. Page 115, line 20 *an Ilande named Di∫cor∫ia*

This is, of course, Socotra, written Scotra in the best MSS. The name as written by F. is a corruption of the *Dioscorides* of the Greeks

448. Page 115, line 22 *great abundãce of Amber*

R gives us more details: "The inhabitants find much ambergris upon their coasts, which is voided from the entrails of whales. Being an article of merchandise in great demand, they make it a business to take these fish; and this they do by means of a barbed iron, which they strike into the whale so firmly that it cannot be drawn out. To the iron a long line is fastened, with a buoy at the end, for the purpose of discovering the place where the fish, when dead, is to be found. They then drag it to the shore, and proceed to extract the ambergris from its belly, whilst from its head they procure several casks of oil"

A very much fuller account will be found in the *Z* text See B p 204.

449. Page 115, line 23 *clothes of Cottenwooll*

F. omits to mention salt fish, meat, milk and rice. Merchants purchase gold at the island, and vessels bound for Aden touch at it.

The Archbishop is subject, not to the Pope, but to the "grant prelais" at Aden. All these details are found in the leading texts.

CHAPTER 128

450. Page 116, line 8 *foure Moores*

Fr. 1116 reads "esceque," sheikh.

451. Page 116, line 9

a thousand four hundred myles

Read "about four thousand miles."

452 Page 116, lines 11, 12

but of Elephants, and of Cammels

This is a mistake. The best texts only say camel's flesh was eaten, but R. and Z add that although the flesh of camels is preferred, that of other cattle is also eaten.

F. omits to mention Madagascar as a great *entrepôt* for ships from all parts of the world, or to speak of the high seas and strong currents of the Indian Ocean.

453. Page 116, lines 16, 17

the Iawe of a wilde Boare which wayghed twentie fiue poundes

Here again is a mistake. Apparently the reference is to the boar's tusks [? hippopotami teeth] mentioned later in the French texts. Their weight, however, is given as fourteen pounds.

454. Page 116, line 18

a certaine foule named Nichas

This is, of course, the roc or rukh, for which see *Ocean of Story*, Vol. I. pp. 103–5, where I have collected numerous references. The best bibliography on the subject is to be found in Chauvin, *Bib des Ouvrages Arabes*, v. p. 228 and VII. pp. 10–14.

I have no idea how F. arrived at his "Nichas."

455 Page 116, line 23 *at his pleasure*

So ends F.'s chapter; but he makes an omission that is of more importance than appears at first sight—he fails to mention asses and giraffes among the "wild beasts of strange aspect" found in Madagascar The point is that giraffes do not, and never did, exist in the island. So also with regard to elephants, camels, lions, leopards and bears. This and other facts led Yule to suspect some confusion between Makdashau (Magadoxo) and Madagascar. We must not forget, of course, that Polo is only speaking by hearsay, and that after coasting past Mekran his next port of call would be Ormuz.

For the giraffes see further, Note 459.

CHAPTER 129

456. Page 117, lines 1, 2

tenne thousand myles in compasse

Read "two thousand miles...."

457. Page 117, lines 4, 5

as fixe of our men may beare

Read "they can carry enough for four and eat enough for five men."

458. Page 117, line 7 *The women are filthy*

F. gives no details as to their huge mouths, big eyes, thick noses and enormous drooping breasts.

Modesty also makes him ignore the old myth about the human method of copulation adopted by the elephants.

459. Page 117, line 10

The wilde beastes of thys Iland

I.e. elephants, lions, and giraffes. Sheep are also mentioned. For an interesting work on giraffes see B. Laufer, *The Giraffe in History and Art*, Field Museum of Natural History, Chicago, 1928. In dealing with the Middle Ages the author points out (p. 74) that Polo is the first to recognize the wider distribution of the giraffe and to look for it beyond the limits of Abyssinia. Clavijo (1403) gives us a very good account. See Broadway Travellers edition, p. 149.

F. omits to tell us of the equipment of the warriors, and of the howdahs from which they fight, and of the curious custom of making the elephants intoxicated before a battle.

CHAPTER 130

460. Page 117, line 12

all whyche I haue declared of India

All the best texts speak more especially of the islands of the Sea of India, and the divisions of India the Greater and India the Lesser. See Y. Vol. II. pp. 424–7.

CHAPTER 131

461. Page 118, line 1

prouince named Abashia

This is the Italianized Ḥabash, i.e. Abyssinia. The text is muddled here. There were six kings, three of whom were Christians and three Saracens. The Christians bear *three* facial marks, the Jews two, and the Saracens only one.

462. Page 118, line 8

Cadamy

I.e. Aden.

463. Page 118, line 13

war against the Souldan of Aden

F. omits the story of the bishop who visited the Holy Sepulchre at Jerusalem on behalf of the Christian King of Abash. It tells how he was subsequently circumcized by the Soldan of Aden because he refused to renounce his faith. On hearing of the insult offered to the bishop the King of Abash attacked Aden and fully avenged himself by a severe defeat of the Saracens.

464 Page 118, line 14

fleshe, milke, and Rice,...

Sesame is omitted, while both R. and Z add corn. All the best texts also mention elephants, not bred in the country, giraffes (see Note 459), bears, leopards and lions, as well as a variety of other animals such as wild asses, cocks, hens, ostriches, parrots, and monkeys with faces like men.

R. alone adds that Abyssinia is extremely rich in gold, and much frequented by merchants who obtain large profits.

CHAPTER 132

465. Page 118, line 16

Adem. named the Sowdan

In the previous chapter F. speaks correctly of the Souldan of Aden! This chapter is greatly abbreviated. The last four lines really belong to the chapter on the city of Esher, or Escier (Es-Sheḥr, 330 miles east of Aden) omitted by F. together with those on Dufar (Dhofar), Calatu or Kalayati (Ḳalhāt), and Ormus or Curmos (Hormuz). These chapters will be found in full in Appendix II. No. 31, pp. 321–5.

F.'s abbreviation of the description of Aden has caused the interesting account of the over-land trade-route to Alexandria to be omitted. The texts have somewhat varying readings, due in all probability to subsequent editing as increased knowledge prompted.

R tells us that the merchants unlade their cargoes and put them on smaller vessels "with which they navigate a gulf of the sea for twenty days, more or less, according to the weather they experience Having reached their port, they then load their goods upon the backs of camels, and transport them overland (thirty days' journey) to the river Nile, where they are again put into small vessels, called *jerms*, in which they are conveyed by the stream of that river to Kairo, and from thence, by an artificial canal, named Kalizene, at length to Alexandria "

Mention is also made of the large trade in horses done at Aden, and of the great wealth of the Soldan arising from the duty thereon

The *Z* text contains an interesting passage on the precautions taken to ensure as far as possible against loss of the more valuable part of the cargoes (pearls, precious stones, etc.) due to the numerous shipwrecks which are encountered in this region. Bags of skins are filled with the valuables as well as necessary food and clothing, and, after being joined together, are fastened to rafts. See B p 213 note f.

CHAPTER 133

466. Page 119

This chapter on Siberia, and half of the next, being much abbreviated, are included in full in the "Ormus" chapter of R already referred to in the last note. See pp 324, 325

CHAPTER 134

467. Page 120, line 10

for the Sunne ſhyneth not there, and is not ſéene

Only in R. do we get an intelligent account of the phenomena of the arctic circle. As Yule says, all other versions imply a belief in the perpetuity of the darkness. The following extract from R. makes this clear

" ..during most of the winter months the sun is invisible, and the atmosphere is obscured to the same degree as that in which we find it just about the dawn of day, when we may be said to see and not to see ...The inhabitants of this region take advantage of the summer season, when they enjoy continual daylight, to catch vast multitudes of ermines, martens, arcolini, foxes and other animals of that kind, the furs of which are more delicate, and consequently more valuable, than those found in the districts inhabited by the Tartars, who, on that account, are induced to undertake the plundering expeditions that have been described .. "

CHAPTER 135

468. Page 121, line 1

the King of Tartarie of the Occident

Fr. 1116 reads "un roi dou ponent qui est Tartars, que a a non Toctai" Toktai was a son of Mangku-Temur, and ascended the throne of Kipchak in 1291

We meet him again in some of the quasi-historical chapters (e g Chs 69, 70 and 71) reprinted in Appendix II. pp 339–41.

469 Page 121, line 2

noble furres for apparell

Here again F. does not enumerate them. "Car il ont gebellines assez et ermin et vair et ercolin et voupes en abondance," says fr. 1116 R. also mentions the export of wax.

470 Page 121, lines 3, 4

ſuch feruent colde, that the people can ſcarce liue

At this point Z has a most interesting and important passage of over 50 lines dealing with the intimate social customs of the Russians, necessitated chiefly on account of the intense cold. See B. pp. 233, 234.

So Frampton's translation ends. There remains only the chapters dealing with Great Turkey, Kaidu, Abaga, Argon, Acomat, Baidu, Alau, Nogai and Toktai.

All these will be found reprinted from Wright's edition of Marsden in Appendix II. No 32, pp. 325–41.

THE TRAVELS OF NICOLO DE' CONTI IN THE EAST

Although it is not within the scope of the present work to annotate the Travels of Nicolò de' Conti, I feel that it is impossible to reprint them at all without saying a few words on the great need for a new edition of what is undoubtedly the best account of Southern Asia by any European traveller of the fifteenth century.

In the first place the present English translation by Frampton from the Castilian of Santaella seems to have escaped notice. Thus when J. Winter Jones translated the Latin version from Poggio's *De Varietate Fortunæ libri quatuor* for the Hakluyt Society in 1857, he imagined he was making the first English translation Nor can I find any mention of Frampton's work in recent years. This is to be regretted, because a comparison of the present version with that of Jones at once shows the superiority of the former. The numerous misreadings in Jones are largely corrected in Frampton, and although many of the place-names are hopelessly corrupted by Poggio, the excellence of the translation as a whole is undeniable. The narrative of Conti is short, and I believe that this is why it has not attracted more notice than it has. As in the present case, owing to its brevity it has continually been included with other works. Its value as a kind of commentary to, and proof of the veracity of the wonders related in *Marco Polo* has constantly been recognized. Thus we find it in the Portuguese edition of 1502, and in the Dutch version of the *Novus Orbis* of 1664, etc. For a full account of all versions and translations of Conti, see Cordier, "Deux Voyageurs dans l'Extrême-Orient au XVe et XVIe Siècles," *T'oung-pao*, Vol x. 1899, No. 4.

Nicolò de' Conti started on his travels in 1419, for on his return to Venice in 1444 he tells us he has been absent 25 years. He passed through Damascus, Baghdad, Basra, and Hormuz to India, and on to Ceylon, Java, Sumatra, and the south of China. On his return journey he visited Burma, and ascended the Irrawaddy to Ava, and touching at Cochin, Calicut, Cambay, Socotra and Aden arrived at Jidda, the port of Mecca, whence he reached Cairo At Mount Sinai he met the Spanish traveller Pero Tafur to whom he related many of his experiences Among other things he told him how on his arrival in India he was taken to see Prester John who received him graciously and married him to a woman by whom he had several children. On reaching Mecca he was ordered to abjure his Faith or be killed He chose the former course for the sake of his family [1] On his arrival in Venice in 1444 he sought absolution for his apostasy from

[1] An excellent and complete translation of Pero Tafur has recently been made by Malcolm Letts (Broadway Travellers Series, 1926) Conti's account of his travels appears on pp. 84–95 As the translator states, a reference to the volume establishes the fact that Conti told Tafur much that he did not relate to Poggio.

Eugenius IV. This was granted on condition that he would truthfully relate his travels to the papal secretary Poggio Bracciolini. This was accordingly done, being written by Poggio in Latin. Copies, if they existed at all, must have been very scarce, for Ramusio, after vainly attempting to find one, had to use a Portuguese translation, which is of little value. It found its way later into Purchas. (See Vol. XI p. 394 *et seq.* of the MacLehose edition, 1906)

Frampton's translation has only one important omission, and that is with regard to the "sonalia," or little bells of gold, silver or copper worn on the members of the men to excite the women to lechery. The practice has been noted by many travellers[1], yet in his recent edition of Duarte Barbosa, Longworth Dames doubts if the custom really existed, and suggest it to be "a mere figment of the imagination such as sailors picked up from the loose talk at seaports." There is no doubt whatever as to the existence of the custom, although the objects used vary[2]. See, for instance, the curious articles known as *ampallang* at the Wellcome Historical Medical Museum in London. The point of interest as raised by Briffault is whether they represented mere voluptuary ingenuity, or must be regarded as amulets used solely to guard the opening of the body against evil spirits. (See Briffault, *op cit.* p. 280)

The latter portion of Conti's travels consists of an account of the manners and customs of the Indians given in direct answer to questions asked by Poggio himself. Thus we have the heading "Pogio" on p 137 of the present edition

[1] Linschoten, *Voyage to the East Indies*, Vol I p 99; Mandelslo, 1669 edit p 97, Barbosa, Dames' edit. 1921, Vol. II. p 154 See also Yule's note in his *Mission to Ava*, p 208, F. Carletti, *Viaggi da lui racontati in dodici ragionamenti*, pp 347 *et seq* ; V Bellemo, *I Viaggi di Nicolò De' Conti*, Milano, 1883, pp 132, 140, Briffault, *The Mothers*, Vol III. p 329

[2] See O Hovorka, *Mitt. Anthrop. Gesell. in Wien*, XXIV. Band, 1894, pp. 131–43; *id Vergleichende Volksmedizin*, II Band, 1909, pp 179–84, and E J. Dingwall, *Male Infibulation*, London, 1925, p 53

APPENDIX II

SELECTED PASSAGES FROM RAMUSIO &c.

APPENDIX II

SELECTED PASSAGES FROM RAMUSIO, &c

The following more lengthy passages are, with but few exceptions, from Ramusio as translated by Marsden, 1818.

The pagination of the revised edition of Marsden, edited by Thomas Wright for Bohn's Antiquarian Library in 1854 (reprinted 1886, 1890, etc) is also added. This is followed by references, in square brackets, to the chapter and page of Frampton's text where the particular passage fits in, together with the number and page of its corresponding Note in Appendix I.

1. *Ramusio* I, Ch. VIII. *Marsden*, pp. 66, 67 *Wright*, pp. 40, 41.
[*Frampton*, Ch 10, p 29, Appendix I Note 78, p 169]

Concerning the capture and death of the *Khalif of Baldach* (Mosta'sim Billah)

THE above-mentioned khalif, who is understood to have amassed greater treasures than had ever been possessed by any other sovereign, perished miserably under the following circumstances At the period when the Tartar princes began to extend their dominion, there were amongst them four brothers, of whom the eldest, named Mangu, reigned in the royal seat of the family. Having subdued the country of Cathay, and other districts in that quarter, they were not satisfied, but coveting further territory, they conceived the idea of universal empire, and proposed that they should divide the world amongst them With this object in view. it was agreed that one of them should proceed to the east, that another should make conquests in the south, and that the other two should direct their operations against the remaining quarters. The southern portion fell to the lot of Ulaù, who assembled a vast army, and having subdued the provinces through which his route lay, proceeded in the year 1255 to the attack of this city of Baldach. Being aware, however, of its great strength and the prodigious number of its inhabitants, he trusted rather to stratagem than to force for its reduction, and in order to deceive the enemy with regard to the number of his troops, which consisted of a hundred thousand horse, besides foot soldiers, he posted one division of his army on the one side, another division on the other side of the approach to the city, in such a manner as to be concealed by a wood, and placing himself at the head of the third, advanced boldly to within a short distance of the gate. The khalif made light of a force apparently so inconsiderable, and confident in the efficacy of the usual Mahometan ejaculation, thought of nothing less than its entire destruction, and for that purpose marched out of the city with his guards; but as soon as Ulaù perceived his approach, he feigned to retreat before him, until by this means he had drawn him beyond the wood where the other divisions were posted. By the closing of these from both sides, the army of the khalif was surrounded and broken, himself was made prisoner, and the city surrendered to the conqueror Upon entering it, Ulaù discovered, to his great astonishment, a tower filled with gold He called the khalif before him, and after reproaching him with his avarice, that prevented him from employing his treasures in the formation of an army for the defence of his capital against the powerful invasion with which it had long been threatened, gave orders for his being shut up in this same tower, without sustenance, and there, in the midst of his wealth, he soon finished a miserable existence.

2. *Ramusio* I, Chs. xv, xvi *Marsden*, pp. 95, 96; 100–102. *Wright*, Chs xvi, xvii, pp. 63–68.
[*Frampton*, Ch 16, pp 36, 37, Appendix I. Note 109, p 175]

Of the City of *Ormus* (Hormuz), its hot wind, shipping, &c.

During the summer season, the inhabitants do not remain in the city, on account of the excessive heat, which renders the air unwholesome, but retire to their gardens along the shore or on the banks of the rivers, where with a kind of ozier-work they construct huts over the water. These they enclose with stakes, driven in the water on the one side, and on the other upon the shore, making a covering of leaves to shelter them from the sun. Here they reside during the period in which there blows, every day, from about the hour of nine until noon, a land-wind so intensely hot as to impede respiration, and to occasion death by suffocating the person exposed to it None can escape from its effects who are overtaken by it on the sandy plain As soon as the approach of this wind is perceived by the inhabitants, they immerge themselves to the chin in water, and continue in that situation until it ceases to blow In proof of the extraordinary degree of this heat, Marco Polo says that he happened to be in these parts when the following circumstance occurred The ruler of Ormus having neglected to pay his tribute to the king of Kierman, the latter took the resolution of enforcing it at the season when the principal inhabitants reside out of the city, upon the main land, and for this purpose despatched a body of troops, consisting of sixteen hundred horse and five thousand foot, through the country of Reobarle, in order to seize them by surprise In consequence, however, of their being misled by the guides, they failed to arrive at the place intended before the approach of night, and halted to take repose in a grove not far distant from Ormus, but upon recommencing their march in the morning, they were assailed by this hot wind, and were all suffocated; not one escaping to carry the fatal intelligence to his master. When the people of Ormus became acquainted with the event, and proceeded to bury the carcases, in order that their stench might not infect the air, they found them so baked by the intenseness of the heat, that the limbs, upon being handled, separated from the trunks, and it became necessary to dig the graves close to the spot where the bodies lay.

The vessels built at Ormus are of the worst kind, and dangerous for navigation, exposing the merchants and others who make use of them to great hazards Their defects proceed from the circumstance of nails not being employed in the construction; the wood being of too hard a quality, and liable to split or to crack like earthenware. When an attempt is made to drive a nail, it rebounds, and is frequently broken The planks are bored, as carefully as possible, with an iron auger, near the extremities, and wooden pins or trenails being driven into them, they are in this manner fastened (to the stem and stern) After this they are bound, or rather sewed together, with a kind of rope-yarn stripped from the husk of the Indian (cocoa) nuts, which are of a large size, and covered with a fibrous stuff like horse-hair. This being steeped in water until the softer parts putrefy, the threads or strings remain clean, and of these they make twine for sewing the planks, which lasts long under water Pitch is not used for preserving the bottoms of vessels, but they are smeared with an oil made from the fat of fish, and then caulked with oakum. The vessel has no more than one mast, one helm, and one deck. When she has taken in her lading, it is covered over with hides, and upon these hides they place the horses which they carry to India They have no iron anchors, but in their stead employ another kind of ground-tackle; the consequence of which is, that in bad weather, (and these seas are very tempestuous,) they are frequently driven on shore and lost.

The inhabitants of the place are of a dark colour, and are Mahometans. They sow their wheat, rice, and other grain in the month of November, and reap their harvest in March. The fruits also they gather in that month, with the exception of the dates, which are collected in May Of these, with other ingredients, they make a good kind of wine. When it is drunk, however, by persons not accustomed to the beverage, it occasions an immediate flux; but upon their recovering from its first effects, it proves beneficial to them, and contributes to render them fat. The food of the natives is different from ours; for were they to eat wheaten bread and flesh meat their health would be injured. They live chiefly upon dates and salted fish, such as the thunnus, cepole (*cepola tania*), and others which from experience they know to be wholesome. Excepting in marshy places, the soil of this country is not covered with grass, in consequence of the extreme heat, which burns up everything. Upon the death of men of rank, their wives loudly bewail them, once in the course of each day, during four successive weeks [all the French MSS. read "years"]; and there are also people to be found here who make such lamentations a profession, and are paid for uttering them over the corpses of persons to whom they are not related.

3. *Ramusio* I, Chs. xviii–xxi. *Marsden*, pp. 105, 106; 107, 108; 109, 110; 112–114. *Wright*, Chs. xix–xxii, pp 68–76
[*Frampton*, Ch 17, p 37, Appendix I Note 112, p 176]

Of Kobiam, Timochain, and of the Old *Man of the Mountain* (Kuh-Banān and Tun-and-Kain)

UPON leaving Kierman and travelling three days, you reach the borders of a desert extending to the distance of seven days' journey, at the end of which you arrive at Kobiam [not mentioned till later in the French MSS]. During the first three days (of these seven) but little water is to be met with, and that little is impregnated with salt, green as grass, and so nauseous that none can use it as drink. Should even a drop of it be swallowed, frequent calls of nature will be occasioned; and the effect is the same from eating a grain of the salt made from this water. In consequence of this, persons who travel over the desert are obliged to carry a provision of water along with them. The cattle, however, are compelled by thirst to drink such as they find, and a flux immediately ensues In the course of these three days not one habitation is to be seen The whole is arid and desolate. Cattle are not found there, because there is no subsistence for them. *On the fourth day you come to a river of fresh water, but which has its channel for the most part under ground. In some parts however there are abrupt openings, caused by the force of the current, through which the stream becomes visible for a short space, and water is to be had in abundance. Here the wearied traveller stops to refresh himself and his cattle after the fatigues of the preceding journey* [the passage in italics is unique to R.] The circumstances of the latter three days resemble those of the former, and conduct him at length to the town of Kobiam. [The French texts mention that asses are found during these three days]

Kobiam is a large town, the inhabitants of which observe the law of Mahomet. They have plenty of iron, steel, and ondanique. Here they make mirrors of highly polished steel, of a large size and very handsome. Much antimony or zinc is found in the country, and they procure tutty which makes an excellent collyrium, together with spodium, by the following process They take the crude ore from a vein that is known to yield such as is fit for the purpose, and put it into a heated furnace. Over the furnace they place an iron grating formed of small bars set close together. The smoke or vapour ascending from the ore in burning attaches itself to the bars, and as it cools becomes hard. This is the tutty, whilst the gross and heavy part, which does not ascend, but remains as a cinder in the furnace, becomes the spodium. [For an article on kohl and collyrium, see my *Ocean of Story*, vol I, pp. 211–18.]

Leaving Kobiam you proceed over a desert of eight days' journey exposed to great drought, neither fruits nor any kind of trees are met with, and what water is found has a bitter taste. Travellers are therefore obliged to carry with them so much as may be necessary for their sustenance. Their cattle are constrained by thirst to drink such as the desert affords, *which their owners endeavour to render palatable to them by mixing it with flour* [unique to R. and Z]. At the end of eight days you reach the province of Timochain [Tonocain] situated towards the north, on the borders of Persia, in which are many towns and strong places. There is here an extensive plain remarkable for the production of a species of tree called the tree of the sun, and by Christians *arbor secco,* the dry or fruitless tree. Its nature and qualities are these.—It is lofty, with a large stem, having its leaves green on the upper surface, but white or glaucous on the under. It produces husks or capsules like those in which the chestnut is enclosed, but these contain no fruit. The wood is solid and strong, and of a yellow colour resembling the box There is no other species of tree near it for the space of a hundred miles, excepting in one quarter, where trees are found within the distance of about ten miles. It is reported by the inhabitants of this district that a battle was fought there between Alexander, king of Macedonia, and Darius. The towns are well supplied with every necessary and convenience of life, the climate being temperate, and not subject to extremes either of heat or cold. The people are of the Mahometan religion. They are in general a handsome race, especially the women, who, in my opinion, are the most beautiful in the world

Having spoken of this country, mention shall now be made of the old man of the mountain. The district in which his residence lay obtained the name of Mulehet [fr. 1116 has "Muleete"] signifying in the language of the Saracens, the place of heretics, and his people that of Mulehetites, or holders of heretical tenets, as we apply the term of Patharini to certain heretics amongst Christians The following account of this chief, Marco Polo testifies to having heard from sundry persons. He was named Alo-eddin, and his religion was that of Mahomet. In a beautiful valley enclosed between two lofty mountains, he had formed a luxurious garden, stored with every delicious fruit and every fragrant shrub that could be procured. Palaces of various sizes and forms were erected in different parts of the grounds, ornamented with works in gold, with paintings, and with furniture of rich silks. By means of small conduits contrived in these buildings, streams of wine, milk, honey, and some of pure water, were seen to flow in every direction The inhabitants of these palaces were elegant and beautiful damsels, accomplished in the arts of singing, playing upon all sorts of musical instruments, dancing, and especially those of dalliance and amorous allurement. Clothed in rich dresses they were seen continually sporting and amusing themselves in the garden and pavilions, their female guardians being confined within doors and never suffered to appear The object which the chief had in view in forming a garden of this fascinating kind, was this: that Mahomet having promised to those who should obey his will the enjoyments of Paradise, where every species of sensual gratification should be found, in the society of beautiful nymphs, he was desirous of its being understood by his followers that he also was a prophet and the compeer of Mahomet, and had the power of admitting to Paradise such as he should choose to favour In order that none without his licence might find their way into this delicious valley ["except those whom he intended to be his Assassins," add the French MSS.], he caused a strong and inexpugnable castle to be erected at the opening of it, through which the entry was by a secret passage. At his court, likewise, this chief entertained a number of youths, from the age of twelve to twenty years, selected from the inhabitants of the surrounding mountains, who showed a disposition for martial exercises, and appeared to possess the quality of daring courage. To them he was in the daily practice of discoursing on the subject of the paradise announced by the prophet, and of his own power of granting admission; and at certain times he caused opium to be administed to ten or a dozen of the youths; and when half dead with sleep he had them conveyed to the several apartments of the palaces in the garden Upon awakening from this state of lethargy, their senses were struck with all the delightful objects that have been described, and each perceived himself surrounded by loving damsels, singing, playing, and attracting his regards by the most fascinating caresses, serving him also with delicate viands and exquisite wines; until intoxicated with excess of enjoyment amidst actual rivulets of milk and wine, he believed himself assuredly in Paradise, and felt an unwillingness to relinquish its delights When four or five days had thus been passed, they were thrown once more into a state of somnolency, and carried out of the garden Upon their being introduced to his presence, and questioned by him as to where they had been, their answer was, "In Paradise, through the favour of your highness": and then before the whole court, who listened to them with eager curiosity and astonishment, they gave a circumstantial account of the scenes to which they had been witnesses The chief thereupon addressing them, said· "We have the assurances of our prophet that he who defends his lord shall inherit Paradise, and if you show yourselves devoted to the obedience of my orders, that happy lot awaits you " Animated to enthusiasm by words of this nature, all deemed themselves happy to receive the commands of their master, and were forward to die in his service. The consequence of this system was, that when any of the neighbouring princes, or others, gave umbrage to this chief, they were put to death by these his disciplined Assassins, none of whom felt terror at the risk of losing their own lives, which they held in little estimation, provided they could execute their master's will. On this account his tyranny became the subject of dread in all the surrounding countries He had also constituted two deputies or representatives of himself, of whom one had his residence in the vicinity of Damascus, and the other in Kurdistan, and these pursued the plan he had established for training their young dependants Thus there was no person, however powerful, who, having become exposed to the enmity of the old man of the mountain, could escape assassination.

4. *Ramusio* I, Ch. xxv. *Marsden*, pp. 130, 131. *Wright*, Ch. xxvi, pp. 84–86.

[*Frampton*, Ch 23, p. 40, Appendix I Note 132, p 179]

Of the province of Balashan (Badakhshan)

The mines of silver, copper, and lead, are likewise very productive. It is a cold country. The horses bred here are of a superior quality, and have great speed. Their hoofs are so hard that they do not require shoeing The natives are in the practice of galloping them on declivities where other cattle could not or would not venture to run They asserted that not long since there were still found in this province horses of the breed of Alexander's celebrated Bucephalus, which were all foaled with a particular mark in the forehead The whole of the breed was in the possession of one of the king's uncles, who, upon his refusal to yield them to his nephew, was put to death; whereupon his widow, exasperated at the murder, caused them all to be destroyed; and thus the race was lost to the world. In the mountains there are falcons of the species called saker (*falco sacer*), which are excellent birds, and of strong flight; as well as of that called laner, (*falco lanarius*) There are also goshawks of a perfect kind (*falco astur*, or *palumbarius*), and sparrow-hawks (*falco nisus*). The people of the country are expert at the chase both of beasts and birds Good wheat is grown there, and a species of barley without the husk. There is no oil of olives, but they express it from certain nuts, and from the grain called sesame, which resembles the seed of flax, excepting that it is light-coloured; and the oil this yields is better, and has more flavour than any other. It is used by the Tartars and other inhabitants of these parts.

In this kingdom there are many narrow defiles, and strong situations, which diminish the apprehension of any foreign power entering it with a hostile intention The men are good archers and excellent sportsmen; generally clothing themselves with the skins of wild animals; other materials for the purpose being scarce The mountains afford pasture for an innumerable quantity of sheep, which ramble about in flocks of four, five, and six hundred, all wild, and although many are taken and killed, there does not appear to be any diminution. These mountains are exceedingly lofty, insomuch that it employs a man from morning till night to ascend to the top of them Between them there are wide plains clothed with grass and with trees, and large streams of the purest water precipitating themselves through the fissures of the rocks In these streams are trout and many other delicate sorts of fish On the summits of the mountains the air is so pure and so salubrious, that when those who dwell in the towns, and in the plains and valleys below, find themselves attacked with fevers or other inflammatory complaints, they immediately remove thither, and remaining for three or four days in that situation, recover their health. Marco Polo affirms that he had experience in his own person of its excellent effects, for having been confined by sickness, in this country, for nearly a year, he was advised to change the air by ascending the hills; when he presently became convalescent.

5. *Ramusio* II, Chs. II, III. *Marsden*, pp. 274, 276; 278, 279. *Wright*, pp 167–171

[Omitted in *Frampton* See Appendix I Note 231, p. 198]

The reason of the Khan's not becoming a Christian—his various kinds of rewards

THE Grand Khan, having obtained this signal victory, returned with great pomp and triumph to the capital city of Kanbalu. This took place in the month of November, and he continued to reside there during the months of February and March, in which latter was our festival of Easter Being aware that this was one of our principal solemnities, he commanded all the Christians to attend him, and to bring with them their Book, which contains the four Gospels of the Evangelists. After causing it to be repeatedly perfumed with incense, in a ceremonious manner, he devoutly kissed it, and directed that the same should be done by all his nobles who were present This was his usual practice upon each of the principal Christian festivals, such as Easter and Christmas; and he observed the same at the festivals of the Saracens, Jews, and idolaters. Upon being asked his motive for this conduct, he said: "There are four great Prophets who are reverenced and worshipped by the different classes of mankind. The Christians regard Jesus Christ as their divinity; the Saracens, Mahomet; the Jews, Moses; and the idolaters, Sogomombar-kan, the most eminent amongst their idols. I do honour and show respect to all the four, and invoke to my aid whichever amongst them is in truth supreme in heaven." But from the manner in which his majesty acted towards them, it is evident that he regarded the faith of the Christians as the truest and the best; nothing, as he observed, being enjoined to its professors that was not replete with virtue and holiness. By no means, however, would he permit them to bear the cross before them in their processions, because upon it so exalted a personage as Christ had been scourged and (ignominiously) put to death. It may perhaps be asked by some, why, if he showed such a preference to the faith of Christ, he did not conform to it, and become a Christian? His reason for not so doing, he assigned to Nicolo and Maffio Polo, when, upon the occasion of his sending them as his ambassadors to the Pope, they ventured to address a few words to him on the subject of Christianity. "Wherefore," he said, "should I become a Christian? You yourselves must perceive that the Christians of these countries are ignorant, inefficient persons, who do not possess the faculty of performing anything (miraculous), whereas you see that the idolaters can do whatever they will When

I sit at table the cups that were in the middle of the hall come to me filled with wine and other beverage, spontaneously and without being touched by human hand, and I drink from them They have the power of controlling bad weather and obliging it to retire to any quarter of the heavens, with many other wonderful gifts of that nature. You are witnesses that their idols have the faculty of speech, and predict to them whatever is required Should I become a convert to the faith of Christ, and profess myself a Christian, the nobles of my court and other persons who do not incline to that religion will ask me what sufficient motives have caused me to receive baptism, and to embrace Christianity. 'What extraordinary powers,' they will say, 'what miracles have been displayed by its ministers? Whereas the idolaters declare that what they exhibit is performed through their own sanctity, and the influence of their idols.' To this I shall not know what answer to make, and I shall be considered by them as labouring under a grievous error; whilst the idolaters, who by means of their profound art can effect such wonders, may without difficulty compass my death. But return you to your pontiff, and request of him, in my name, to send hither a hundred persons well skilled in your law, who being confronted with the idolaters shall have power to coerce them, and showing that they themselves are endowed with similar art, but which they refrain from exercising, because it is derived from the agency of evil spirits, shall compel them to desist from practices of such a nature in their presence. When I am witness of this, I shall place them and their religion under an interdict, and shall allow myself to be baptized. Following my example, all my nobility will then in like manner receive baptism, and this will be imitated by my subjects in general; so that the Christians of these parts will exceed in number those who inhabit your own country " From this discourse it must be evident that if the Pope had sent out persons duly qualified to preach the gospel, the Grand Khan would have embraced Christianity, for which, it is certainly known, he had a strong predilection. But, to return to our subject, we shall now speak of the rewards and honours he bestows on such as distinguish themselves by their valour in battle.

The Grand Khan appoints twelve of the most intelligent amongst his nobles, whose duty it is to make themselves acquainted with the conduct of the officers and men of his army, particularly upon expeditions and in battles, and to present their reports to him, and he, upon being apprised of their respective merits, advances them in his service, raising those who commanded an hundred men to the command of a thousand, and presenting many with vessels of silver, as well as the customary tablets or warrants of command and of government. The tablets given to those commanding a hundred men are of silver; to those commanding a thousand, of gold or of silver gilt; and those who command ten thousand receive tablets of gold, bearing the head of a lion; the former being of the weight of a hundred and twenty *saggi*, and these with the lion's head, two hundred and twenty At the top of the inscription on the tablet is a sentence to this effect. "By the power and might of the great God, and through the grace which he vouchsafes to our empire, be the name of the Kaan blessed, and let all such as disobey (what is herein directed) suffer death and be utterly destroyed." The officers who hold these tablets have privileges attached to them, and in the inscription is specified what are the duties and the powers of their respective commands. He who is at the head of a hundred thousand men, or the commander in chief of a grand army, has a golden tablet weighing three hundred *saggi*, with the sentence above mentioned, and at the bottom is engraved the figure of a lion, together with representations of the sun and moon. He exercises also the privileges of his high command, as set forth in this magnificent tablet. Whenever he rides in public, an umbrella is carried over his head, denoting the rank and authority he holds; and when he is seated, it is always upon a silver chair. The Grand Khan confers likewise upon certain of his nobles tablets on which are represented figures of the gerfalcon, in virtue of which they are authorised to take with them as their guard of honour the whole army of any great prince They can also make use of the horses of the imperial stud at their pleasure, and can appropriate the horses of any officers inferior to themselves in rank

6. *Ramusio* II, Ch. IV. *Marsden*, pp. 281–283. *Wright*, pp. 172–174.

[*Frampton*, Ch. 54, p. 64; Appendix I Note 233, p 199]

THITHER the Grand Khan sends his officers every second year, or oftener, as it may happen to be his pleasure, who collect for him, to the number of four or five hundred, or more, of the handsomest of the young women, according to the estimation of beauty communicated to them in their instructions. The mode of their appreciation is as follows. Upon the arrival of these commissioners, they give orders for assembling all the young women of the province, and appoint qualified persons to examine them, who, upon careful inspection of each of them separately, that is to say, of the hair, the countenance, the eyebrows, the mouth, the lips, and other features, as well as the symmetry of these with each other, estimate their value at sixteen, seventeen, eighteen, or twenty, or more carats, according to the greater or less degree of beauty. The number required by the Grand Khan, at the rates, perhaps, of twenty or twenty-one carats, to which their commission was limited, is then selected from the rest, and they are conveyed to his court. Upon their arrival in his presence, he causes a new examination to be made by a different set of inspectors, and from amongst them a further selection takes place, when thirty or forty are retained for his own chamber, at a higher valuation These, in the first instance, are committed separately to the care of the wives of certain of the nobles, whose duty it is to observe them attentively during the course of the night, in order to ascertain that they have not any concealed imperfections, that they sleep tranquilly, do not snore, have sweet breath, and are free from unpleasant scent in any part of the body. Having undergone this rigorous scrutiny, they are divided into parties of five, one of which parties attends during three days and three nights, in His Majesty's interior apartment, where they are to perform every service that is required of them, and he does with them as he likes. When this term is completed, they are relieved by another party, and in this manner successively, until the whole number have taken their turn; when the first five recommence their attendance. But whilst the one party officiates in the inner chamber, another is stationed in the outer apartment adjoining; in order that if his majesty should have occasion for anything, such as drink or victuals, the former may signify his commands to the latter, by which the article required is immediately procured· and thus the duty of waiting upon his majesty's person is exclusively performed by these young females. The remainder of them, whose value had been estimated at an inferior rate, are assigned to the different lords of the household; under whom they are instructed in cookery, in dressmaking, and other suitable works; and upon any person belonging to the court expressing an inclination to take a wife, the Grand Khan bestows upon him one of these damsels, with a handsome portion. In this manner he provides for them all amongst his nobility. It may be asked whether the people of the province do not feel themselves aggrieved in having their daughters thus forcibly taken from them by the sovereign? Certainly not; but, on the contrary, they regard it as a favour and an honour done to them; and those who are the fathers of handsome children feel highly gratified by his condescending to make choice of their daughters. "If," say they, "my daughter is born under an auspicious planet and to good fortune, His Majesty can best fulfil her destinies, by matching her nobly; which it would not be in my power to do." If, on the other hand, the daughter misconducts herself, or any mischance befalls her (by which she becomes disqualified), the father attributes the disappointment to the malign influence of her stars.

7. *Ramusio* II, Ch. VII. *Marsden*, pp. 297–299. *Wright*, pp. 182–186.
[*Frampton*, Ch 55, p. 65, Appendix I. Note 234, p 199.]

Of the New City of Tai-du

SOME of the inhabitants, however, of whose loyalty he did not entertain suspicion, were suffered to remain, especially because the latter, although of the dimensions that shall presently be described, was not capable of containing the same number as the former, which was of vast extent.

This new city is of a form perfectly square, and twenty-four miles in extent, each of its sides being neither more nor less than six miles. It is enclosed with walls of earth, that at the base are about ten paces thick, but gradually diminish to the top, where the thickness is not more than three paces. In all parts the battlements are white. The whole plan of the city was regularly laid out by line, and the streets in general are consequently so straight, that when a person ascends the wall over one of the gates, and looks right forward, he can see the gate opposite to him on the other side of the city. In the public streets there are, on each side, booths and shops of every description. All the allotments of ground upon which the habitations throughout the city were constructed are square, and exactly on a line with each other; each allotment being sufficiently spacious for handsome buildings, with corresponding courts and gardens. One of these was assigned to each head of a family; that is to say, such a person of such a tribe had one square allotted to him, and so of the rest. Afterwards the property passed from hand to hand. In this manner the whole interior of the city is disposed in squares, so as to resemble a chess-board, and planned out with a degree of precision and beauty impossible to describe. The wall of the city has twelve gates, three on each side of the square, and over each gate and compartment of the wall there is a handsome building; so that on each side of the square there are five such buildings, containing large rooms, in which are disposed the arms of those who form the garrison of the city, every gate being guarded by a thousand men. It is not to be understood that such a force is stationed there in consequence of the apprehension of danger from any hostile power whatever, but as a guard suitable to the honour and dignity of the sovereign. Yet it must be allowed that the declaration of the astrologers has excited in his mind a degree of suspicion with regard to the Cathaians. In the centre of the city there is a great bell suspended in a lofty building, which is sounded every night, and after the third stroke no person dares to be found in the streets, unless upon some urgent occasion, such as to call assistance to a woman in labour, or a man attacked with sickness, and even in such necessary cases the person is required to carry a light.

Withoutside of each of the gates is a suburb so wide that it reaches to and unites with those of the other nearest gates on both sides, and in length extends to the distance of three or four miles, so that the number of inhabitants in these suburbs exceeds that of the city itself. Within each suburb there are, at intervals, as far perhaps as a mile from the city, many hotels, or caravanserais, in which the merchants arriving from various parts take up their abode; and to each description of people a separate building is assigned, as we should say, one to the Lombards, another to the Germans, and a third to the French The number of public women who prostitute themselves for money, reckoning those in the new city as well as those in the suburbs of the old, is twenty-five thousand To each hundred and to each thousand of these there are superintending officers appointed, who are under the orders of a captain-general. The motive for placing them under such command is this: when ambassadors arrive charged with any business in which the interests of the Grand Khan are concerned, it is customary to maintain them at His Majesty's expense, and in order that they may be treated in the most honourable manner, the captain is ordered to furnish nightly to each individual of the embassy one of these courtezans, who is likewise to be changed every night, for which service, as it is considered in the light of a tribute they owe to the sovereign, they do not receive any remuneration. Guards, in parties of thirty or forty, continually patrol the streets during the course of the night, and make diligent search for persons who may be from their homes at an unseasonable hour, that is, after the third stroke of the great bell. When any are met with under such circumstances, they immediately apprehend and confine them, and take them in the morning for examination before officers appointed for that purpose, who, upon the proof of any delinquency, sentence them, acording to the nature of the offence, to a severer or lighter infliction of the bastinade, which sometimes, however, occasions their death. It is in this manner that crimes are usually punished amongst these people, from a disinclination to the shedding of blood, which their *baksis* or learned astrologers instruct them to avoid. Having thus described the interior of the city of Tai-du, we shall now speak of the disposition to rebellion shown by its Cathaian inhabitants.

8. *Ramusio* II, Ch. VIII. *Marsden*, pp. 309–313 *Wright*, pp. 187–192.
[Omitted by *Frampton* See Appendix I. Note 247, p 202]

Of the Oppressions of Achmac (Ahmad)

PARTICULAR mention will hereafter be made of the establishment of a council of twelve persons, who had the power of disposing, at their pleasure, of the lands, the governments, and everything belonging to the state Amongst these was a Saracen, named Achmac, a crafty and bold man, whose influence with the Grand Khan surpassed that of the other members. To such a degree was his master infatuated with him that he indulged him in every liberty It was discovered, indeed, after his death, that he had by means of spells so fascinated His Majesty as to oblige him to give ear and credit to whatever he represented, and by these means was enabled to act in all matters according to his own arbitrary will. He gave away all the governments and public offices, pronounced judgment upon all offenders, and when he was disposed to sacrifice any man to whom he bore ill-will, he had only to go to the emperor and say to him, "Such a person has committed an offence against Your Majesty, and is deserving of death," when the emperor was accustomed to reply, "Do as you judge best", upon which he caused him to be immediately executed. So evident were the proofs of the authority he possessed, and of His Majesty's implicit faith in his representations, that none had the hardiness to contradict him in any matter; nor was there a person, however high in rank or office, who did not stand in awe of him If any one was accused by him of capital crime, however anxious he might be to exculpate himself, he had not the means of refuting the charge, because he could not procure an advocate, none daring to oppose the will of Achmac. By these means he occasioned many to die unjustly. Besides this, there was no handsome female who became an object of his sensuality that he did not contrive to possess, taking her as a wife if she was unmarried, or otherwise compelling her to yield to his desires When he obtained information of any man having a beautiful daughter, he despatched his emissaries to the father of the girl, with instructions to say to him: "What are your views with regard to this handsome daughter of yours? You cannot do better than give her in marriage to the Lord Deputy or Vicegerent" (that is, to Achmac, for so they termed him, as implying that he was His Majesty's representative). "We shall prevail upon him to appoint you to such a government or to such an office for three years." Thus tempted, he is prevailed upon to part with his child; and the matter being so far arranged, Achmac repairs to the emperor and informs His Majesty that a certain government is vacant, or that the period for which it is held will expire on such a day, and recommends the father as a person well qualified to perform the duties. To this His Majesty gives his consent, and the appointment is immediately carried into effect. By such means as these, either from the ambition of holding high offices or the apprehension of his power, he obtained the sacrifice of all the most beautiful young women, either under the denomination of wives, or as the slaves of his pleasure. He had sons to the number of twenty-five, who held the highest offices of the state, and some of them, availing themselves of the authority of their father, formed adulterous connexions, and committed many other unlawful and atrocious acts. Achmac had likewise accumulated great wealth, for every person who obtained an appointment found it necessary to make him a considerable present.

During a period of twenty-two years he exercised this uncontrolled sway At length the natives of the country, that is, the Cathaians, no longer able to endure his multiplied acts of injustice or the flagrant wickedness committed against their families, held meetings in order to devise means of putting him to death and raising a rebellion against the government. Amongst the persons principally concerned in this plot was a Cathaian, named Chen-ku, a chief of six thousand men, who, burning with resentment on account of the violation of his mother, his wife, and his daughter, proposed the measure to one of his countrymen, named Van-ku, who was at the head of ten thousand men, and recommended its being carried into execution at the time when the Grand Khan, having completed his three months' residence in Kanbalu, had departed for his palace of Shan-du, and when his son Chingis also had retired to the place he was accustomed to visit at that season; because the charge of the city was then entrusted to Achmac, who communicated to his master whatever matters occurred during his absence, and received in return the signification of his pleasure Van-ku and Chen-ku, having held this consultation together, imparted their designs to some of the leading persons of the Cathaians, and through them to their friends in many other cities. It was accordingly determined amongst them that, on a certain day, immediately upon their perceiving the signal of a fire, they should rise and put to death all those who wore beards; and should extend the signal to other places, in order that the same might be carried into effect throughout the country. The meaning of the distinction with regard to beards was this; that whereas the Cathaians themselves are naturally beardless, the Tartars, the Saracens, and the Christians wear beards. It should be understood that the Grand Khan not having obtained the sovereignty of Cathay by any legal right, but only by force of arms, had no confidence in the inhabitants, and therefore bestowed all the provincial governments and magistracies upon Tartars, Saracens, Christians, and other foreigners, who belonged to his household, and in whom he could trust. In consequence of this, his government was universally hated by the natives, who found themselves treated as slaves by these Tartars, and still worse by the Saracens.

Their plans being thus arranged, Van-ku and Chen-ku contrived to enter the palace at night, where the former, taking his place on one of the royal seats, caused the apartment to be lighted up, and sent a messenger to Achmac, who resided in the old city, requiring his immediate attendance upon Chingis, the emperor's son, who (he should say) had unexpectedly arrived that night. Achmac was much astonished at the intelligence, but, being greatly in awe of the prince, instantly obeyed. Upon passing the gate of the (new) city, he met a Tartar officer named Kogatai, the commandant of the guard of twelve thousand men, who asked him whither he was going at that late hour. He replied that he was proceeding to wait upon Chingis, of whose arrival he had just heard. "How is it possible," said the officer, "that he can have arrived in so secret a manner, that I should not have been aware of his approach in time to order a party of his guards to attend him?" In the meanwhile the two Cathaians felt assured that if they could but succeed in despatching Achmac they had nothing further to apprehend. Upon his entering the palace and seeing so many lights burning, he made his prostrations before Van-ku, supposing him to be the prince, when Chen-ku, who stood there provided with a sword, severed his head from his body. Kogatai had stopped at the door, but upon observing what had taken place, exclaimed that there was treason going forward, and instantly let fly an arrow at Van-ku as he sat upon the throne, which slew him. He then called to his men, who seized Chen-ku, and despatched an order into the city, that every person found out of doors should be put to death. The Cathaians perceiving, however, that the Tartars had discovered the conspiracy, and being deprived of their leaders, one of whom was killed and the other a prisoner, kept within their houses, and were unable to make the signals to the other towns, as had been concerted Kogatai immediately sent messengers to the Grand Khan, with a circumstantial relation of all that had passed, who, in return, directed him to make a diligent investigation of the treason, and to punish, according to the degree of their guilt, those whom he should find to have been concerned. On the following day, Kogatai examined all the Cathaians, and upon such as were principals in the conspiracy he inflicted capital punishment. The same was done with respect to the other cities that were known to have participated in the guilt.

When the Grand Khan returned to Kanbalu, he was desirous of knowing the causes of what had happened, and then learned that the infamous Achmac and seven of his sons (for all were not equally culpable) had committed those enormities which have been described. He gave orders for removing the treasure which had been accumulated by the deceased to an incredible amount, from the place of his residence in the old city to the new, where it was deposited in his own treasury. He likewise directed that his body should be taken from the tomb, and thrown into the street to be torn to pieces by the dogs The sons who had followed the steps of their father in his iniquities he caused to be flayed alive. Reflecting also upon the principles of the accursed sect of the Saracens, which indulge them in the commission of every crime, and allow them to murder those who differ from them on points of faith, so that even the nefarious Achmac and his sons might have supposed themselves guiltless, he held them in contempt and abomination. Summoning, therefore, these people to his presence, he forbade them to continue many practices enjoined to them by their law, commanding that in future their marriages should be regulated by the custom of the Tartars, and that instead of the mode of killing animals for food, by cutting their throats, they should be obliged to open the belly. At the time that these events took place Marco Polo was on the spot.

9. *Ramusio* II, Chs xx–xxvi. *Marsden*, pp. 362–366; 370–378; 381, 382. *Wright*, pp. 221–237.

[Omitted by *Frampton*, save for the few lines that constitute Ch 62 See Appendix I. Note 251, p 203]

Owing to the number of chapters omitted by Frampton at this point, the Ramusian chapter-headings are given consecutively:

CHAPTER XX

Of the places established on all the great roads for supplying post-horses—of the couriers on foot—and of the mode in which the expense is defrayed

FROM the city of Kanbalu there are many roads leading to the different provinces, and upon each of these, that is to say, upon every great high road, at the distance of twenty-five or thirty miles, accordingly as the towns happen to be situated, there are stations, with houses of accommodation for travellers, called *yamb* or post-houses. These are large and handsome buildings, having several well-furnished apartments, hung with silk, and provided with everything suitable to persons of rank. Even kings may be lodged at these stations in a becoming manner, as every article required may be obtained from the towns and strong places in the vicinity; and for some of them the court makes regular provision. At each station four hundred good horses are kept in constant readiness, in order that all messengers going and coming upon the business of the Grand Khan, and all ambassadors, may have relays, and, leaving their jaded horses, be supplied with fresh ones. Even in mountainous districts, remote from the great roads, where there were no villages, and the towns are far distant from each other, His Majesty has equally caused buildings of the same kind to be erected, furnished with everything necessary, and provided with the usual establishment of horses. He sends people to dwell upon the spot, in order to cultivate the land, and attend to the service of the post; by which means large villages are formed. In consequence of these regulations, ambassadors to the court, and the royal messengers, go and return through every province and kingdom of the empire with the greatest convenience and facility; in all which the Grand Khan exhibits a superiority over every other emperor, king, or human being. In his dominions no fewer than two hundred thousand horses are thus employed in the department of the post, and ten thousand buildings, with suitable furniture, are kept up. It is indeed so wonderful a system, and so effective in its operation, as it is scarcely possible to describe. If it be questioned how the population of the country can supply sufficient numbers for these duties, and by what means they can be victualled, we may answer, that all the idolaters, and likewise the Saracens, keep six, eight, or ten women, according to their circumstances, by whom they have a prodigious number of children; some of them as many as thirty sons capable of following their fathers in arms; whereas with us a man has only one wife, and even although she should prove barren, he is obliged to pass his life with her, and is by that means deprived of the chance of raising a family. Hence it is that our population is so much inferior to theirs. With regard to food, there is no deficiency of it, for these people, especially the Tartars, Cathaians, and inhabitants of the province of Manji (or Southern China), subsist, for the most part, upon rice, panicum, and millet; which three grains yield, in their soil, an hundred measures for one Wheat, indeed, does not yield a similar increase, and bread not being in use with them, it is eaten only in the form of vermicelli or of pastry. The former grains they boil in milk or stew with their meat. With them no spot of earth is suffered to lie idle, that can possibly be cultivated; and their cattle of different kinds multiply exceedingly, insomuch that when they take the field, there is scarcely an individual that does not carry with him six, eight, or more horses, for his own personal use. From all this may be seen the causes of so large a population, and the circumstances that enable them to provide so abundantly for their subsistence.

In the intermediate space between the post-houses, there are small villages settled at the distance of every three miles, which may contain, one with another, about forty cottages. In these are stationed the foot-messengers, likewise employed in the service of His Majesty. They wear girdles round their waists, to which several small bells are attached, in order that their coming may be perceived at a distance; and as they run only three miles, that is, from one of these foot-stations to another next adjoining, the noise serves to give notice of their approach, and preparation is accordingly made by a fresh courier to proceed with the packet instantly upon the arrival of the former. Thus it is so expeditiously conveyed from station to station, that in the course of two days and two nights His Majesty receives distant intelligence that in the ordinary mode could not be obtained in less than ten days; and it often happens that in the fruit season, what is gathered in the morning at Kanbalu (Cambaluc) is conveyed to the Grand Khan, at Shan-du (Chandu), by the evening of the following day; although the distance is generally considered as ten days' journey. At each of these three-mile stations there is a clerk, whose business it is to note the day and hour at which the one courier arrives and the other departs; which is likewise done at all the post-houses. Besides this, officers are directed to pay monthly visits to every station, in order to examine into the management of them, and to punish those couriers who have neglected to use proper diligence All these couriers are not only exempt from the (capitation) tax, but also receive from His Majesty good allowances. The horses employed in this service are not attended with any (direct) expense; the cities, towns, and villages in the neighbourhood being obliged to furnish, and also to maintain them. By His Majesty's command the governors of the cities cause examination to be made by well informed persons, as to the number of horses the inhabitants, individually, are capable of supplying. The same is done with respect to the towns and villages; and according to their means the requisition is enforced, those on each side of the station contributing their due proportion. The charge of the maintenance of the horses is afterwards deducted by the cities out of the revenue payable to the Grand Khan; inasmuch as the sum for which each inhabitant would be liable is commuted for an equivalent of horses or share of horses, which he maintains at the nearest adjoining station

It must be understood, however, that of the four hundred horses the whole are not constantly on service at the station, but only two hundred, which are kept there for the space of a month, during which period the other half are at pasture; and at the beginning of the month, these in their turn take the duty, whilst the former have time to recover their flesh; each alternately relieving the other. Where it happens that there is a river or a lake which the couriers on foot, or the horsemen, are under the necessity of passing, the neighbouring cities are obliged to keep three or four boats in continual readiness for that purpose; and where there is a desert of several days' journey, that does not admit of any habitation, the city on its borders is obliged to furnish horses to such persons as ambassadors to and from the court, that they may be enabled to pass the desert, and also to supply provisions to them and their suite; but cities so circumstanced have a remuneration from His Majesty. Where the post stations lie at a distance from the great road, the horses are partly those of his majesty, and are only in part furnished by the cities and towns of the district.

When it is necessary that the messengers should proceed with extraordinary despatch, as in the cases of giving information of disturbance in any part of the country, the rebellion of a chief, or other important matter, they ride two hundred, or sometimes two hundred and fifty miles in the course of a day. On such occasions they carry with them the tablet of the gerfalcon as a signal of the urgency of their business and the necessity for despatch. And when there are two messengers, they take their departure together from the same place, mounted upon good fleet horses; and they gird their bodies tight, bind a cloth round their heads, and push their horses to the greatest speed. They continue thus till they come to the next post-house, at twenty-five miles distant, where they find two other horses, fresh and in a state for work, they spring upon them without taking any repose, and changing in the same manner at every stage, until the day closes, they perform a journey of two hundred and fifty miles. In cases of great emergency they continue their course during the night, and if there should be no moon, they are accompanied to the next station by persons on foot, who run before them with lights; when of course they do not make the same expedition as in the day-time, the light-bearers not being able to exceed a certain pace Messengers qualified to undergo this extraordinary degree of fatigue are held in high estimation Now we will leave this subject, and I will tell you of a great act of benevolence which the Grand Khan performs twice a-year.

CHAPTER XXI

Of the relief afforded by the Grand Khan to all the provinces of his empire, in times of dearth or mortality of cattle

The Grand Khan sends every year his commissioners to ascertain whether any of his subjects have suffered in their crops of corn from unfavourable weather, from storms of wind or violent rains, or by locusts, worms, or any other plague; and in such cases he not only refrains from exacting the usual tribute of that year, but furnishes them from his granaries with so much corn as is necessary for their subsistence, as well as for sowing their land. With this view, in times of great plenty, he causes large purchases to be made of such kinds of grain as are most serviceable to them, which is stored in granaries provided for the purpose in the several provinces, and managed with such care as to ensure its keeping for three or four years without damage. It is his command, that these granaries be always kept full, in order to provide against times of scarcity; and when, in such seasons, he disposes of the grain for money, he requires for four measures no more than the purchaser would pay for one measure in the market. In like manner where there has been a mortality of cattle in any district, he makes good the loss to the sufferers from those belonging to himself, which he has received as his tenth of produce in other provinces. All his thoughts, indeed, are directed to the important object of assisting the people whom he governs, that they may be enabled to live by their labour and improve their substance We must not omit to notice a peculiarity of the Grand Khan, that where an accident has happened by lightning to any herd of cattle, flock of sheep, or other domestic animals, whether the property of one or more persons, and however large the herd may be, he does not demand the tenth of the increase of such cattle during three years, and so also if a ship laden with merchandise has been struck by lightning, he does not collect from her any custom or share of her cargo, considering the accident as an ill omen. God, he says, has shown himself to be displeased with the owner of the goods, and he is unwilling that property bearing the mark of divine wrath should enter his treasury.

CHAPTER XXII

Of the trees which he causes to be planted at the sides of the roads, and of the order in which they are kept

There is another regulation adopted by the Grand Khan, equally ornamental and useful. At both sides of the public roads he causes trees to be planted, of a kind that become large and tall, and being only two paces asunder, they serve (besides the advantage of their shade in summer) to point out the road (when the ground is covered with snow); which is of great assistance and affords much comfort to travellers This is done along all the high roads, where the nature of the soil admits of plantation, but when the way lies through sandy deserts or over rocky mountains, where it is impossible to have trees, he orders stones to be placed and columns to be erected, as marks for guidance. He also appoints officers of rank, whose duty it is to see that all these are properly arranged and the roads constantly kept in good order Besides the motives that have been assigned for these plantations, it may be added that the Grand Khan is the more disposed to make them, from the circumstance of his diviners and astrologers having declared that those who plant trees are rewarded with long life.

CHAPTER XXIII

Of the kind of wine made in the province of Cathay—and of the stones used there for burning in the manner of charcoal

The greater part of the inhabitants of the province of Cathay drink a sort of wine made from rice mixed with a variety of spices and drugs. This beverage, or wine as it may be termed, is so good and well flavoured that they do not wish for better. It is clear, bright, and pleasant to the taste, and being (made) very hot, has the quality of inebriating sooner than any other

Throughout this province there is found a sort of black stone, which they dig out of the mountains, where it runs in veins When lighted, it burns like charcoal, and retains the fire much better than wood, insomuch that it may be preserved during the night, and in the morning be found still burning. These stones do not flame, excepting a little when first lighted, but during their ignition give out a considerable heat. It is true there is no scarcity of wood in the country, but the multitude of inhabitants is so immense, and their stoves and baths, which they are continually heating, so numerous, that the quantity could not supply the demand; for there is no person who does not frequent the warm bath at least three times in the week, and during the winter daily, if it is in their power. Every man of rank or wealth has one in his house for his own use, and the stock of wood must soon prove inadequate to such consumption; whereas these stones may be had in the greatest abundance, and at a cheap rate.

CHAPTER XXIV

Of the great and admirable liberality exercised by the Grand Khan towards the poor of Kanbalu, and other persons who apply for relief at his court

It has been already stated that the Grand Khan distributes large quantities of grain to his subjects (in the provinces). We shall now speak of his great charity to and provident care of the poor in the city of Kanbalu. Upon his being apprised of any respectable family, that had lived in easy circumstances, being by misfortunes reduced to poverty, or who, in consequence of infirmities, are unable to work for their living or to raise a supply of any kind of grain to a family in that situation he gives what is necessary for their year's consumption, and at the customary period they present themselves before the officers who manage the department of His Majesty's expenses and who reside in a palace where that business is transacted, to whom they deliver a statement in writing of the quantity furnished to them in the preceding year, according to which they receive also for the present He provides in like manner for their clothing, which he has the means of doing from his tenths of wool, silk, and hemp. These materials he has woven into the different sorts of cloth, in a house erected for that purpose, where every artisan is obliged to work one day in the week for his majesty's service. Garments made of the stuffs thus manufactured he orders to be given to the poor families above described, as they are wanted for their winter and their summer dresses He also has clothing prepared for his armies, and in every city has a quantity of woollen cloth woven, which is paid for from the amount of the tenths levied at the place.

It should be known that the Tartars, when they followed their original customs, and had not yet adopted the religion of the idolaters, were not in the practice of bestowing alms, and when a necessitous man applied to them, they drove him away with injurious expressions, saying, "Begone with your complaint of a bad season which God has sent you; had he loved you, as it appears he loves me, you would have prospered as I do." But since the wise men of the idolaters, and especially the baksis, already mentioned, have represented to His Majesty that providing for the poor is a good work and highly acceptable to their deities, he has relieved their wants in the manner stated, and at his court none are denied food who come to ask it. Not a day passes in which there are not distributed, by the regular officers, twenty thousand vessels of rice, millet, and panicum. By reason of this admirable and astonishing liberality which the Grand Khan exercises towards the poor, the people all adore him as a divinity.

CHAPTER XXV

Of the astrologers of the city of Kanbalu

There are in the city of Kanbalu, amongst Christians, Saracens, and Cathaians, about five thousand astrologers and prognosticators, for whose food and clothing the Grand Khan provides in the same manner as he does for the poor families above mentioned, and who are in the constant exercise of their art. They have their astrolabes, upon which are described the planetary signs, the hours (at which they pass the meridian), and their several aspects for the whole year The astrologers (or almanac-makers) of each distinct sect annually proceed to the examination of their respective tables, in order to ascertain from thence the course of the heavenly bodies, and their relative positions for every lunation. They discover therein what the state of the weather shall be, from the paths and configurations of the planets in the different signs, and thence foretell the peculiar phenomena of each month: that in such a month, for instance, there shall be thunder and storms, in such another, earthquakes; in another, strokes of lightning and violent rains; in another, diseases, mortality, wars, discords, conspiracies. As they find the matter in their astrolabes, so they declare it will come to pass; adding, however, that God, according to his good pleasure, may do more or less than they have set down. They write their predictions for the year upon certain small squares, which are called *takuini*, and these they sell, for a groat apiece, to all persons who are desirous of peeping into futurity. Those whose predictions are found to be the more generally correct are esteemed the most perfect masters of their art, and are consequently the most honoured. When any person forms the design of executing some great work, of performing a distant journey in the way of commerce, or of commencing any other undertaking, and is desirous of knowing what success may be likely to attend it, he has recourse to one of these astrologers, and, informing him that he is about to proceed on such an expedition, inquires in what disposition the heavens appear to be at the time. The latter thereupon tells him, that before he can answer, it is necessary he should be informed of the year, the month, and the hour in which he was born; and that, having learned these particulars, he will then proceed to ascertain in what respects the constellation that was in the ascendant at his nativity corresponds with the aspect of the celestial bodies at the time of making the inquiry. Upon this comparison he grounds his prediction of the favourable or unfavourable termination of the adventure

It should be observed that the Tartars compute their time by a cycle of twelve years; to the first of which they give the name of the lion; to the second year, that of the ox; to the third, the dragon; to the fourth, the dog; and so of the rest, until the whole of the twelve have elapsed. When a person, therefore, is asked in what year he was born, he replies, In the course of the year of the lion, upon such a day, at such an hour and minute; all of which has been carefully noted by his parents in a book. Upon the completion of the twelve years of the cycle, they return to the first, and continually repeat the same series.

CHAPTER XXVI

Of the religion of the Tartars—of the opinions they hold respecting the soul—and of some of their customs

As has already been observed, these people are idolaters, and for deities, each person has a tablet fixed up against a high part of the wall of his chamber, upon which is written a name, that serves to denote the high, celestial, and sublime God; and to this they pay daily adoration, with incense burning. Lifting up their hands and then striking their faces against the floor three times, they implore from him the blessings of sound intellect and health of body; without any further petition. Below this, on the floor, they have a statue which they name *Natigai*, which they consider as the God of all terrestrial things, or whatever is produced from the earth. They give him a wife and children, and worship him in a similar manner, burning incense, raising their hands, and bending to the floor To him they pray for seasonable weather, abundant crops, increase of family, and the like They believe the soul to be immortal, in this sense, that immediately upon the death of a man, it enters into another body, and that accordingly as he has acted virtuously or wickedly during his life, his future state will become, progressively, better or worse. If he be a poor man, and has conducted himself worthily and decently, he will be re-born, in the first instance, from the womb of a gentlewoman, and become, himself, a gentleman; next, from the womb of a lady of rank, and become a nobleman; thus continually ascending in the scale of existence until he be united to the divinity. But if, on the contrary, being the son of a gentleman, he has behaved unworthily, he will, in his next state, be a clown, at length a dog, continually descending to a condition more vile than the preceding

Their style of conversation is courteous; they salute each other politely, with countenances expressive of satisfaction, have an air of good breeding, and eat their victuals with particular cleanliness. To their parents they show the utmost reverence; but should it happen that a child acts disrespectfully to or neglects to assist his parents in their necessity, there is a public tribunal, whose especial duty it is to punish with severity the crime of filial ingratitude, when the circumstance is known. Malefactors guilty of various crimes, who are apprehended and thrown into prison, are executed by strangling, but such as remain till the expiration of three years, being the time appointed by His Majesty for a general gaol delivery, and are then liberated, have a mark imprinted upon one of their cheeks, that they may be recognised.

The present Grand Khan has probibited all species of gambling and other modes of cheating, to which the people of this country are addicted more than any others upon earth; and as an argument for deterring them from the practice, he says to them (in his edict), "I subdued you by the power of my sword, and consequently whatever you possess belongs of right to me· if you gamble, therefore, you are sporting with my property." He does not, however, take anything arbitrarily in virtue of this right. The order and regularity observed by all ranks of people, when they present themselves before His Majesty, ought not to pass unnoticed. When they approach within half a mile of the place where he happens to be, they show their respect for his exalted character by assuming a humble, placid, and quiet demeanour, insomuch that not the least noise, nor the voice of any person calling out, or even speaking aloud, is heard. Every man of rank carries with him a small vessel, into which he spits, so long as he continues in the hall of audience, no one daring to spit on the floor, and this being done, he replaces the cover, and makes a salutation. They are accustomed likewise to take with them handsome buskins made of white leather, and when they reach the court, but before they enter the hall (for which they wait a summons from the Grand Khan), they put on these white buskins, and give those in which they had walked to the care of the servants This practice is observed that they may not soil the beautiful carpets, which are curiously wrought with silk and gold, and exhibit a variety of colours

10. *Ramusio* II, Ch. xxxviii. *Marsden*, pp. 420, 421. *Wright*, pp. 259, 260.
[*Frampton*, Ch 77, p 81, Appendix I Note 291, p. 213]

Of the Province of Kain-du (the Kien-ch'ang valley)

THE money or currency they make use of is thus prepared. Their gold is formed into small rods, and (being cut into certain lengths) passes according to its weight, without any stamp This is their greater money: the smaller is of the following description. In this country there are salt-springs, from which they manufacture salt by boiling it in small pans. When the water has boiled for an hour, it becomes a kind of paste, which is formed into cakes of the value of twopence each. These, which are flat on the lower, and convex on the upper side, are placed upon hot tiles, near a fire, in order to dry and harden On this latter species of money the stamp of His Majesty is impressed, and it cannot be prepared by any other than his own officers. Eighty of the cakes are made to pass for a saggio of gold But when these are carried by the traders amongst the inhabitants of the mountains and other parts little frequented, they obtain a saggio of gold for sixty, fifty, or even forty of the salt-cakes, in proportion as they find the natives less civilised, further removed from the towns, and more accustomed to remain on the same spot; inasmuch as people so circumstanced cannot always have a vend for their gold, musk, and other commodities. And yet even at this rate it answers well to them who collect the gold-dust from the beds of the rivers, as has been mentioned. The same merchants travel in like manner through the mountainous and other parts of the province of Tebeth, last spoken of, where the money of salt has equal currency. Their profits are considerable, because these country people consume the salt with their food, and regard it as an indispensable necessary; whereas the inhabitants of the cities use for the same purpose only the broken fragments of the cakes, putting the whole cakes into circulation as money. Here also the animals, which yield the musk, are taken in great numbers, and the article is proportionably abundant Many fish, of good kinds, are caught in the lake In the country are found tigers, bears, deer, stags, and antelopes. There are numerous birds also, of various sorts The wine is not made from grapes, but from wheat and rice, with a mixture of spices, which is an excellent beverage

This province likewise produces cloves. The tree is small, the branches and leaves resemble those of the laurel, but are somewhat longer and narrower. Its flowers are white and small, as are the cloves themselves, but as they ripen they become dark-coloured. Ginger grows there and also cassia in abundance, besides many other drugs, of which no quantity is ever brought to Europe

11. *Ramusio* II, Ch. XL. *Marsden*, pp. 429–431. *Wright*, pp. 265–267.
[*Frampton*, Ch. 81, p 82, Appendix I. Note 305, p 216]

Of the province named Karazan (Ta-lifu)

HERE are seen huge serpents, ten paces in length, and ten spans in the girt of the body. At the fore-part, near the head, they have two short legs, having three claws like those of a tiger, with eyes larger than a fourpenny loaf (*pane da quattro denari*) and very glaring The jaws are wide enough to swallow a man, the teeth are large and sharp, and their whole appearance is so formidable, that neither man, nor any kind of animal, can approach them without terror. Others are met with of a smaller size, being eight, six, or five paces long; and the following method is used for taking them In the day-time, by reason of the great heat, they lurk in caverns, from whence, at night, they issue to seek their food, and whatever beast they meet with and can lay hold of, whether tiger, wolf, or any other, they devour; after which they drag themselves towards some lake, spring of water, or river, in order to drink. By their motion in this way along the shore, and their vast weight, they make a deep impression, as if a heavy beam had been drawn along the sands. Those whose employment it is to hunt them observe the track by which they are most frequently accustomed to go, and fix into the ground several pieces of wood, armed with sharp iron spikes, which they cover with the sand in such a manner as not to be perceptible. When therefore the animals make their way towards the places they usually haunt, they are wounded by these instruments, and speedily killed. The crows, as soon as they perceive them to be dead, set up their scream; and this serves as a signal to the hunters, who advance to the spot, and proceed to separate the skin from the flesh, taking care immediately to secure the gall, which is most highly esteemed in medicine In cases of the bite of a mad dog, a pennyweight of it, dissolved in wine, is administered. It is also useful in accelerating parturition, when the labour pains of women have come on. A small quantity of it being applied to carbuncles, pustules, or other eruptions on the body, they are presently dispersed; and it is efficacious in many other complaints. The flesh also of the animal is sold at a dear rate, being thought to have a higher flavour than other kinds of meat, and by all persons it is esteemed a delicacy. In this province the horses are of a large size, and whilst young, are carried for sale to India It is the practice to deprive them of one joint of the tail, in order to prevent them from lashing it from side to side, and to occasion its remaining pendent; as the whisking it about, in riding, appears to them a vile habit. These people ride with long stirrups, as the French do in our part of the world; whereas the Tartars, and almost all other people, wear them short, for the more conveniently using the bow; as they rise in their stirrups above the horse, when they shoot their arrows. They have complete armour of buffalo-leather, and carry lances, shields, and cross-bows. All their arrows are poisoned I was assured, as a certain fact, that many persons, and especially those who harbour bad designs, always carry poison about them, with the intention of swallowing it, in the event of their being apprehended for any delinquency, and exposed to the torture, that, rather than suffer it, they may effect their own destruction. But their rulers, who are aware of this practice, are always provided with the dung of dogs, which they oblige the accused to swallow immediately after, as it occasions their vomiting up the poison, and thus an antidote is ready against the arts of these wretches. Before the time of their becoming subject to the dominion of the Grand Khan, these people were addicted to the following brutal custom. When any stranger of superior quality, who united personal beauty with distinguished valour, happened to take up his abode at the house of one of them, he was murdered during the night; not for the sake of his money, but in order that the spirit of the deceased, endowed with his accomplishments and intelligence, might remain with the family, and that through the efficacy of such an acquisition, all their concerns might prosper. Accordingly the individual was accounted fortunate who possessed in this manner the soul of any noble personage; and many lost their lives in consequence. But from the time of His Majesty's beginning to rule the country, he has taken measures for suppressing the horrid practice, and from the effect of severe punishments that have been inflicted, it has ceased to exist.

12. *Ramusio* II, Ch. XLII *Marsden*, pp. 441–445. *Wright*, pp. 271–276.
[*Frampton*, Chs. 82-83, pp 84, 85, Appendix I Note 312, p. 218]

Of the manner in which the Grand Khan effected the conquest of the Kingdom of Mien and Bangala (Burma and Bengal)

BEFORE we proceed further (in describing the country), we shall speak of a memorable battle that was fought in this kingdom of Vochang (Unchang, or Yun-chang). It happened that in the year 1272 the Grand Khan sent an army into the countries of Vochang and Karazan, for their protection and defence against any attack that foreigners might attempt to make; for at this period he had not as yet appointed his own sons to the governments, which it was afterwards his policy to do; as in the instance of Cen-temur, for whom those places were erected into a principality. When the king of Mien and Bangala, in India, who was powerful in the number of his subjects, in extent of territory, and in wealth, heard that an army of Tartars had arrived at Vochang, he took the resolution of advancing immediately to attack it, in order that by its destruction the Grand Khan should be deterred from again attempting to station a force upon the borders of his dominions. For this purpose he assembled a very large army, including a multitude of elephants (an animal with which his country abounds), upon whose backs were placed battlements or castles, of wood, capable of containing to the number of twelve or sixteen in each. With these, and a numerous army of horse and foot, he took the road to Vochang, where the Grand Khan's army lay, and encamping at no great distance from it, intended to give his troops a few days of rest. As soon as the approach of the king of Mien, with so great a force, was known to Nestardín, who commanded the troops of the Grand Khan, although a brave and able officer, he felt much alarmed, not having under his orders more than twelve thousand men (veterans, indeed, and valiant soldiers); whereas the enemy had sixty thousand, besides the elephants armed as has been described. He did not, however, betray any signs of apprehension, but descending into the plain of Vochang, took a position in which his flank was covered by a thick wood of large trees, whither, in case of a furious charge by the elephants, which his troops might not be able to sustain, they could retire, and from thence, in security, annoy them with their arrows. Calling together the principal officers of his army, he exhorted them not to display less valour on the present occasion than they had done in all their preceding engagements, reminding them that victory did not depend upon the number of men, but upon courage and discipline. He represented to them that the troops of the king of Mien and Bangala were raw and unpractised in the art of war, not having had the opportunities of acquiring experience that had fallen to their lot; that instead of being discouraged by the superior number of their foes, they ought to feel confidence in their own valour so often put to the test; that their very name was a subject of terror, not merely to the enemy before them, but to the whole world; and he concluded by promising to lead them to certain victory. Upon the king of Mien's learning that the Tartars had descended into the plain, he immediately put his army in motion, took up his ground at the distance of about a mile from the enemy, and made a disposition of his force, placing the elephants in the front, and the cavalry and infantry, in two extended wings, in their rear, but leaving between them a considerable interval. Here he took his own station, and proceeded to animate his men and encourage them to fight valiantly, assuring them of victory, as well from the superiority of their numbers, being four to one, as from their formidable body of armed elephants, whose shock the enemy, who had never before been engaged with such combatants, could by no means resist. Then giving orders for sounding a prodigious number of warlike instruments, he advanced boldly with his whole army towards that of the Tartars, which remained firm, making no movement, but suffering them to approach their entrenchments. They then rushed out with great spirit and the utmost eagerness to engage; but it was soon found that the Tartar horses, unused to the sight of such huge animals, with their castles, were terrified, and wheeling about endeavoured to fly, nor could their riders by any exertions restrain them, whilst the king, with the whole of his forces, was every moment gaining ground. As soon as the prudent commander perceived this unexpected disorder, without losing his presence of mind, he instantly adopted the measure of ordering his men to dismount and their horses to be taken into the wood, where they were fastened to the trees. When dismounted, the men, without loss of time, advanced on foot towards the line of elephants, and commenced a brisk discharge of arrows, whilst, on the other side, those who were stationed in the castles, and the rest of the king's army, shot volleys in return with great activity; but their arrows did not make the same impression as those of the Tartars, whose bows were drawn with a stronger arm. So incessant were the discharges of the latter, and all their weapons (according to the instructions of their commander) being directed against the elephants, these were soon covered with arrows, and, suddenly giving way, fell back upon their own people in the rear, who were thereby thrown into confusion. It soon became impossible for their drivers to manage them, either by force or address. Smarting under the pain of their wounds, and terrified by the shouting of the assailants, they were no longer governable, but without guidance or control ran about in all directions, until at length, impelled by rage and fear, they rushed into a part of the wood not occupied by the Tartars. The consequence of this was, that from the closeness of the branches of large trees, they broke, with loud crashes, the battlements or castles that were upon their backs, and involved in the destruction those who sat upon them Upon seeing the rout of the elephants the Tartars acquired fresh courage, and filing off by detachments, with perfect order and regularity, they remounted their horses, and joined their several divisions, when a sanguinary and dreadful combat was renewed. On the part of the king's troops there was no want of valour, and he himself went amongst the ranks entreating them to stand firm, and not to be alarmed by the accident that had befallen the elephants. But the Tartars, by their consummate skill in archery, were too powerful for them, and galled them the more exceedingly, from their not being provided with such armour as was worn by the former. The arrows having been expended on both sides, the men grasped their swords and iron maces, and violently encountered each other. Then in an instant were to be seen many horrible wounds, limbs dismembered, and multitudes falling to the ground, maimed and dying; with such effusion of blood as was dreadful to behold. So great also was the clangour of arms, and such the shoutings and the shrieks, that the noise seemed to ascend to the skies. The king of Mien, acting as became a valiant chief, was present wherever the greatest danger appeared, animating his soldiers, and beseeching them to maintain their ground with resolution. He ordered fresh squadrons from the reserve to advance to the support of those that were exhausted, but perceiving at length that it was impossible any longer to sustain the conflict or to withstand the impetuosity of the Tartars, the greater part of his troops being either killed or wounded, and all the field covered with the carcases of men and horses, whilst those who survived were beginning to give way, he also found himself compelled to take to flight with the wreck of his army, numbers of whom were afterwards slain in the pursuit.

The losses in this battle, which lasted from the morning till noon, were severely felt on both sides; but the Tartars were finally victorious; a result that was materially to be attributed to the troops of the king of Mien and Bangala not wearing armour as the Tartars did, and to their elephants, especially those of the foremost line, being equally without that kind of defence, which, by enabling them to sustain the first discharges of the enemy's arrows, would have allowed them to break his ranks and throw him into disorder. A point perhaps of still greater importance is, that the king ought not to have made his attack on the Tartars in a position where their flank was supported by a wood, but should have endeavoured to draw them into the open country, where they could not have resisted the first impetuous onset of the armed elephants, and where, by extending the cavalry of his two wings, he might have surrounded them The Tartars having collected their force after the slaughter of the enemy, returned towards the wood into which the elephants had fled for shelter, in order to take possession of them, where they found that the men who had escaped from the overthrow were employed in cutting down trees and barricading the passages, with the intent of defending themselves But their ramparts were soon demolished by the Tartars, who slew many of them, and with the assistance of the persons accustomed to the management of the elephants, they possessed themselves of these to the number of two hundred or more From the period of this battle the Grand Khan has always chosen to employ elephants in his armies, which before that time he had not done. The consequences of the victory were, that the Grand Khan acquired possession of the whole of the territories of the king of Bangala and Mien, and annexed them to his dominions.

13. *Ramusio* II, Ch. XLIV. *Marsden*, pp. 448, 449. *Wright*, pp. 277–279.
[Omitted in *Frampton* See Appendix I. Note 317, p. 219]

Of the city of Mien, and of a great sepulchre of its king. (The city of Pagan in Burma)

AFTER the journey of fifteen days that has been mentioned, you reach the city of Mien, which is large, magnificent, and the capital of the kingdom. The inhabitants are idolaters, and have a language peculiar to themselves. It is related that there formerly reigned in this country a rich and powerful monarch, who, when his death was drawing near, gave orders for erecting on the place of his interment, at the head and foot of the sepulchre, two pyramidal towers, entirely of marble, ten paces in height, of a proportionate bulk, and each terminating with a ball. One of these pyramids was covered with a plate of gold an inch in thickness, so that nothing besides the gold was visible; and the other with a plate of silver, of the same thickness Around the balls were suspended small bells of gold and of silver, which sounded when put in motion by the wind. The whole formed a splendid object. The tomb was in like manner covered with a plate, partly of gold and partly of silver. This the king commanded to be prepared for the honour of his soul, and in order that his memory might not perish. The Grand Khan, having resolved upon taking possession of this city, sent thither a valiant officer to effect it, and the army, at its own desire, was accompanied by some of the jugglers or sorcerers, of whom there were always a great number about the court. When these entered the city, they observed the two pyramids so richly ornamented, but would not meddle with them until his majesty's pleasure respecting them should be known. The Grand Khan, upon being informed that they had been erected in pious memory of a former king, would not suffer them to be violated nor injured in the smallest degree; the Tartars being accustomed to consider as a heinous sin the removal of any article appertaining to the dead. [The *Z* text has a unique passage here. See B. p. 124 note d.] In this country were found many elephants, large and handsome wild oxen, with stags, fallow deer, and other animals in great abundance.

14. *Ramusio* II, Ch. XLV. *Marsden*, pp. 451, 452. *Wright*, pp. 279–281.
[Omitted in *Frampton* See Appendix I. Note 317, p. 219]

Of the province of Bangala (Bengal)

THE province of Bangala is situated on the southern confines of India, and was (not yet) brought under the dominion of the Grand Khan at the time of Marco Polo's residence at his court [the date is given in the French texts as 1290]; (although) the operations against it occupied his army for a considerable period, the country being strong and its king powerful, as has been related. It has its peculiar language. The people are worshippers of idols, and amongst them there are teachers [this is a corruption, eunuchs are intended; see Y. Vol. II, p. 115 note] at the head of schools for instruction in the principles of their idolatrous religion and of necromancy, whose doctrine prevails amongst all ranks, including the nobles and chiefs of the country. Oxen are found here almost as tall as elephants, but not equal to them in bulk The inhabitants live upon flesh, milk, and rice, of which they have abundance. Much cotton is grown in the country, and trade flourishes. Spikenard, galangal, ginger, sugar, and many sorts of drugs are amongst the productions of the soil; to purchase which the merchants from various parts of India resort thither. They likewise make purchases of eunuchs, of whom there are numbers in the country, as slaves; for all the prisoners taken in war are presently emasculated, and as every prince and person of rank is desirous of having them for the custody of their women, the merchants obtain a large profit by carrying them to other kingdoms, and there disposing of them. This province is thirty days' journey in extent, and at the eastern extremity of it lies a country named Kangigu

15. *Ramusio* II, Ch. XLVI. *Marsden*, p. 455. *Wright*, pp. 281, 282.
[Omitted in *Frampton*. See Appendix I. Note 317, p 219]

Of the province of Kangigu ("Ciugiu" or "Cangigu," Upper Laos)

KANGIGU is a province situated towards the east, and is governed by a king. The people are idolaters, have a peculiar language, and made a voluntary submission to the Grand Khan, to whom they pay an annual tribute. The king is devoted to sensual pleasures. He has about three hundred wives; and when he hears of any handsome woman, he sends for her, and adds her to the number. Gold is found here in large quantities, and also many kinds of drugs; but, being an inland country, distant from the sea, there is little opportunity of vending them. There are elephants in abundance, and other beasts. The inhabitants live upon flesh, rice, and milk They have no wine made from grapes, but prepare it from rice and a mixture of drugs. Both men and women have their bodies punctured all over, in figures of beasts and birds; and there are among them practitioners whose sole employment it is to trace out these ornaments with the point of a needle, upon the hands, the legs, and the breast When a black colouring stuff has been rubbed over these punctures, it is impossible, either by water or otherwise, to efface the marks. The man or woman who exhibits the greatest profusion of these figures, is esteemed the most handsome. [A fuller account of the tattooing is found in *Z* See B p. 126 note b.]

16. *Ramusio* II, Ch. XLVII. *Marsden*, p. 456. *Wright*, p. 282.
[Omitted in *Frampton* See Appendix I. Note 317, p 219.]

Of the province of Amu ("Aniu" or "Anin," the S.E. corner of Yunnan)

AMU, also, is situated towards the east, and its inhabitants are subjects of the Grand Khan. They are idolaters. and live upon the flesh of their cattle and the fruits of the earth. They have a peculiar language The country produces many horses and oxen, which are sold to the itinerant merchants, and conveyed to India. Buffaloes also, as well as oxen, are numerous, in consequence of the extent and excellence of the pastures. Both men and women wear rings, of gold and silver, upon their wrists, arms, and legs, but those of the females are the more costly The distance between this province and that of Kangigu is twenty-five [the French texts read fifteen] days' journey, and thence to Bangala is twenty days' journey. We shall now speak of a province named Tholoman, situated eight days' journey from the former.

17. *Ramusio* II, Ch. XLVIII. *Marsden*, p. 457. *Wright*, p. 283.
[Omitted in *Frampton* See Appendix I Note 317, p. 219]

Of Tholoman ("Toloman" or "Coloman," the western frontier of Kwei-chau)

THE province of Tholoman lies towards the east, and its inhabitants are idolaters. They have a peculiar language, and are subjects of the Grand Khan. The people are tall and good-looking, their complexions inclining rather to brown than fair.

They are just in their dealings, and brave in war. Many of their towns and castles are situated upon lofty mountains. They burn the bodies of their dead, and the bones that are not reduced to ashes, they put into wooden boxes, and carry them to the mountains, where they conceal them in caverns of the rocks, in order that no wild animal may disturb them. Abundance of gold is found here. For the ordinary small currency they use the porcelain shells that come from India; and this sort of money prevails also in the two before-mentioned provinces of Kangigu and Amu. Their food and drink are the same that has been already mentioned.

18. *Ramusio* II, Ch. XLIX. *Marsden*, pp. 458–460. *Wright*, pp. 284–287.
[Omitted in *Frampton* See Appendix I Note 317, p. 219]

Of the cities of Chintigui, Sidin-fu, Gin-gui and Pazan-fu (Kwei-chau, Ch'êng-tu fu, Chochau and Ho-kien fu)

LEAVING the province of Tholoman, and pursuing a course towards the east, you travel for twelve days by a river, on each side of which lie many towns and castles; when at length you reach the large and handsome city of Chintigui, the inhabitants of which are idolaters, and are the subjects of the Grand Khan. They are traders and artisans. They make cloth of the bark of certain trees, which looks well, and is the ordinary summer clothing of both sexes. The men are brave warriors. They have no other kind of money than the stamped paper of the Grand Khan.

In this province the tigers are so numerous, that the inhabitants, from apprehension of their ravages, cannot venture to sleep at night out of their towns; and those who navigate the river dare not go to rest with their boats moored near the banks; for these animals have been known to plunge into the water, swim to the vessel, and drag the men from thence, but find it necessary to anchor in the middle of the stream, where, in consequence of its great width, they are in safety. In this country are likewise found the largest and fiercest dogs that can be met with: so courageous and powerful are they, that a man, with a couple of them, may be an overmatch for a tiger. Armed with a bow and arrows, and thus attended, should he meet a tiger, he sets on his intrepid dogs, who instantly advance to the attack. The animal instinctively seeks a tree, against which to place himself, in order that the dogs may not be able to get behind him, and that he may have his enemies in front. With this intent, as soon as he perceives the dogs, he makes towards the tree, but with a slow pace, and by no means running, that he may not show any signs of fear, which his pride would not allow. During this deliberate movement, the dogs fasten upon him, and the man plies him with his arrows. He, in his turn, endeavours to seize the dogs, but they are too nimble for him, and draw back, when he resumes his slow march; but before he can gain his position, he has been wounded by so many arrows, and so often bitten by the dogs, that he falls through weakness and from loss of blood. By these means it is that he is at length taken. [F. gives the above description as part of his Ch. 83, but makes the animal an elephant instead of a tiger.]

There is here an extensive manufacture of silks, which are exported in large quantities to other parts by the navigation of the river, which continues to pass amongst towns and castles, and the people subsist entirely by trade. At the end of twelve days, you arrive at the city of Sidin-fu, of which an account has been already given. From thence, in twenty days, you reach Gin-gui, and in four days more the city of Pazan-fu, which belongs to Cathay, and lies towards the south, in returning by the other side of the province. The inhabitants worship idols, and burn the bodies of their dead. There are here also certain Christians, who have a church. They are subjects of the Grand Khan, and his paper money is current among them. They gain their living by trade and manufacture, having silk in abundance, of which they weave tissues mixed with gold, and also very fine scarfs. This city has many towns and castles under its jurisdiction. A great river flows beside it, by means of which large quantities of merchandise are conveyed to the city of Kanbalu; for by the digging of many canals it is made to communicate with the capital. But we shall take our leave of this, ...

19. *Ramusio* II, Ch. LII. *Marsden*, pp. 466, 467. *Wright*, pp. 289–291.
[*Frampton*, Ch. 86, p. 86; Appendix I. Note 324, p. 221.]

Of the city of Tudin-fu ("Candrafra," "Tandinfu," Yen-chau)

WHEN you depart from Chan-gli, and travel southwards six days' [so also B., but Y. reads "five"] journey, you pass many towns and castles of great importance and grandeur, whose inhabitants worship idols, and burn the bodies of their dead. They are the subjects of the Grand Khan, and receive his paper money as currency. They subsist by trade and manufactures, and have provisions in abundance. At the end of these six days you arrive at a city named Tudin-fu, which was formerly a magnificent capital, but the Grand Khan reduced it to his subjection by force of arms. It is rendered a delightful residence by the gardens which surround it, stored as they are with handsome shrubs and excellent fruits. Silk is produced here in large quantities. It has under its jurisdiction eleven cities and considerable towns of the empire, all places of great trade, and having abundance of silk. It was the seat of government of its own king, before the period of its reduction by the Grand Khan In 1272 [Y. reads 1273] the latter appointed one of his officers of the highest rank, named Lucansor [corrupted from "Lutam Sangon" of fr. 1116 and "Liyton Sangon" of Y. See Pelliot, *Journ. As.* 1912, p. 584 n., and Moule, *T'oung Pao*, July 1915, p. 417] to the government of this city, with a command of eighty thousand horse, for the protection of that part of the country. This man upon finding himself master of a rich and highly productive district, and at the head of so powerful a force, became intoxicated with pride, and formed schemes of rebellion against his sovereign. With this view he tampered with the principal persons of the city, persuaded them to become partakers in his evil designs, and by their means succeeded in producing a revolt throughout all the towns and fortified places of the province As soon as the Grand Khan became acquainted with these traitorous proceedings, he despatched to that quarter an army of a hundred thousand men, under the orders of two others of his nobles, one of whom was named Angul [read Aguil] and the other Mongatai. When the approach of this force was known to Lucansor, he lost no time in assembling an army no less numerous than that of his opponents, and brought them as speedily as possible to action. There was much slaughter on both sides, when at length, Lucansor being killed, his troops betook themselves to flight. Many were slain in the pursuit, and many were made prisoners. These were conducted to the presence of the Grand Khan, who caused the principals to be put to death, and pardoning the others took them into his own service, to which they ever afterwards continued faithful. [At this point the newly found *Z* text adds 64 entirely new lines. See B. pp. 130–132.]

20. *Ramusio* II, Ch. LIII. *Marsden*, pp. 469, 470. *Wright*, pp. 291, 292.
[*Frampton*, Ch. 86, p 86, Appendix I. Note 326, p. 222.]

Of the city of Singui-matu (Tsi-ning-chau)

TRAVELLING from Tudin-fu three [M originally had "seven"] days, in a southerly direction, you pass many considerable towns and strong places, where commerce and manufactures flourish. The inhabitants are idolaters, and are subjects of the Grand Khan. The country abounds with game, both beasts and birds, and produces an ample supply of the necessaries of life. At the end of three days you arrive at the city of Singui-matu [see above], within which, but on the southern side, passes a large and deep river, which the inhabitants divided into two branches, one of which, taking its course to the east, runs through Kataia, whilst the other, taking a westerly course, passes towards the province of Manji. This river is navigated by so many vessels that the number might seem incredible, and serves to convey from both provinces, that is, from the one province to the other, every requisite article of consumption. It is indeed surprising to observe the multitude and the size of the vessels that are continually passing and repassing, laden with merchandise of the greatest value. On leaving Singui-matu and travelling towards the south for sixteen days, you unceasingly meet with commercial towns and with castles. The people throughout the country are idolaters, and subjects of the Grand Khan.

21. As both Frampton and Ramusio omit the chapters on the cities of "Linju," "Piju," and "Siju" they are given below from the translation by Yule, corrected, however, by fr. 1116.

In Frampton's text they should come at the end of Ch. 86. See Appendix I, Note 326, p. 222.

Yule, Bk. II, Ch. LXIII; *Ben.*, Chs. CXXXVII and CXXXVIII

Concerning the cities of Linju and Piju (? Süchaufu and Pei-chau)

ON leaving the city [fr. 1116 has "ceste ville de Singiu"] of Sinju-matu you travel for eight days towards the south, always coming to great and rich towns and villages flourishing with trade and manufactures. The people are all subjects of the Great Khan, use paper-money, and burn their dead. At the end of those eight days you come to the city of Linju, in the province of the same name of which it is the capital. It is a rich and noble city, and the men are good soldiers, natheless they carry on great trade and manufactures [fr. 1116 also has "Il sunt ydres..."]. There is a great abundance of game in both beasts and birds, and all the necessaries of life are in profusion The place stands on the river of which I told you above. And they have here great numbers of vessels, even greater than those of which I spoke before, and these transport a great amount of costly merchandize. [Here the next chapter of B. commences.]

So, quitting this province and city [fr. 1116 has only "cité .."] of Linju, you travel three days more towards the south, constantly finding numbers of rich towns and villages These still belong to Cathay; and the people are all idolaters, burning their dead, and using paper-money, that I mean of their Lord the Great Khan, whose subjects they are [fr 1116 simply says, "et sunt au grant Kaan," but adds, "Et ausint sunt (com) les autres que je vos ai contés en ariere"]. This is the finest country for game, whether in beasts or birds, that is anywhere to be found, and all the necessaries of life are in profusion

At the end of those three days you find the city of Piju, a great, rich, and noble city, with large trade and manufactures, and a great production of silk. This city stands at the entrance to the great province of Manzi, and there reside at it a great number of merchants who despatch carts from this place loaded with great quantities of goods to the different towns ["towns and villages" in fr. 1116] of Manzi.

The city brings in a great revenue to the Great Khan [fr. 1116 adds "Il n'i a autre couse que a mentovoir face, et par ce nos en partiron et vos conteron de un autre cité, qui est apellé Cingiu que est encore a midi"].

Yule, Bk. II, Ch. LXIV; *Ben.*, Ch. CXXXIX

Concerning the city of Siju, and the Great River Caramoran (Su-t'sien and the old bed of the Hwang ho, or Yellow River)

When you leave Piju [fr. 1116 has "la cité de Pingiu"] you travel towards the south for two days, through beautiful districts abounding in everything, and in which you find quantities of all kinds of game. At the end of those two days you reach the city of Siju ["Cingiu"], a great, rich, and noble city, flourishing with trade and manufactures. The people are idolaters, burn their dead, use paper-money, and are subjects of the Great Khan [Y. omits the last part of this now usual formula, found in fr. 1116: "and use paper-money"].

This ends the portions omitted in F. The rest of Y. Bk. II. Ch. LXIV and B. Ch. CXXXIX corresponds roughly to F. Ch. 87, p. 87 of this volume.

22. *Ramusio* II, Ch. LV. *Marsden*, pp. 474–476. *Wright*, pp. 294–298.
[*Frampton*, Ch 88, p 88, Appendix I Note 339, p 225]

Of the most noble province of Manji, and of the manner in which it was subdued by the Grand Khan

THE province of Manji is the most magnificent and the richest that is known in the eastern world. About the year 1269 it was subject to a prince who was styled Fanfur, and who surpassed in power and wealth any other that for a century had reigned in that country His disposition was pacific, and his actions benevolent. So much was he beloved by his people, and such the strength of his kingdom, enclosed by rivers of the largest size, that his being molested by any power upon earth was regarded as an impossible event The effect of this opinion was, that he neither paid any attention himself to military affairs, nor encouraged his people to become acquainted with military exercises. The cities of his dominions were remarkably well fortified, being surrounded by deep ditches, a bow-shot in width, and full of water. He did not keep up any force in cavalry, because he was not apprehensive of attack. The means of increasing his enjoyments and multiplying his pleasures were the chief employment of his thoughts. He maintained at his court, and kept near his person, about a thousand beautiful women, in whose society he took delight He was a friend to peace and to justice, which he administered strictly. The smallest act of oppression, or injury of any kind, committed by one man against another, was punished in an exemplary manner, without respect of persons. Such indeed was the impression of his justice, that when shops, filled with goods, happened, through the negligence of the owners, to be left open, no person dared to enter them, or to rob them of the smallest article. Travellers of all descriptions might pass through every part of the kingdom, by night as well as by day, freely and without apprehension of danger. He was religious, and charitable to the poor and needy. Children whom their wretched mothers exposed in consequence of their inability to rear them, he caused to be saved and taken care of, to the number of twenty thousand annually. When the boys attained a sufficient age, he had them instructed in some handicraft, and afterwards married them to young women who were brought up in the same manner.

Very different from the temper and habits of Fanfur were those of Kublai-khan, emperor of the Tartars, whose whole delight consisted in thoughts of a warlike nature, of the conquest of countries, and of extending his renown. After having annexed to his dominions a number of provinces and kingdoms, he now directed his views to the subduing that of Manji, and for this purpose assembled a numerous army of horse and foot, the command of which he gave to a general named Chin-san Bay-an, which signifies in our language, the "Hundred-eyed." A number of vessels were likewise put under his orders, with which he proceeded to the invasion of Manji. Upon landing there, he immediately summoned the inhabitants of the city of Koi-gan-zu to surrender to the authority of his sovereign. Upon their refusal to comply, instead of giving orders for an assault, he advanced to the next city, and when he there received a similar answer, proceeded to a third and a fourth, with the same result Deeming it no longer prudent to leave so many cities in his rear, whilst not only his army was strong, but he expected to be soon joined by another of equal force, which the Grand Khan was to send to him from the interior, he resolved upon the attack of one of these cities; and having, by great exertions and consummate skill, succeeded in carrying the place, he put every individual found in it to the sword As soon as the intelligence of this event reached the other cities, it struck their inhabitants with such consternation and terror, that of their own accord they hastened to declare their submission. This being effected, he advanced, with the united force of his two armies, against the royal city of Kinsai, the residence of king Fanfur, who felt all the agitation and dread of a person who had never seen a battle, nor been engaged in any sort of warfare. Alarmed for the safety of his person, he made his escape to a fleet of vessels that lay in readiness for the purpose, and embarking all his treasure and valuable effects, left the charge of the city to his queen, with directions for its being defended to the utmost; feeling assured that her sex would be a protection to her, in the event of her falling into the hands of the enemy. He from thence proceeded to sea, and reaching certain islands, where were some strongly fortified posts, he continued there till his death. After the queen had been left in the manner related, it is said to have come to her knowledge that the king had been told by his astrologers that he could never be deprived of his sovereignty by any other than a chief who should have a hundred eyes. On the strength of this declaration she felt confident, notwithstanding that the city became daily more and more straitened, that it could not be lost, because it seemed a thing impossible that any mortal could have that number of eyes. Inquiring, however, the name of the general who commanded the enemy's troops, and being told it was Chin-san Bay-an, which means a hundred eyes, she was seized with horror at hearing it pronounced, as she felt a conviction that this must be the person who, according to the saying of the astrologers, might drive her husband from his throne. Overcome by womanish fear, she no longer attempted to make resistance, but immediately surrendered. Being thus in possession of the capital, the Tartars soon brought the remainder of the province under their subjection. The queen was sent to the presence of Kublai-khan, where she was honourably received by him, and an allowance was by his orders assigned, that enabled her to support the dignity of her rank. Having stated the manner in which the conquest of Manji was effected, we shall now speak of the different cities of that province, ...

23. *Ramusio* II, Ch. LXII *Marsden*, pp. 488, 489. *Wright*, pp. 302–304.
[*Frampton*, Ch 93, pp 90, 91, Appendix I. Note 353, p 228]

Of the City of Sa-yan-fu, that was taken by the means of MM. Nicolo and Maffeo Polo (Siege of Siang-Yang)

SA-YAN-FU is a considerable city of the province of Manji, having under its jurisdiction twelve wealthy and large towns. It is a place of great commerce and extensive manufactures. The inhabitants burn the bodies of their dead, and are idolaters. They are the subjects of the Grand Khan, and use his paper currency. Raw silk is there produced in great quantity, and the finest silks, intermixed with gold, are woven. Game of all kinds abounds. The place is amply furnished with everything that belongs to a great city, and by its uncommon strength it was enabled to stand a siege of three years; refusing to surrender to the Grand Khan, even after he had obtained possession of the province of Manji. The difficulties experienced in the reduction of it were chiefly occasioned by the army's not being able to approach it, excepting on the northern side; the others being surrounded with water, by means of which the place continually received supplies, which it was not in the power of the besiegers to prevent. When the operations were reported to His Majesty, he felt extremely hurt that this place alone should obstinately hold out, after all the rest of the country had been reduced to obedience. The circumstance having come to the knowledge of the brothers Nicolo and Maffeo, who were then resident at the imperial court, they immediately presented themselves to the Grand Khan, and proposed to him that they should be allowed to construct machines, such as were made use of in the West, capable of throwing stones of three hundred pounds weight, by which the buildings of the city might be destroyed and the inhabitants killed. Their memorial was attended to by the Grand Khan, who, warmly approving of the scheme, gave orders that the ablest smiths and carpenters should be placed under their direction, amongst whom were some Nestorian Christians, who proved to be most able mechanics In a few days they completed their mangonels, according to the instructions furnished by the two brothers, and a trial being made of them in the presence of the Grand Khan, and of his whole court, an opportunity was afforded of seeing them cast stones, each of which weighed three hundred pounds. They were then put on board of vessels, and conveyed to the army. When set up in front of the city of Sa-van-fu, the first stone projected by one of them fell with such weight and violence upon a building, that a great part of it was crushed, and fell to the ground. So terrified were the inhabitants by this mischief, which to them seemed to be the effect of a thunderbolt from heaven, that they immediately deliberated upon the expediency of surrendering. Persons authorised to treat were accordingly sent from the place, and their submission was accepted on the same terms and conditions as had been granted to the rest of the province. This prompt result of their ingenuity increased the reputation and credit of these two Venetian brothers in the opinion of the Grand Khan and of all his courtiers.

24. *Ramusio* II, Ch. LXIII. *Marsden*, pp. 494, 495. *Wright*, pp. 305–307.
[*Frampton*, Ch 94, p 91; Appendix I Note 356, p. 230]

Of the city of Sin-gui, and of the very great river Kiang (Tcheng on the Yangtze)

LEAVING the city of Sa-yan-fu, and proceeding fifteen days' journey towards the south-east, you reach the city of Sin-gui, which, although not large, is a place of great commerce. The number of vessels that belong to it is prodigious, in consequence of its being situated near the Kiang, which is the largest river in the world, its width being in some places ten, in others eight, and in others six miles. Its length, to the place where it discharges itself into the sea, is upwards of one hundred days' journey It is indebted for its great size to the vast number of other navigable rivers that empty their waters into it, which have their sources in distant countries. A great number of cities and large towns are situated upon its banks, and more than two hundred, with sixteen provinces, partake of the advantages of its navigation, by which the transport of merchandise is to an extent that might appear incredible to those who have not had an opportunity of witnessing it. When we consider, indeed, the length of its course, and the multitude of rivers that communicate with it (as has been observed), it is not surprising that the quantity and value of articles for the supply of so many places, lying in all directions, should be incalculable. The principal commodity, however, is salt, which is not only conveyed by means of the Kiang, and the rivers connected with it, to the towns upon their banks, but afterwards from thence to all places in the interior of the country. On one occasion, when Marco Polo was at the city of Sin-gui, he saw there not fewer than five thousand vessels; and yet there are other towns along the river where the number is still more considerable. All these vessels are covered with a kind of deck, and have a mast with one sail. Their burthen is in general about four thousand *cantari*, or quintals, of Venice, and from that upwards to twelve thousand cantari, which some of them are capable of loading. They do not employ hempen cordage, excepting for the masts and sails (standing and running rigging). They have canes of the length of fifteen paces, such as have been already described, which they split, in their whole length, into very thin pieces, and these, by twisting them together, they form into ropes three hundred paces long. So skilfully are they manufactured, that they are equal in strength to cordage made of hemp. With these ropes the vessels are tracked along the rivers, by means of ten or twelve horses to each, as well upwards, against the current, as in the opposite direction. At many places near the banks of this river there are hills and small rocky eminences, upon which are erected idol temples and other edifices, and you find a continual succession of villages and inhabited places.

25. *Ramusio* II, Ch. LXVIII. *Marsden*, pp. 508–542. *Wright*, pp. 313–335.
[*Frampton*, Chs. 98–100, pp. 93–95, Appendix I. Note 368, p. 233]

Of the noble and magnificent city of Kin-sai (Hang-chau)

UPON leaving Va-giu you pass, in the course of three days' [F. has "five"; possibly an attempt to make up for the omission of "Vuju," etc.] journey, many towns, castles, and villages, all of them well inhabited and opulent. The people are idolaters, and the subjects of the Grand Khan, and they use paper money and have abundance of provisions. At the end of three days you reach the noble and magnificent city of Kin-sai, a name that signifies "the celestial city," and which it merits from its preeminence to all others in the world, in point of grandeur and beauty, as well as from its abundant delights, which might lead an inhabitant to imagine himself in paradise. This city was frequently visited by Marco Polo, who carefully and diligently observed and inquired into every circumstance respecting it, all of which he entered in his notes, from whence the following particulars are briefly stated. According to common estimation, this city is an hundred miles in circuit Its streets and canals are extensive, and there are squares, or market-places, which, being necessarily proportioned in size to the prodigious concourse of people by whom they are frequented, are exceedingly spacious. It is situated between a lake of fresh and very clear water on the one side, and a river of great magnitude on the other, the waters of which, by a number of canals, large and small, are made to run through every quarter of the city, carrying with them all the filth into the lake, and ultimately to the sea. This, whilst it contributes much to the purity of the air, furnishes a communication by water, in addition to that by land, to all parts of the town; the canals and the streets being of sufficient width to allow of boats on the one, and carriages in the other, conveniently passing, with articles necessary for the consumption of the inhabitants. It is commonly said that the number of bridges, of all sizes, amounts to twelve thousand. Those which are thrown over the principal canals and are connected with the main streets, have arches so high, and built with so much skill, that vessels with [M had "without"] their masts can pass under them, whilst, at the same time, carts and horses are passing over their heads,—so well is the slope from the street adapted to the height of the arch. If they were not in fact so numerous, there would be no convenience of crossing from one place to another.

Beyond the city, and enclosing it on that side, there is a fosse about forty miles in length, very wide, and full of water that comes from the river before mentioned. This was excavated by the ancient kings of the province, in order that when the river should overflow its banks, the superfluous water might be diverted into this channel, and to serve at the same time as a measure of defence. The earth dug out from thence was thrown to the inner side, and has the appearance of many hillocks surrounding the place. There are within the city ten principal squares or market-places, besides innumerable shops along the streets. Each side of these squares is half a mile in length, and in front of them is the main street, forty paces in width, and running in a direct line from one extremity of the city to the other. It is crossed by many low and convenient bridges. These market-squares (two miles in their whole dimension) are at the distance of four miles from each other. In a direction parallel to that of the main street, but on the opposite side of the squares, runs a very large canal, on the nearer bank of which capacious warehouses are built of stone, for the accommodation of the merchants who arrive from India and other parts, together with their goods and effects, in order that they may be conveniently situated with respect to the market-places. In each of these, upon three days in every week, there is an assemblage of from forty to fifty thousand persons, who attend the markets and supply them with every article of provision that can be desired. There is an abundant quantity of game of all kinds, such as roebucks, stags, fallow deer, hares, and rabbits, together with partridges, pheasants, francolins, quails, common fowls, capons, and such numbers of ducks and geese as can scarcely be expressed, for so easily are they bred and reared on the lake, that, for the value of a Venetian silver groat, you may purchase a couple of geese and two couple of ducks There, also, are the shambles, where they slaughter cattle for food, such as oxen, calves, kids, and lambs, to furnish the tables of rich persons and of the great magistrates. As to people of the lower classes, they do not scruple to eat every other kind of flesh, however unclean, without any discrimination At all seasons there is in the markets a great variety of herbs and fruits, and especially pears of an extraordinary size, weighing ten pounds each, that are white in the inside, like paste, and have a very fragrant smell. There are peaches also, in their season, both of the yellow and the white kind, and of a delicious flavour. Grapes are not produced there, but are brought in a dried state, and very good, from other parts. This applies also to wine, which the natives do not hold in estimation, being accustomed to their own liquor prepared from rice and spices. From the sea, which is fifteen miles distant, there is daily brought up the river, to the city, a vast quantity of fish, and in the lake also there is abundance, which gives employment at all times to persons whose sole occupation it is to catch them The sorts are various according to the season of the year, and, in consequence of the offal carried thither from the town, they become large and rich. At the sight of such an importation of fish, you would think it impossible that it could be sold, and yet, in the course of a few hours, it is all taken off, so great is the number of inhabitants, even of those classes which can afford to indulge in such luxuries, for fish and flesh are eaten at the same meal. Each of the ten market-squares is surrounded with high dwelling-houses, in the lower part of which are shops, where every kind of manufacture is carried on, and every article of trade is sold; such, amongst others, as spices, drugs, trinkets, and pearls In certain shops nothing is vended but the wine of the country, which they are continually brewing, and serve out fresh to their customers at a moderate price. The streets connected with the market-squares are numerous, and in some of them are many cold baths, attended by servants of both sexes, to perform the offices of ablution for the men and women who frequent them, and who from their childhood have been accustomed at all times to wash in cold water, which they reckon highly conducive to health. At these bathing places, however, they have apartments provided with warm water, for the use of strangers, who, from not being habituated to it, cannot bear the shock of the cold All are in the daily practice of washing their persons, and especially before their meals. [See Y. Vol. II. p 189, and note 8 on p 198.]

In other streets are the habitations of the courtesans, who are here in such numbers as I dare not venture to report: and not only near the squares, which is he situation usually appropriated for their residence, but in every part of the city they are to be found, adorned with much finery, highly perfumed, occupying well-furnished houses, and attended by many female domestics. These women are accomplished, and are perfect in the arts of blandishment and dalliance, which they accompany with expressions adapted to every description of person, insomuch that strangers who have once tasted of their charms, remain in a state of fascination, and become so enchanted by their meretricious arts, that they can never divest themselves of the impression. Thus intoxicated with sensual pleasures, when they return to their homes they report that they have been in Kin-sai, or the celestial city, and pant for the time when they may be enabled to revisit paradise. In other streets are the dwellings of the physicians and the astrologers, who also give instructions in reading and writing, as well as in many other arts. They have apartments also amongst those which surround the market-squares. On opposite sides of each of these squares there are two large edifices, where officers appointed by the Grand Khan are stationed, to take immediate cognisance of any differences that may happen to arise between the foreign merchants, or amongst the inhabitants of the place. It is their duty likewise to see that the guards upon the several bridges in their respective vicinities (of whom mention shall be made hereafter) are duly placed, and in cases of neglect, to punish the delinquents at their discretion.

On each side of the principal street, already mentioned as extending from one end of the city to the other, there are houses and mansions of great size, with their gardens, and near to these, the dwellings of the artisans, who work in shops, at their several trades; and at all hours you see such multitudes of people passing and repassing, on their various avocations, that the providing food in sufficiency for their maintenance might be deemed an impossibility; but other ideas will be formed when it is observed that, on every market-day, the squares are crowded with tradespeople, who cover the whole space with the articles brought by carts and boats, for all of which they find a sale. By instancing the single article of pepper, some notion may be formed of the whole quantity of provisions, meat, wine, groceries, and the like, required for the consumption of the inhabitants of Kin-sai; and of this, Marco Polo learned from an officer employed in the Grand Khan's customs, the daily amount was forty-three loads, each load being two hundred and forty-three pounds. [R. has "dugento, & ventitre," so we must alter M.'s translation to "223" This is supported by *Z* as well as Y.]

The inhabitants of the city are idolaters, and they use paper money as currency. The men as well as the women have fair complexions, and are handsome. The greater part of them are always clothed in silk, in consequence of the vast quantity of that material produced in the territory of Kin-sai, exclusively of what the merchants import from other provinces. Amongst the handicraft trades exercised in the place, there are twelve considered to be superior to the rest, as being more generally useful; for each of which there are a thousand workshops, and each shop furnishes employment for ten, fifteen, or twenty workmen, and in a few instances as many as forty, under their respective masters The opulent principals in these manufactories do not labour with their own hands, but, on the contrary, assume airs of gentility and affect parade Their wives equally abstain from work. They have much beauty, as has been remarked, and are brought up with delicate and languid habits The costliness of their dresses, in silks and jewellery, can scarcely be imagined Although the laws of their ancient kings ordained that each citizen should exercise the profession of his father, yet they were allowed, when they acquired wealth, to discontinue the manual labour, provided they kept up the establishment, and employed persons to work at their paternal trades Their houses are well built and richly adorned with carved work So much do they delight in ornaments of this kind, in paintings, and fancy buildings, that the sums they lavish on such objects are enormous. The natural disposition of the native inhabitants of Kin-sai is pacific, and by the example of their former kings, who were themselves unwarlike, they have been accustomed to habits of tranquillity. The management of arms is unknown to them, nor do they keep any in their houses. Contentious broils are never heard among them They conduct their mercantile and manufacturing concerns with perfect candour and probity. They are friendly towards each other, and persons who inhabit the same street, both men and women, from the mere circumstance of neighbourhood, appear like one family. In their domestic manners they are free from jealousy or suspicion of their wives, to whom great respect is shown, and any man would be accounted infamous who should presume to use indecent expressions to a married woman. To strangers also, who visit their city in the way of commerce, they give proofs of cordiality, inviting them freely to their houses, showing them hospitable attention, and furnishing them with the best advice and assistance in their mercantile transactions. On the other hand, they dislike the sight of soldiery, not excepting the guards of the Grand Khan, as they preserve the recollection that by them they were deprived of the government of their native kings and rulers

On the borders of the lake [F. has "in" instead of "on" the lake] are many handsome and spacious edifices belonging to men of rank and great magistrates. There are likewise many idol temples, with their monasteries, occupied by a number of monks, who perform the service of the idols. Near the central part are two islands, upon each of which stands a superb building, with an incredible number of apartments and separate pavilions ["Pallaces" in F] When the inhabitants of the city have occasion to celebrate a wedding, or to give a sumptuous entertainment, they resort to one of these islands, where they find ready for their purpose every article that can be required, such as vessels, napkins, table-linen, and the like, which are provided and kept there at the common expense of the citizens, by whom also the buildings were erected It may happen that at one time there are a hundred parties assembled there, at wedding or other feasts, all of whom, notwithstanding, are accommodated with separate rooms or pavilions, so judiciously arranged that they do not interfere with or incommode each other. In addition to this, there are upon the lake a great number of pleasure-vessels or barges, calculated for holding ten, fifteen, to twenty persons, being from fifteen to twenty paces in length, with a wide and flat flooring, and not liable to heel to either side in passing through the water Such persons as take delight in the amusement, and mean to enjoy it, either in the company of their women or that of their male companions, engage one of these barges, which are always kept in the nicest order, with proper seats and tables, together with every other kind of furniture necessary for giving an entertainment. The cabins have a flat roof or upper deck, where the boatmen take their place, and by means of long poles, which they thrust to the bottom of the lake (not more than one or two fathoms in depth), they shove the barges along, until they reach the intended spot. These cabins are painted withinside of various colours and with a variety of figures; all parts of the vessel are likewise adorned with painting. There are windows on each side, which may either be kept shut, or opened, to give an opportunity to the company, as they sit at table, of looking out in every direction and feasting their eyes on the variety and beauty of the scenes as they pass them And truly the gratification afforded in this manner, upon the water, exceeds any that can be derived from the amusements on the land; for as the lake extends the whole length of the city, on one side, you have a view, as you stand in the boat, at a certain distance from the shore, of all its grandeur and beauty, its palaces, temples, convents, and gardens, with trees of the largest size growing down to the water's edge, whilst at the same time you enjoy the sight of other boats of the same description, continually passing you, filled in like manner with parties in pursuit of amusement. In fact, the inhabitants of this place, as soon as the labours of the day have ceased, or their mercantile transactions are closed, think of nothing else than of passing the remaining hours in parties of pleasure, with their wives or their mistresses, either in these barges, or about the city in carriages, of which it will here be proper to give some account, as constituting one of the amusements of these people.

It must be observed, in the first place, that the streets of Kin-sai are all paved with stones and bricks, and so likewise are all the principal roads extending from thence through the province of Manji, by means of which passengers can travel to every part without soiling their feet, but as the couriers of His Majesty, who go on horseback with great speed, cannot make use of the pavement, a part of the road, on one side, is on their account left unpaved The main street of the city, of which we have before spoken, as leading from one extremity to the other, is paved with stone and brick to the width of ten paces on each side, the intermediate part being filled up with small gravel, and provided with arched drains for carrying off the rain-water that falls, into the neighbouring canals, so that it remains always dry. On this gravel it is that the carriages are continually passing and repassing They are of a long shape, covered at top, have curtains and cushions of silk, and are capable of holding six persons. Both men and women who feel disposed to take their pleasure, are in the daily practice of hiring them for that purpose, and accordingly at every hour you may see vast numbers of them driven along the middle part of the street Some of them proceed to visit certain gardens, where the company are introduced, by those who have the management of the place, to shady recesses contrived by the gardeners for that purpose, and here the men indulge themselves all day in the society of their women, returning home, when it becomes late, in the manner they came

It is the custom of the people of Kin-sai, upon the birth of a child, for the parents to make a note, immediately, of the day, hour, and minute at which the delivery took place They then inquire of an astrologer under what sign or aspect of the heavens the child was born; and his answer is likewise committed carefully to writing. When therefore he is grown up, and is about to engage in any mercantile adventure, voyage, or treaty of marriage, this document is carried to the astrologer, who, having examined it, and weighed all the circumstances, pronounces certain oracular words, in which these people, who sometimes find them justified by the event, place great confidence. Of these astrologers, or rather magicians, great numbers are to be met with in every market-place, and no marriage is ever celebrated until an opinion has been pronounced upon it by one of that profession.

It is also their custom, upon the death of any great and rich personage, to observe the following ceremonies. The relations, male and female, clothe themselves in coarse dresses, and accompany the body to the place appointed for burning it. The procession is likewise attended by performers on various musical instruments, which are sounded as it moves along, and prayers to their idols are chanted in a loud voice. When arrived at the spot, they throw into the flame many pieces of cotton-paper, upon which are painted representations of male and female servants, horses, camels, silk wrought with gold, as well as of gold and silver money. This is done, in consequence of their belief that the deceased will possess in the other world all these conveniences, the former in their natural state of flesh and bones, together with the money and the silks. As soon as the pile has been consumed, they sound all the instruments of music at the same time, producing a loud and long-continued noise; and they imagine that by these ceremonies their idols are induced to receive the soul of the man whose corpse has been reduced to ashes, in order to its being regenerated in the other world, and entering again into life.

In every street of this city there are stone buildings or towers, to which, in case of a fire breaking out in any quarter (an accident by no means unusual, as the houses are mostly constructed of wood), the inhabitants may remove their effects for security. By a regulation which His Majesty has established, there is a guard of ten watchmen stationed, under cover, upon all the principal bridges, of whom five do duty by day and five by night Each of these guard-rooms is provided with a sonorous wooden instrument as well as one of metal, together with a *clepsydra* (*horiuolo*), by means of which latter the hours of the day and night are ascertained. As soon as the first hour of the night is expired, one of the watchmen gives a single stroke upon the wooden instrument, and also upon the metal *gong* (*bacino*), which announces to the people of the neighbouring streets that it is the first hour. At the expiration of the second, two strokes are given, and so on progressively, increasing the number of strokes as the hours advance The guard is not allowed to sleep, and must be always on the alert In the morning, as soon as the sun begins to appear, a single stroke is again struck, as in the evening, and so onwards from hour to hour. Some of these watchmen patrol the streets, to observe whether any person has a light or fire burning after the hour appointed for extinguishing them. Upon making the discovery, they affix a mark to the door, and in the morning the owner of the house is taken before the magistrates, by whom, if he cannot assign a legitimate excuse for his offence, he is condemned to punishment Should they find any person abroad at an unseasonable hour, they arrest and confine him, and in the morning he is carried before the same tribunal If, in the course of the day, they notice any person who from lameness or other infirmity is unable to work, they place him in one of the hospitals, of which there are several in every part of the city, founded by the ancient kings, and liberally endowed When cured, he is obliged to work at some trade. Immediately upon the appearance of fire breaking out in a house, they give the alarm by beating on the wooden machine, when the watchmen from all the bridges within a certain distance assemble to extinguish it, as well as to save the effects of the merchants and others, by removing them to the stone towers that have been mentioned The goods are also sometimes put into boats, and conveyed to the islands in the lake Even on such occasions the inhabitants dare not stir out of their houses, when the fire happens in the night-time, and only those can be present whose goods are actually removing, together with the guard collected to assist, which seldom amounts to a smaller number than from one to two thousand men In cases also of tumult or insurrection amongst the citizens, the services of this police guard are necessary; but, independently of them, His Majesty always keeps on foot a large body of troops, both infantry and cavalry, in the city and its vicinity, the command of which he gives to his ablest officers, and those in whom he can place the greatest confidence, on account of the extreme importance of this province, and especially its noble capital, which surpasses in grandeur and wealth every other city in the world For the purposes of nightly watch, there are mounds of earth thrown up, at the distance of above a mile from each other, on the top of which a wooden frame is constructed, with a sounding board, which being struck with a mallet by the guard stationed there, the noise is heard to a great distance If precautions of this nature were not taken upon occasions of fire, there would be danger of half the city being consumed; and their use is obvious also in the event of popular commotion, as, upon the signal being given, the guards at the several bridges arm themselves, and repair to the spot where their presence is required.

When the Grand Khan reduced to his obedience the province of Manji, which until that time had been one kingdom, he thought proper to divide it into nine parts [F. has "8 kingdomes"], over each of which he appointed a king or viceroy, who should act as supreme governor of that division, and administer justice to the people. These make a yearly report to commissioners acting for His Majesty, of the amount of the revenue, as well as of every other matter pertaining to their jurisdiction. Upon the third year they are changed, as are all other public officers. One of these nine viceroys resides and holds his court in the city of Kin-sai, and has authority over more than a hundred and forty cities and towns, all large and rich. Nor is this number to be wondered at, considering that in the whole of the province of Manji there are no fewer than twelve hundred [F. has "1202"], containing a large population of industrious and wealthy inhabitants. In each of these, according to its size and other circumstances, His Majesty keeps a garrison, consisting, in some places, of a thousand, in others of ten or twenty thousand men, accordingly as he judges the city to be, in its own population, more or less powerful. It is not to be understood that all these troops are Tartars On the contrary, they are chiefly natives of the province of Cathay. The Tartars are universally horsemen, and cavalry cannot be quartered about those cities which stand in the low, marshy parts of the province, but only in firm, dry situations, where such troops can be properly exercised. To the former he sends Cathaians, and such men of the province of Manji as appear to have a military turn, for it is his practice to make an annual selection amongst all his subjects of such as are best qualified to bear arms, and these he enrolls to serve in his numerous garrisons, that may be considered as so many armies. But the soldiers drawn from the province of Manji he does not employ in the duty of their native cities, on the contrary, he marches them to others at the distance of perhaps twenty days' journey, where they are continued for four or five years, at the expiration of which they are allowed to return to their homes, and others are sent to replace them. This regulation applies equally to the Cathaians The greater part of the revenues of the cities, paid into the treasury of the Grand Khan, is appropriated to the maintenance of these garrisons When it happens that a city is in a state of rebellion (and it is not an uncommon occurrence for these people, actuated by some sudden exasperation, or when intoxicated, to murder their governors), a part of the garrison of a neighbouring city is immediately despatched with orders to destroy the place where such guilty excesses have been committed; whereas it would be a tedious operation to send an army from another province, that might be two months on its march. For such purposes, the city of Kin-sai constantly supports a garrison of thirty thousand soldiers; and the smallest number stationed at any place is one thousand.

It now remains to speak of a very fine palace that was formerly the residence of king Fanfur, whose ancestors enclosed with high walls an extent of ground ten miles in compass, and divided it into three parts. That in the centre was entered by a lofty portal, on each side of which was a magnificent colonnade, on a flat terrace, the roofs of which were supported by rows of pillars, highly ornamented with the most beautiful azure and gold. The colonnade opposite to the entrance, at the further side of the court, was still grander than the others, its roof being richly adorned, the pillars gilt, and the walls on the inner side ornamented with exquisite paintings, representing the histories of former kings. Here, annually, upon certain days consecrated to the service of their idols, king Fanfur was accustomed to hold his court, and to entertain at a feast his principal nobles, the chief magistrates, and the opulent citizens of Kin-sai. Under these colonnades might be seen, at one time, ten thousand persons suitably accommodated at table. This festival lasted ten or twelve days, and the magnificence displayed on the occasion, in silks, gold, and precious stones, exceeded all imagination; for every guest, with a spirit of emulation, endeavoured to exhibit as much finery as his circumstances would possibly allow. Behind the colonnade last mentioned, or that which fronted the grand portal, there was a wall, with a passage, that divided this exterior court of the palace from an interior court, which formed a kind of large cloister, with its rows of pillars sustaining a portico that surrounded it, and led to various apartments for the use of the king and queen. These pillars were ornamented in a similar manner, as were also the walls. From this cloister you entered a covered passage or corridor, six paces in width, and of such a length as to reach to the margin of the lake. On each side of this there were corresponding entrances to ten courts, in the form of long cloisters, surrounded by their porticoes, and each cloister or court had fifty apartments, with their respective gardens, the residence of a thousand young women, whom the king retained in his service. Accompanied sometimes by his queen, and on other occasions by a party of these females, it was his custom to take amusement on the lake, in barges covered with silk, and to visit the idol temples on its borders. The other two divisions of this seraglio were laid out in groves, pieces of water, beautiful gardens stored with fruit-trees, and also enclosures for all sorts of animals that are the objects of sport, such as antelopes, deer, stags, hares, and rabbits. Here likewise the king amused himself, in company with his damsels, some in carriages and some on horseback. No male person was allowed to be of these parties, but on the other hand, the females were practised in the art of coursing with dogs, and pursuing the animals that have been mentioned. When fatigued with these exercises, they retired into the groves on the banks of the lake, and there quitting their dresses, rushed into the water in a state of nudity, sportively swimming about, some in one direction and some in another, whilst the king remained a spectator of the exhibition. After this they returned to the palace. Sometimes he ordered his repast to be provided in one of these groves, where the foliage of lofty trees afforded a thick shade, and was there waited upon by the same damsels. Thus was his time consumed amidst the enervating charms of his women, and in profound ignorance of whatever related to martial concerns, the consequence of which was, that his depraved habits and his pusillanimity enabled the Grand Khan to deprive him of his splendid possessions, and to expel him with ignominy from his throne, as has been already stated All these particulars were communicated to me, when I was in that city, by a rich merchant of Kin-sai, then very old, who had been a confidential servant of king Fanfur, and was acquainted with every circumstance of his life. Having known the palace in its original state, he was desirous of conducting me to view it. Being at present the residence of the Grand Khan's viceroy, the colonnades are preserved in the style in which they had formerly subsisted, but the chambers of the females had been suffered to go to ruin, and the foundations only were visible. The wall likewise that enclosed the park and gardens was fallen to decay, and neither animals nor trees were any longer to be found there.

At the distance of twenty-five miles [F. has "fifteene myles"] from this city, in a direction to the northward of east, lies the sea, near to which is a town named Gan-pu [F. has "Ganfu" and Y. "Ganfu"], where there is an extremely fine port, frequented by all the ships that bring merchandise from India The river that flows past the city of Kin-sai forms this port, at the place where it falls into the sea. Boats are continually employed in the conveyance of goods up and down the river, and those intended for exportation are there put on board of ships bound to various parts of India and of Cathay.

Marco Polo, happening to be in the city of Kin-sai at the time of making the annual report to His Majesty's commissioners of the amount of revenue and the number of inhabitants, had an opportunity of observing that the latter were registered at one hundred and sixty *tomans* of fire-places, that is to say, of families dwelling under the same roof, and as a *toman* is ten thousand, it follows that the whole city must have contained one million six hundred thousand families, amongst which multitude of people there was only one church of Nestorian Christians. Every father of a family, or housekeeper, is required to affix a writing to the door of his house, specifying the name of each individual of his family, whether male or female, as well as the number of his horses. When any person dies, or leaves the dwelling, the name is struck out, and upon the occasion of a birth, it is added to the list. By these means the great officers of the province and governors of the cities are at all times acquainted with the exact number of the inhabitants. The same regulation is observed throughout the province of Cathay as well as of Manji. In like manner, all the keepers of inns and public hotels inscribe in a book the names of those who take up their occasional abode with them, particularising the day and the hour of their arrival and departure; a copy of which is transmitted daily to those magistrates who have been spoken of as stationed in the market-squares It is a custom in the province of Manji, with the indigent class of the people, who are unable to support their families, to sell their children to the rich, in order that they may be fed and brought up in a better manner that their own poverty would admit.

26. *Ramusio* III, Ch. I. *Marsden*, pp. 565–567. *Wright*, pp. 347–349.
[Omitted in *Frampton* See Appendix I Note 380, p 237]

Of the Merchant Ships of the Indian Seas

HAVING treated, in the preceding parts of our work, of various provinces and regions, we shall now take leave of them, and proceed to the account of India, the admirable circumstances of which shall be related. We shall commence with a description of the ships employed by the merchants, which are built of fir-timber They have a single deck, and below this the space is divided into about sixty small cabins, fewer or more, according to the size of the vessels, each of them affording accommodation for one merchant They are provided with a good helm They have four masts, with as many sails, and some of them have two masts which can be set up and lowered again, as may be found necessary. Some ships of the larger class have, besides (the cabins), to the number of thirteen bulk-heads or divisions in the hold, formed of thick planks let into each other (*incastrati*, mortised or rabbeted) The object of these is to guard against accidents which may occasion the vessel to spring a leak, such as striking on a rock or receiving a stroke from a whale, a circumstance that not unfrequently occurs; for, when sailing at night, the motion through the waves causes a white foam that attracts the notice of the hungry animal. In expectation of meeting with food, it rushes violently to the spot, strikes the ship, and often forces in some part of the bottom. The water, running in at the place where the injury has been sustained, makes its way to the well, which is always kept clear. The crew, upon discovering the situation of the leak, immediately remove the goods from the division affected by the water, which, in consequence of the boards being so well fitted, cannot pass from one division to another. They then repair the damage, and return the goods to that place in the hold from whence they had been taken. The ships are all double-planked; that is, they have a course of sheathing-boards laid over the planking in every part These are caulked with oakum both withinside and without, and are fastened with iron nails. They are not coated with pitch, as the country does not produce that article, but the bottoms are smeared over with the following preparation. The people take quick-lime and hemp, which latter they cut small, and with these, when pounded together, they mix oil procured from a certain tree, making of the whole a kind of unguent, which retains its viscous properties more firmly, and is a better material than pitch.

Ships of the largest size require a crew of three hundred men; others, two hundred; and some, one hundred and fifty only, according to their greater or less bulk They carry from five to six thousand baskets (or mat bags) of pepper. In former times they were of greater burthen than they are at present; but the violence of the sea having in many places broken up the islands, and especially in some of the principal ports, there is a want of depth of water for vessels of such draught, and they have on that account been built, in latter times, of a smaller size. The vessels are likewise moved with oars or sweeps, each of which requires four men to work it Those of the larger class are accompanied by two or three large barks, capable of containing about one thousand baskets of pepper, and are manned with sixty, eighty, or one hundred sailors These small craft are often employed to tow the larger, when working their oars, or even under sail, provided the wind be on the quarter, but not when right aft, because, in that case, the sails of the larger vessel must becalm those of the smaller, which would, in consequence, be run down The ships also carry with them as many as ten small boats, for the purpose of carrying out anchors, for fishing, and a variety of other services They are slung over the sides, and lowered into the water when there is occasion to use them. The barks are in like manner provided with their small boats When a ship, having been on a voyage for a year or more, stands in need of repair, the practice is to give her a course of sheathing over the original boarding, forming a third course, which is caulked and paid in the same manner as the others; and this, when she needs further repairs, is repeated, even to the number of six layers, after which she is condemned as unserviceable and not sea-worthy. Having thus described the shipping, we shall proceed to the account of India; but in the first instance we shall speak of certain islands in the part of the ocean where we are at present, and shall commence with the island named Zipangu . . .

27. *Ramusio* III, Ch. XVI. *Marsden*, pp. 614, 615 *Wright*, pp. 374, 375.
[*Frampton*, Ch 113 (second half), p 105, Appendix I Note 410, p 244]

Of the Kingdom of Fanfur

FANFUR is a kingdom of the same island, governed by its own prince, where the people likewise worship idols, and profess obedience to the Grand Khan. In this part of the country a species of camphor, much superior in quality to any other, is produced It is named the camphor of Fanfur, and is sold for its weight in gold. There is not any wheat nor other corn, but the food of the inhabitants is rice, with milk, and the wine extracted from trees in the manner that has been described in the chapter respecting Samara. They have also a tree from which, by a singular process, they obtain a kind of meal. The stem is lofty, and as thick as can be grasped by two men. When from this the outer bark is stripped, the ligneous substance is found to be about three inches in thickness, and the central part is filled with pith, which yields a meal or flour, resembling that procured from the acorn. The pith is put into vessels filled with water, and is stirred about with a stick, in order that the fibres and other impurities may rise to the top, and the pure farinaceous part subside to the bottom When this has been done, the water is poured off, and the flour which remains, divested of all extraneous matter, is applied to use, by making it into cakes and various kinds of pastry Of this, which resembles barley bread in appearance and taste, Marco Polo has frequently eaten, and some of it he brought home with him to Venice. The wood of the tree, in thickness about three inches (as has been mentioned), may be compared to iron in this respect, that when thrown into water it immediately sinks It admits of being split in an even direction from one end to the other, like the bamboo cane Of this the natives make short lances: were they to be of any considerable length, their weight would render it impossible to carry or to use them. They are sharpened at one end, and rendered so hard by fire that they are capable of penetrating any sort of armour, and in many respects are preferable to iron. What we have said on the subject of this kingdom (one of the divisions of the island) is sufficient Of the other kingdoms composing the remaining part we shall not speak, because Marco Polo did not visit them.

28. *Ramusio* III, Ch. xx. *Marsden*, pp. 638–640, 647–648. *Wright*, pp 388–395.

[*Frampton*, Chs 116, 117, pp 107–109, Appendix I. Note 426, p 248]

Of the Province of Maabar

THE country produces no other grain than rice and sesamé. The people go to battle with lances and shields, but without clothing, and are a despicable unwar-like race. They do not kill cattle nor any kind of animals for food, but when desirous of eating the flesh of sheep or other beasts, or of birds, they procure the Saracens, who are not under the influence of the same laws and customs, to perform the office Both men and women wash their whole bodies in water twice every day, that is, in the morning and the evening. Until this ablution has taken place they neither eat nor drink; and the person who should neglect this observance, would be regarded as a heretic It ought to be noticed, that in eating they make use of the right hand only, nor do they ever touch their food with the left. For every cleanly and delicate work they employ the former, and reserve the latter for the base uses of personal abstersion, and other offices connected with the animal functions. They drink out of a particular kind of vessel, and each individual from his own, never making use of the drinking pot of another person. When they drink they do not apply the vessel to the mouth, but hold it above the head, and pour the liquor into the mouth, not suffering the vessel on any account to touch the lips In giving drink to a stranger, they do not hand their vessel to him, but, if he is not provided with one of his own, pour the wine or other liquor into his hands, from which he drinks it, as from a cup.

Offences in this country are punished with strict and exemplary justice, and with regard to debtors the following customs prevail. If application for payment shall have been repeatedly made by a creditor, and the debtor puts him off from time to time with fallacious promises, the former may attach his person by drawing a circle round him, from whence he dares not depart until he has satisfied his creditor, either by payment, or by giving adequate security. Should he attempt to make his escape, he renders himself liable to the punishment of death, as a violator of the rules of justice. Messer Marco, when he was in this country on his return homeward, happened to be an eye-witness of a remarkable transaction of this nature. The king was indebted in a sum of money to a certain foreign merchant, and although frequently importuned for payment, amused him for a long time with vain assurances. One day when the king was riding on horseback, the merchant took the opportunity of describing a circle round him and his horse. As soon as the king perceived what had been done, he immediately ceased to proceed, nor did he move from the spot until the demand of the merchant was fully satisfied. The bystanders beheld what passed with admiration, and pronounced that king to merit the title of most just, who himself submitted to the laws of justice. [See *Ocean of Story,* Vol. III. p. 201 *et seq.*]

These people abstain from drinking wine made from grapes, and should a person be detected in the practice, so disreputable would it be held, that his evidence would not be received in court. A similar prejudice exists against persons frequenting the sea, who, they observe, can only be people of desperate fortunes, and whose testimony, as such, ought not to be admitted They do not hold fornication to be a crime. The heat of the country is excessive, and the inhabitants on that account go naked. There is no rain excepting in the months of June, July, and August, and if it was not for the coolness imparted to the air during these three months by the rain, it would be impossible to support life.

In this country there are many adepts in the science denominated physiognomy, which teaches the knowledge of the nature and qualities of men, and whether they tend to good or evil. These qualities are immediately discerned upon the appearance of the man or woman. They also know what events are portended by meeting certain beasts or birds. More attention is paid by these people to the flight of birds than by any others in the world, and from thence they predict good or bad fortune. In every day of the week there is one hour which they regard as unlucky, and this they name *choiach*; thus, for example, on Monday the (canonical) hour of *mi-tierce*, on Tuesday the hour of *tierce*, on Wednesday the hour of *none*; and on these hours they do not make purchases, nor transact any kind of business, being persuaded that it would not be attended with success. In like manner they ascertain the qualities of every day throughout the year, which are described and noted in their books. They judge of the hour of the day by the length of a man's shadow when he stands erect When an infant is born, be it a boy or a girl, the father or the mother makes a memorandum in writing of the day of the week on which the birth took place, also of the age of the moon, the name of the month, and the hour. This is done because every future act of their lives is regulated by astrology As soon as a son attains the age of thirteen years, they set him at liberty, and no longer suffer him to be an inmate in his father's house; giving him to the amount in their money, of twenty to twenty-four groats. Thus provided, they consider him as capable of gaining his own livelihood, by engaging in some kind of trade and thence deriving a profit. These boys never cease to run about in all directions during the whole course of the day, buying an article in one place, and selling it in another. At the season when the pearl fishery is going on, they frequent the beach, and make purchases from the fishermen or others, of five, six, or more (small) pearls, according to their means, carrying them afterwards to the merchants, who, on account of the heat of the sun, remain sitting in their houses, and to whom they say: "These pearls have cost us so much; pray allow such a profit on them as you may judge reasonable." The merchants then give something beyond the price at which they had been obtained In this way likewise they deal in many other articles, and become excellent and most acute traders. When business is over for the day, they carry to their mothers the provisions necessary for their dinners, which they prepare and dress for them; but these never eat anything at their fathers' expense.

Not only in this kingdom, but throughout India in general, all the beasts and birds are unlike those of our own country, excepting the quails, which perfectly resemble ours; the others are all different There are bats as large as vultures, and vultures as black as crows, and much larger than ours Their flight is rapid, and they do not fail to seize their bird.

In their temples there are many idols, the forms of which represent them of the male and the female sex, and to these, fathers and mothers dedicate their daughters Having been so dedicated, they are expected to attend whenever the priests of the convent require them to contribute to the gratification of the idol; and on such occasions they repair thither, singing and playing on instruments, and adding by their presence to the festivity .These young women are very numerous, and form large bands. Several times in the week they carry an offering of victuals to the idol to whose service they are devoted, and of this food they say the idol partakes. A table for the purpose is placed before it, and upon this the victuals are suffered to remain for the space of a full hour, during which the damsels never cease to sing, and play, and exhibit wanton gestures This lasts as long as a person of condition would require for making a convenient meal. They then declare that the spirit of the idol is content with its share of the entertainment provided, and, ranging themselves around it, they proceed to eat in their turn, after which they repair to their respective homes The reason given for assembling the young women, and performing the ceremonies that have been described, is this.—The priests declare that the male divinity is out of humour with and incensed against the female, refusing to have connexion or even to converse with her, and that if some measure were not adopted to restore peace and harmony between them, all the concerns of the monastery would go to ruin, as the grace and blessing of the divinities would be withheld from them. For this purpose it is, they expect the votaries to appear in a state of nudity, with only a cloth round their waists, and in that state to chaunt hymns to the god and goddess. These people believe that the former often solaces himself with the latter. [See *Ocean of Story*, Vol. 1 pp. 231–269]

The natives make use of a kind of bedstead, or cot, of very light cane-work, so ingeniously contrived that when they repose on them, and are inclined to sleep, they can draw close the curtains about them by pulling a string This they do in order to exclude the tarantulas, which bite grievously, as well as to prevent their being annoyed by fleas and other small vermin; whilst at the same time the air, so necessary for mitigating the excessive heat, is not excluded Indulgences of this nature, however, are enjoyed only by persons of rank and fortune; others of the inferior class lie in the open streets. [A large portion of the above is also found in *Z*; see B. pp. 182–185.]

29. *Ramusio* III, Ch. XXIII. *Marsden*, pp. 669, 670. *Wright*, pp. 405–408. [*Frampton*, Ch 115, p 106 See Appendix I Note 435, p 250]

Of the island of Zeilan (Ceylon)

I AM unwilling to pass over certain particulars which I omitted when before speaking of the island of Zeilan, and which I learned when I visited that country in my homeward voyage In this island there is a very high mountain, so rocky and precipitous that the ascent to the top is impracticable, as it is said, excepting by the assistance of iron chains employed for that purpose By means of these some persons attain the summit, where the tomb of Adam, our first parent, is reported to be found Such is the account given by the Saracens. But the idolaters assert that it contains the body of Sogomon-barchan, the founder of their religious system, and whom they revere as a holy personage He was the son of a king of the island, who devoted himself to an ascetic life, refusing to accept of kingdoms or any other worldly possessions, although his father endeavoured, by the allurements of women, and every other imaginable gratification, to divert him from the resolution he had adopted. Every attempt to dissuade him was in vain, and the young man fled privately to this lofty mountain, where, in the observance of celibacy and strict abstinence, he at length terminated his mortal career. By the idolaters he is regarded as a saint. The father, distracted with the most poignant grief, caused an image to be formed of gold and precious stones, bearing the resemblance of his son, and required that all the inhabitants of the island should honour and worship it as a deity Such was the origin of the worship of idols in that country; but Sogomon-barchan is still regarded as superior to every other. In consequence of this belief, people flock from various distant parts in pilgrimage to the mountain on which he was buried. Some of his hair, his teeth, and the basin he made use of, are still preserved, and shown with much ceremony. The Saracens, on the other hand, maintain that these belonged to the prophet Adam, and are in like manner led by devotion to visit the mountain.

It happened that, in the year 1281, the Grand Khan heard from certain Saracens who had been upon the spot, the fame of these relics belonging to our first parent, and felt so strong a desire to possess them, that he was induced to send an embassy to demand them of the king of Zeilan After a long and tedious journey, his ambassadors at length reached the place of their destination, and obtained from the king two large back-teeth, together with some of the hair, and a handsome vessel of porphyry. When the Grand Khan received intelligence of the approach of the messengers, on their return with such valuable curiosities, he ordered all the people of Kanbalu to march out of the city to meet them, and they were conducted to his presence with great pomp and solemnity Having mentioned these particulars respecting the mountain of Zeilan, we shall return to the kingdom of Maabar, and speak of the city of Kael.

30. *Ramusio* III, Ch. xxiv. *Marsden*, pp. 674, 675. *Wright*, pp. 408–410.
[Omitted in *Frampton* See Appendix I Note 435, p. 250]

Of the City of Kael

KAEL is a considerable city, governed by Astiar, one of the four brothers, kings of the country of Maabar, who is rich in gold and jewels, and preserves his country in a state of profound peace. On this account it is a favourite place of resort for foreign merchants, who are well received and treated by the king Accordingly all the ships coming from the west—as from Ormus, Chisti, Adem, and various parts of Arabia—laden with merchandise and horses, make this port, which is besides well situated for commerce. The prince maintains in the most splendid manner not fewer than three hundred women.

All the people of this city, as well as the natives of India in general, are addicted to the custom of having continually in their mouths the leaf called *tembul*; which they do, partly from habit, and partly from the gratification it affords. Upon chewing it, they spit out the saliva to which it gives occasion. Persons of rank have the leaf prepared with camphor and other aromatic drugs, and also with a mixture of quick lime. I have been told that it is extremely conducive to health. If it is an object with any man to affront another in the grossest and most contemptuous manner, he spits the juice of this masticated leaf in his face. Thus insulted, the injured party hastens to the presence of the king, states the circumstances of his grievance, and declares his willingness to decide the quarrel by combat. The king thereupon furnishes them with arms, consisting of a sword and small shield; and all the people assemble to be spectators of the conflict, which lasts till one of them remains dead on the field They are, however, forbidden to wound with the point of the sword. [See *Ocean of Story*, Vol. VIII, pp. 237–319.]

31. *Ramusio* III, Chs. XLI, XLII, XLIII and XLIV. *Marsden*, pp. 728, 729, 733, 734, 735, 737–739. *Wright*, pp. 440–449.
[Omitted in *Frampton*. See Appendix I Note 465, p 256]

The Ramusian chapter-headings are given consecutively:

CHAPTER 41

Of the City of Escier

THE ruler of this city is a Mahometan, who governs it with exemplary justice, under the superior authority of the sultan of Aden. Its distance from thence is about forty miles to the south-east Subordinate to it there are many towns and castles. Its port is good, and it is visited by many trading ships from India, which carry back a number of excellent horses, highly esteemed in that country, and sold there at considerable prices.

This district produces a large quantity of white frankincense of the first quality, which distils, drop by drop, from a certain small tree that resembles the fir. The people occasionally tap the tree, or pare away the bark, and from the incision the frankincense gradually exudes, which afterwards becomes hard Even when an incision is not made, an exudation is perceived to take place, in consequence of the excessive heat of the climate. There are also many palm-trees, which produce good dates in abundance. No grain excepting rice and millet is cultivated in this country, and it becomes necessary to obtain supplies from other parts. There is no wine made from grapes; but they prepare a liquor from rice, sugar, and dates, that is a delicious beverage. They have a small breed of sheep, the ears of which are not situated like those in others of the species, two small horns growing in the place of them, and lower down, towards the nose, there are two orifices that serve the purpose of ears.

These people are great fishermen, and catch the tunny in such numbers, that two may be purchased for a Venetian groat. They dry them in the sun; and as, by reason of the extreme heat, the country is in a manner burnt up, and no sort of vegetable is to be seen, they accustom their cattle, cows, sheep, camels, and horses, to feed upon dried fish, which being regularly served to them, they eat without any signs of dislike. The fish used for this purpose are of a small kind, which they take in vast quantities during the months of March, April, and May; and when dried, they lay up in their houses for the food of their cattle. These will also feed upon the fresh fish, but are more accustomed to eat them in the dried state. In consequence also of the scarcity of grain, the natives make a kind of biscuit of the substance of the larger fish, in the following manner: they chop it into very small particles, and moisten the preparation with a liquor rendered thick and adhesive by a mixture of flour, which gives to the whole the consistence of paste. This they form into a kind of bread, which they dry and harden by exposure to a burning sun. A stock of this biscuit is laid up to serve them for the year's consumption. The frankincense before mentioned is so cheap in the country as to be purchased by the governor at the rate of ten besants (gold ducats) the quintal, who sells it again to the merchants at forty besants. This he does under the direction of the soldan of Aden, who monopolises all that is produced in the district at the above price, and derives a large profit from the re-sale. Nothing further presenting itself at this place, we shall now speak of the city of Dulfar.

CHAPTER 42

Of the City of Dulfar

Dulfar is a large and respectable city or town, at the distance of twenty miles from Escier, in a south-easterly direction. Its inhabitants are Mahometans, and its ruler also is a subject of the soldan of Aden. This place lies near the sea, and has a good port, frequented by many ships Numbers of Arabian horses are collected here from the inland country, which the merchants buy up and carry to India, where they gain considerably by disposing of them. Frankincense is likewise produced here, and purchased by the merchants Dulfar has other towns and castles under its jurisdiction. We shall now speak of the gulf of Kalayati.

CHAPTER 43

Of the City of Kalayati

Kalayati is a large town situated near a gulf which has the name of Kalatu, distant from Dulfar about fifty miles towards the south-east. The people are followers of the law of Mahomet, and are subjects to the melik of Ormus, who, when he is attacked and hard pressed by another power, has recourse to the protection afforded by this city, which is so strong in itself, and so advantageously situated, that it has never yet been taken by an enemy. The country around it not yielding any kind of grain, it is imported from other districts Its harbour is good, and many trading ships arrive there from India, which sell their piece-goods and spiceries to great advantage, the demand being considerable for the supply of towns and castles lying at a distance from the coast. These likewise carry away freights of horses, which they sell advantageously in India.

The fortress is so situated at the entrance of the gulf of Kalatu, that no vessel can come in or depart without its permission. Occasionally it happens that the melik of this city, who is under certain engagements with, and is tributary to the king of Kermain, throws off his allegiance in consequence of the latter's imposing some unusual contribution. Upon his refusing to pay the demand, and an army being sent to compel him, he departs from Ormus, and makes his stand at Kalayati, where he has it in his power to prevent any ship from entering or sailing. By this obstruction of the trade the king of Kermain is deprived of his duties, and being thereby much injured in his revenue, is constrained to accommodate the dispute with the melik The strong castle at this place constitutes, as it were, the key, not only of the gulf, but also of the sea itself, as from thence the ships that pass can at all times be discovered. The inhabitants in general of this country subsist upon dates and upon fish, either fresh or salted, having constantly a large supply of both; but persons of rank, and those who can afford it, obtain corn for their use from other parts. Upon leaving Kalavati, and proceeding three hundred miles towards the north-east, you reach the island of Ormus.

CHAPTER 44

Of Ormus

Upon the island of Ormus there is a handsome and large city, built close to the sea. It is governed by a melik, which is a title equivalent to that of lord of the marches with us, and he has many towns and castles under his authority. The inhabitants are Saracens, all of them professing the faith of Mahomet The heat that reigns here is extreme; but in every house they are provided with ventilators, by means of which they introduce air to the different floors, and into every apartment, at pleasure. Without this resource it would be impossible to live in the place. We shall not now say more of this city, as in a former book we have given an account of it, together with Kisi and Kerman.

Having thus treated sufficiently at length of those provinces and cities of the Greater India which are situated near the sea-coast, as well as of some of the countries of Ethiopia, termed the Middle India, I shall now, before I bring the work to a conclusion, step back, in order to notice some regions lying towards the north, which I omitted to speak of in the preceding books.

It should be known, therefore, that in the northern parts of the world there dwell many Tartars, under a chief of the name of Kaidu, who is of the race of Jengiz-khan, and nearly related to Kublai, the Grand Khan He is not the subject of any other prince. The people observe the usages and manners of their ancestors, and are regarded as genuine Tartars. These Tartars are idolators, and worship a god whom they call Naagai, that is, the god of earth, because they think and believe that this their god has dominion over the earth, and over all things that are born of it; and to this their false god they make idols and images of felt, as is described in a former book Their king and his armies do not shut themselves up in castles or strong places, nor even in towns; but at all times remain in the open plains, the valleys, or the woods, with which this region abounds. They have no corn of any kind, but subsist upon flesh and milk, and live amongst each other in perfect harmony; their king, to whom they all pay implicit obedience, having no object dearer to him than that of preserving peace and union amongst his subjects, which is the essential duty of a sovereign. They possess vast herds of horses, cows, sheep, and other domestic animals. In these northern districts are found bears of a white colour, and of prodigious size, being for the most part about twenty spans in length. There are foxes also whose furs are entirely black, wild asses in great numbers, and certain small animals named rondes, which have most delicate furs, and by our people are called zibelines or sables. Besides these there are various small beasts of the marten or weasel kind, and those which bear the name of Pharaoh's mice The swarms of the latter are incredible; but the Tartars employ such ingenious contrivances for catching them, that none can escape their hands.

In order to reach the country inhabited by these people, it is necessary to perform a journey of fourteen days across a wide plain, entirely uninhabited and desert—a state that is occasioned by innumerable collections of water and springs, that render it an entire marsh. This, in consequence of the long duration of the cold season, is frozen over, excepting for a few months of the year, when the sun dissolves the ice, and turns the soil to mud, over which it is more difficult and fatiguing to travel than when the whole is frozen For the purpose, however, of enabling the merchants to frequent their country, and purchase their furs, in which all their trade consists, these people have exerted themselves to render the marshy desert passable for travellers, by erecting at the end of each day's stage a wooden house, raised some height above the ground, where persons are stationed, whose business it is to receive and accommodate the merchants, and on the following day to conduct them to the next station of this kind; and thus they proceed from stage to stage, until they have effected the passage of the desert. In order to travel over the frozen surface of the ground, they construct a sort of vehicle, not unlike that made use of by the natives of the steep and almost inaccessible mountains in the vicinity of our own country, and which is termed a *tragula* or sledge. It is without wheels, is flat at bottom, but rises with a semi-circular curve in front, by which construction it is fitted for running easily upon the ice For drawing these small carriages they keep in readiness certain animals resembling dogs, and which may be called such, although they approach to the size of asses. They are very strong and inured to the draught. Six of them, in couples, are harnessed to each carriage, which contains only the driver who manages the dogs, and one merchant, with his package of goods. When the day's journey has been performed he quits it, together with that set of dogs, and thus changing both, from day to day, he at length accomplishes his journey across the desert, and afterwards carries with him (in his return) the furs that find their way, for sale, to our part of the world

32. The following chapters are taken from Wright's edition of *Marsden*, pp. 453–471, where they form Chs. XLVII–LXXI. They are not in *Ramusio*, but are found in the French editions.

Pauthier and Yule made them into Book IV, while in *Benedetto* they occupy chapters CC–CCXXXIV. The "Conclusion" is from the Crusca version (see Y. Vol. II, pp. 500, 501).

CHAPTER 47

Of Great Turkey

IN Great Turkey there is a king called Kaidu, who is the nephew of the Grand Khan, for he was son of the son of Ciagatai, who was brother to the Grand Khan He possesses many cities and castles, and is a very great lord. He is Tartar, and his men also are Tartar, and they are good warriors, which is no wonder, for they are all men brought up to war; and I tell you that this Kaidu never gave obedience to the Grand Khan, without first making great war. And you must know that this Great Turkey lies to the north-west when we leave Ormus, by the way already mentioned. Great Turkey is beyond the river Ion, and stretches out northward to the territory of the Grand Khan This Kaidu has already fought many battles with the people of the Grand Khan, and I will relate to you how he came to quarrel with him You must know for a truth that Kaidu sent word one day to the Grand Khan that he wanted his part of what they had obtained by conquest, claiming a part of the province of Cathay and of that of Manji The Grand Khan told him that he was quite willing to give him his share, as he had done to his other sons, if he, on his part, would repair to his court and attend his council as often as he sent for him; and the Grand Khan willed further, that he should obey him like the others his sons and his barons; and on this condition the Grand Khan said that he would give him part of their conquest (of China). Kaidu, who distrusted his uncle the Grand Khan, rejected this condition, saying that he was willing to yield him obedience in his own country, but that he would not go to his court for any consideration, as he feared lest he should be put to death. Thus originated the quarrel between the Grand Khan and Kaidu, which led to a great war, and there were many great battles between them. And the Grand Khan posted an army round the kingdom of Kaidu, to prevent him or his people from committing any injury to his territory or people. But, in spite of all these precautions of the Grand Khan, Kaidu invaded his territory, and fought many times with the forces sent to oppose him. Now king Kaidu, by exerting himself, could bring into the field a hundred thousand horsemen, all good men, and well trained to war and battle. And moreover he has with him many barons of the lineage of the emperor, that is of Jengis-khan, who was the founder of the empire. We will now proceed to narrate certain battles between Kaidu and the Grand Khan's people; but first we will describe their mode of fighting. When they go to war, each is obliged to carry with him sixty arrows, thirty of which are of a smaller size, intended for shooting at a distance, but the other thirty are larger, and have a broad blade; these they use near at hand, and strike their enemies in the faces and arms, and cut the strings of their bows, and do great damage with them. And when they have discharged all their arrows, they take their swords and maces, and give one another heavy blows with them.

In the year 1266, this king Kaidu, with his cousins, one of whom was called Jesudar, assembled a vast number of people, and attacked two of the Grand Khan's barons, who also were cousins of king Kaidu, though they held their lands of the Grand Khan One of these was named Tibai or Ciban They were sons of Ciagatai, who had received Christian baptism, and was own brother to the Grand Khan Kublaï. Well, Kaidu with his people fought with these his two cousins, who also had a great army, for on both sides there were about a hundred thousand horsemen. They fought very hard together, and there were many slain on both sides; but at last king Kaidu gained the victory, and did great damage to the others But the two brothers, the cousins of king Kaidu, escaped without hurt, for they had good horses, which bore them away with great swiftness Having thus gained the victory, Kaidu's pride and arrogance increased; and he returned into his own country, where he remained full two years in peace, without any hostilities between him and the Grand Khan. But at the end of two years Kaidu again assembled a great army. He knew that the Grand Khan's son, named Nomogan, was at Caracorum, and that with him was George the grandson of Prester John, which two barons had also a very great army of horsemen. King Kaidu, having assembled his host, marched from his own country, and, without any occurrence worth mentioning, arrived in the neighbourhood of Caracorum, where the two barons, the son of the Grand Khan and the grandson of Prester John, were with their army. The latter, instead of being frightened, prepared to meet them with the utmost ardour and courage; and having assembled their whole army, which consisted of not less than sixty thousand horsemen, they marched out and established their camp very well and orderly at a distance of about ten miles from king Kaidu, who was encamped with his men in the same plain. Each party remained in their camp till the third day, preparing for battle in the best way they could, for their numbers were about equal, neither exceeding sixty thousand horsemen, well armed with bows and arrows, and a sword, mace, and shield to each. Both armies were divided into six squadrons of ten thousand men each, and each having its commander. And when the two armies were drawn up in the field, and waited only for the signal to be given by sounding the nacar, they sang and sounded their instruments of music in such a manner that it was wonderful to hear For the Tartars are not allowed to commence a battle till they hear the nacars of their lord begin to sound, but the moment it sounds they begin to fight; and it is their custom, while thus waiting the signal of battle, to sing and sound their two-corded instruments very sweetly, and make great solace. As soon as the sound of the nacars was heard, the battle began, and they put their hands to their bows, and placed the arrows to the strings. In an instant the air was filled with arrows like rain, and you might see many a man and many a horse struck down dead, and the shouting and the noise of the battle was so great, that one could hardly have heard God's thunder. In truth, they fought like mortal enemies. And truly, as long as they had any arrows left, those who were able ceased not to shoot, but so many were slain and mortally wounded, that the battle commenced propitiously for neither party. And when they had exhausted their arrows, they placed the bows in their cases, and seized their swords and maces, and, rushing upon each other, began to give terrible blows with them Thus they began a very fierce and dreadful battle, with such execution upon each other, that the ground was soon covered with corpses. Kaidu especially performed great feats of arms, and but for his personal prowess, which restored courage to his followers, they were several times nearly defeated. And on the other side, the son of the Grand Khan and the grandson of Prester John also behaved themselves with great bravery In a word, this was one of the most sanguinary battles that had ever taken place among the Tartars; for it lasted till nightfall; and in spite of all their efforts, neither party could drive the other from the field, which was covered with so many corpses that it was pity to see, and many a lady that day was made a widow, and many a child an orphan. And when the sun set, both parties gave over fighting, and returned to their several camps to repose during the night. Next morning, king Kaidu, who had received information that the Grand Khan had sent a very powerful army against him, put his men under arms at daybreak, and, all having mounted, he ordered them to proceed homewards. Their opponents were so weary with the previous day's battle, that they made no attempt to follow them, but let them go without molestation. Kaidu's men continued their retreat, until they came to Samarcand, in Great Turkey.

CHAPTER 48

What the Grand Khan said of the injuries done to him by Kaidu

Now the Grand Khan was greatly enraged against Kaidu, who was always doing so much injury to his people and his territory, and he said in himself, that if he had not been his nephew, he should not have escaped an evil death But his feelings of relationship hindered him from destroying him and his land, and thus Kaidu escaped from the hands of the Grand Khan. We will now leave this matter, and we will tell you a strange history of king Kaidu's daughter.

CHAPTER 49

Of the daughter of King Kaidu, how strong and valiant she was

You must know, then, that king Kaidu had a daughter named, in the Tartar language, Aigiarm, which means shining moon. This damsel was so strong, that there was no young man in the whole kingdom who could overcome her, but she vanquished them all. Her father the king wished to marry her; but she declined, saying, that she would never take a husband till she met with some gentleman who should conquer her by force, upon which the king, her father, gave her a written promise that she might marry at her own will. She now caused it to be proclaimed in different parts of the world, that if any young man would come and try strength with her, and should overcome her by force, she would accept him for her husband. This proclamation was no sooner made, than many came from all parts to try their fortune. The trial was made with great solemnity. The king took his place in the principal hall of the palace, with a large company of men and women; then came the king's daughter, in a dress of cendal, very richly adorned, into the middle of the hall; and next came the young man, also in a dress of cendal. The agreement was, that if the young man overcame her so as to throw her by force to the ground, he was to have her for wife; but if, on the contrary, he should be overcome by the king's daughter, he was to forfeit to her a hundred horses. In this manner the damsel gained more than ten thousand horses, for she could meet with no one able to conquer her, which was no wonder, for she was so well-made in all her limbs, and so tall and strongly built, that she might almost be taken for a giantess At last, about the year 1280, there came the son of a rich king, who was very beautiful and young; he was accompanied with a very fine retinue, and brought with him a thousand beautiful horses Immediately on his arrival, he announced that he was come to try his strength with the lady. King Kaidu received him very gladly, for he was very desirous to have this youth for his son-in-law, knowing him to be the son of the king of Pamar; on which account, Kaidu privately told his daughter that he wished her on this occasion to let herself be vanquished But she said she would not do so for anything in the world. Thereupon the king and queen took their places in the hall, with a great attendance of both sexes, and the king's daughter presented herself as usual, and also the king's son, who was remarkable no less for his beauty than for his great strength. Now when they were brought into the hall, it was, on account of the superior rank of the claimant, agreed as the conditions of the trial, that if the young prince were conquered, he should forfeit the thousand horses he had brought with him as his stake. This agreement having been made, the wrestling began; and all who were there, including the king and queen, wished heartily that the prince might be the victor, that he might be the husband of the princess. But, contrary to their hopes, after much pulling and tugging, the king's daughter gained the victory, and the young prince was thrown on the pavement of the palace, and lost his thousand horses. There was not one person in the whole hall who did not lament his defeat. After this the king took his daughter with him into many battles, and not a cavalier in the host displayed so much valour; and at last the damsel rushed into the midst of the enemy, and seizing upon a horseman, carried him off to her own people We will now quit this episode, and proceed to relate a great battle which fell out between Kaidu and Argon, the son of Abaga the lord of the east [For the amazon, see Chauvin, *Bib. des Ouvrages Arabes*, VI, p. 112, VIII, p. 55; and Clouston, *Book of Sindibād*, pp. 322 *et seq.*]

CHAPTER 50

How Abaga sent Argon his son with an army

Now Abaga, the lord of the east, held many provinces and many lands, which bordered on the territory of king Kaidu, on the side towards the tree which is called in the book of Alexander, *Arbor Secco*. And Abaga, in consequence of the damages done to his lands by king Kaidu, sent his son Argon with a very great number of horsemen into the country of Arbor Secco, as far as the river Ion, where they remained to protect the country against king Kaidu's people. In this manner Argon and his men remained in the plain of the Arbor Secco, and garrisoned many cities and castles thereabouts Thereupon king Kaidu assembled a great number of horsemen, and gave the command of them to his brother Barac, a prudent and brave man, with orders to fight Argon. Barac promised to fulfil his commandment, and to do his best against Argon and his army; and he marched with his army, which was a very numerous one, and proceeded for many days without meeting with any accident worth mentioning, till he reached the river Ion, where he was only ten miles distant from the army of Argon. Both sides immediately prepared for battle, and in a very fierce engagement, which took place three days afterwards, the army of Barac was overpowered, and pursued with great slaughter over the river.

CHAPTER 51

How Argon succeeded his father in the sovereignty

Soon after this victory, Argon received intelligence that his father Abaga was dead, for which he was very sorrowful, and he set out with all his host on his way to his father's court, a distance of forty days' journey, in order to receive the sovereignty. Now Abaga had a brother named Acomat Soldan, who had become a Saracen, and who no sooner heard of his brother Abaga's death, than he formed the design of seizing the succession for himself, considering that Argon was at too great a distance to prevent him. He therefore collected a powerful army, went direct to the court of his brother Abaga, and seized upon the sovereignty. There he found such an immense quantity of treasure as could hardly be believed, and by distributing this very lavishly among Abaga's barons and knights, he gained so far upon their hearts, that they declared they would have no other lord but him Moreover, Acomat Soldan showed himself a very good lord, and made himself beloved by everybody. But he had not long enjoyed his usurped power, when news came that Argon was approaching with a very great host. Acomat showed no alarm, but courageously summoned his barons and others, and within a week he had assembled a vast number of cavalry, who all declared that they were ready to march against Argon, and that they desired nothing more than to take him and put him to death.

CHAPTER 52

How Acomat went with his host to fight Argon

When Acomat Soldan had collected full sixty thousand horsemen, he set out on his way to encounter Argon and his people, and at the end of ten days' march he halted, having received intelligence that the enemy was only five days' march from him, and equal in number to his own army. Then Acomat established his camp in a very great and fair plain, and announced his intention of awaiting his enemy there, as a favourable place for giving battle. As soon as he arranged his camp, he called together his people, and addressed them as follows: "Lords," said he, "you know well how I ought to be liege lord of all which my brother Abaga held, because I was the son of his father, and I assisted in the conquest of all the lands and territories we possess. It is true that Argon was the son of my brother Abaga, and that some pretend that the succession would go of right to him; but, with all respect to those who hold this opinion, I say that they are in the wrong, for as his father held the whole of so great a lordship, it is but just that I should have it after his death, who ought rightly to have had half of it during his life, though by my generosity he was allowed to retain the whole. But since it is as I tell you, pray, let us defend our right against Argon, that the kingdom and lordship may remain to us all; for I assure you that all I desire for myself is the honour and renown, while you have the profit and the goods and lordships through all our lands and provinces. I will say no more, for I know that you are wise men and love justice, and that you will act for the honour and good of us all." When he had ended, all the barons, and knights, and others who were there, replied with one accord that they would not desert him as long as they had life in their bodies, and that they would aid him against all men whatever, and especially against Argon, adding that they feared not but they should take him and deliver him into his hands. After this, Acomat and his army remained in their camp, waiting the approach of the enemy.

CHAPTER 53

How Argon held council with his Barons before encountering Acomat

To return to Argon, as soon as he received certain intelligence of the movements of Acomat, and knew that he was encamped with so large an army, he was greatly affected, but he thought it wise to show courage and ardour before his men Having called all his barons and wise counsellors into his tent, for he was encamped also in a very far spot, he addressed them as follows: "Fair brothers and friends," said he, "you know well how tenderly my father loved you; while alive he treated you as brothers and sons, and you know in how many battles you were with him, and how you helped him to conquer the land he possessed. You know, too, that I am the son of him who loved you so much, and I myself love you as though you were my own body. It is just and right, therefore, that you aid me against him who comes contrary to justice and right to disinherit us of our land. And you know further how he is not of our law, but that he has abandoned it, and has become a Saracen and worships Mahomet, and it would ill become us to let Saracens have lordship over Tartars. Now, fair brethren and friends, all these reasons ought to give you courage and will to do your utmost to prevent such an occurrence, wherefore I implore each of you to show himself a valiant man, and to put forth all his ardour that we may conquer in the battle, and that the sovereignty may belong to you and not to Saracens. And truly every one ought to reckon on victory, since justice is on our side, and our enemies are in the wrong I will say no more, but again to implore every one of you to do his duty."

CHAPTER 54

How the Barons replied to Argon

When the barons and knights who were present had heard Argon's address, each resolved that he would prefer death in the battle to defeat; and while they stood silent, reflecting on his words, one of the great barons rose and spoke thus "Fair sir Argon, fair sir Argon," said he; "we know well that what you have said to us is the truth, and therefore I will be spokesman for all your men who are with you to fight this battle, and tell you openly that we will not fail you as long as we have life in our bodies, and that we would rather all die than not obtain the victory. We feel confident that we shall vanquish your enemies, on account of the justice of our cause, and the wrong which they have done; and therefore I counsel that we proceed at once against them, and I pray all our companions to acquit themselves in such a manner in this battle, that all the world shall talk of them " When this man had ended, all the others declared that they were of his opinion, and the whole army clamoured to be led against the enemy without delay. Accordingly, early next morning, Argon and his people began their march with very resolute hearts, and when they reached the extensive plain in which Acomat was encamped, they established their camp in good order at a distance of about ten miles from him. As soon as he had encamped, Argon sent two trusty messengers on a mission to his uncle.

CHAPTER 55

How Argon sent his messengers to Acomat

When these two trusty messengers, who were men of very advanced age, arrived at the enemy's camp, they dismounted at Acomat's tent, where he was attended by a great company of his barons, and having entered it, they saluted him courteously. Acomat, who knew them well, received them with the same courtesy, told them they were welcome, and made them sit down before him. After they had remained seated a short space, one of the messengers rose up on his feet and delivered his message as follows. "Fair sir Acomat," said he, "your nephew Argon wonders much at your conduct in taking from him his sovereignty, and now again in coming to engage him in mortal combat; truly this is not well, nor have you acted as a good uncle ought to act towards his nephew. Wherefore he informs you by us that he prays you gently, as that good uncle and father, that you restore him his right, so that there be no battle between you, and he will show you all honour, and you shall be lord of all his land under him This is the message which your nephew sends you by us."

CHAPTER 56

Acomat's reply to the message of Argon

When Acomat Soldan had heard the message of his nephew Argon, he replied as follows: " Sir Messenger," said he, "what my nephew says amounts to nothing, for the land is mine and not his; I conquered it as well as his father; and therefore tell my nephew that if he will, I will make him a great lord, and I will give him land enough, and he shall be as my son, and the highest in rank after me. And if he will not, you may assure him that I will do all in my power to put him to death. Now this is what I will do for my nephew, and no other thing or other arrangement shall you ever have from me." When Acomat had concluded, the messengers asked again, "Is this all the answer which we shall have?" "Yes," said he, "you shall have no other as long as I live." The messengers immediately departed, and riding as fast as they could to Argon's camp, dismounted at his tent and told him all that had passed. When Argon heard his uncle's message, he was so enraged, that he exclaimed in the hearing of all who were near him, "Since I have received such injury and insult from my uncle, I will never live or hold land if I do not take such vengeance that all the world shall talk of it!" After these words, he addressed his barons and knights. "Now we have nothing to do but to go forth as quickly as we can and put these faithless traitors to death; and it is my will that we attack them to-morrow morning, and do our utmost to destroy them." All that night they made preparations for battle; and Acomat Soldan, who knew well by his spies what were Argon's designs, prepared for battle also, and admonished his people to demean themselves with valour

CHAPTER 57

The battle between Argon and Acomat

Next morning, Argon, having called his men to arms and drawn them up skilfully in order of battle, addressed to them an encouraging admonition, after which they advanced towards the enemy. Acomat had done the same, and the two armies met on their way and engaged without further parley The battle began with a shower of arrows so thick that it seemed like rain from heaven, and you might see everywhere the riders cast from the horses, and the cries and groans of those who lay on the earth mortally wounded were dreadful to hear. When they had exhausted their arrows, they took to their swords and clubs, and the battle became so fierce and the noise so great that you could hardly have heard God's thunder. The slaughter was very great on both sides; but at last, though Argon himself displayed extraordinary valour, and set an example to all his men, it was in vain, for fortune turned against him, and his men were compelled to fly, closely pursued by Acomat and his men, who made great havoc of them And in the flight Argon himself was captured, upon which the pursuit was abandoned, and the victors returned to their camp and tents, glad beyond measure. Acomat caused his nephew, Argon, to be confined and closely guarded, and, being a man given to his pleasures, he returned to his court to enjoy the society of the fair ladies who were there, leaving the command of the army to a great melic, or chief, with strict orders to keep Argon closely guarded, and to follow him to court by short marches, so as not to fatigue his men.

CHAPTER 58

How Argon was liberated

Now it happened that a great Tartar baron, who was of great age, took pity on Argon, and said in himself that it was a great wickedness and disloyalty thus to hold their lord a prisoner, and that he would do his best to set him free. He began by persuading many other barons to adopt the same sentiments, and his personal influence, on account of his age and known character for justice and wisdom, was so great, that he easily gained them over to the enterprise, and they promised to be directed by him. The name of the leader of this enterprise was Boga, and the chief of his fellow-conspirators were named Elcidai, Togan, Tegana, Taga, Tiar Oulatai, and Samagar. With these, Boga went to the tent where Argon was confined, and told him that they repented of the part they had taken against him, and that in reparation of their error they had come to set him free and take him for their lord.

CHAPTER 59

How Argon recovered the sovereignty

When Argon heard Boga's words, he thought at first that they came to mock him, and was very angry and cross "Fair sirs," said he, "you sin greatly in making me an object of mockery, and ought to be satisfied with the wrong you have already done me in imprisoning your rightful lord You know that you are behaving wrongfully, and therefore I pray go your way and mock me no more." "Fair Sir Argon," said Boga, "be assured that we are not mocking you at all, but what we say is quite true, and we swear to it upon our faith." Then all the barons took an oath that they would hold him for their lord. And Argon on his side swore that he would never trouble them for what was past, but that he would hold them all as dear as his father Abaga had done. And as soon as these mutual oaths had been taken, they took Argon out of prison, and received him as their lord. Then Argon told them to shoot their arrows at the tent in which the melic who had the command of the army was, and they did so, and thus the melic was slain. This melic was named Soldan, and was the greatest lord after Acomat. Thus Argon recovered the sovereignty.

CHAPTER 60

How Argon caused his Uncle Acomat to be put to death

And when Argon found that he was assured of the sovereignty, he gave orders to the army to commence its march towards the court. It happened one day that Acomat was at court in his principal palace making great festivity, when a messenger came to him and said: "Sir, I bring you news, not such as I would, but very evil. Know that the barons have delivered Argon and raised him to the sovereignty, and have slain Soldan, your dear friend; and I assure you that they are hastening hither to take and slay you; take counsel immediately what is best to be done." When Acomat heard this, he was at first so overcome with astonishment and fear that he knew not what to do or say; but at last, like a brave and prudent man, he told the messenger to mention the news to no one, and hastily ordered his most trusty followers to arm and mount their horses; telling nobody whither he was going, he took the route to go to the Sultan of Babilonia, believing that there his life would be safe. At the end of six days he arrived at a pass which could not be avoided, the keeper of which knew that it was Acomat, and perceived that he was seeking safety by flight. This man determined to take him, which he might easily do, as he was slightly attended. When Acomat was thus arrested, he made great entreaty, and offered great treasure to be allowed to go free; but the keeper of the pass, who was a zealous partizan of Argon, replied that all the treasure in the world should not hinder him from doing his duty towards his rightful lord. He accordingly placed Acomat under a strong guard, and marching with him to the court, arrived there just three days after Argon had taken possession of it, who was greatly mortified that Acomat had escaped. When therefore Acomat was delivered to him a prisoner, he was in the greatest joy imaginable, and commanding the army to be assembled immediately, without consulting with anybody, he ordered one of his men to slay his uncle, and to throw his body into such place as it would never be seen again, which order was immediately executed. Thus ended the affair between Argon and his uncle Acomat.

CHAPTER 61

The death of Argon

When Argon had done all this, and had taken possession of the principal palace with the sovereignty, all the barons who had been in subjection to his father came to perform their homages as to their lord, and obeyed it as such in everything And after this, Argon sent Casan, his son, with full thirty thousand horsemen, to the Arbor Secco, which is in that country, to protect his land and people. Argon thus recovered his sovereignty in the year 1286 of the incarnation of Jesus Christ, and Acomat had held the sovereignty two years. Argon reigned six years, at the end of which he died, as was generally said, by poison.

CHAPTER 62

How Quiacatu seized upon the sovereignty after the death of Argon

When Argon was dead, his uncle, named Quiacatu, seized upon the sovereignty, which he was enabled to do with the more ease in consequence of Casan being so far distant as the Arbor Secco. Casan was greatly angered when he heard of the death of his father and of the usurpation of Quiacatu, but he could not leave his post at that moment for fear of his enemies. He threatened, however, that he would find the occasion to revenge himself as signally as his father had done upon Acomat. Quiacatu held the sovereignty, and all were obedient to him except those who were with Casan; and he took the wife of his nephew Argon and held her as his own, and enjoyed himself much with the ladies, for he was excessively given to his pleasures. Quiacatu held the sovereignty two years, at the end of which he was carried off by poison.

CHAPTER 63

How Baidu seized upon the sovereignty after the death of Quiacatu

When Quiacatu was dead, Baidu, who was his uncle, and a Christian, seized upon the sovereignty, and all obeyed him except Casan and the army with him. This occurred in the year 1294. When Casan learnt what had occurred, he was more furious against Baidu than he had been against Quiacatu, and, threatening to take such vengeance on him as should be talked of by everybody, he resolved that he would delay no longer, but march immediately against him. He accordingly provisioned his army, and commenced his march. When Baidu knew for certain that Casan was coming against him, he assembled a vast number of men, and marched forwards full ten days, and then encamped and waited for him to give battle. On the second day Casan appeared, and immediately there began a fierce battle, which ended in the entire defeat of Baidu, who was slain in the combat. Casan now assumed the sovereignty, and began his reign in the year 1294 of the Incarnation. Thus did the kingdom of the Eastern Tartars descend from Abaga to Casan, who now reigns.

CHAPTER 64

Of the Lords of the Tartars of the West

The first lord of the Tartars of the West was Sain, who was a very great and powerful king. He conquered Russia, and Comania, and Alania, and Lac, and Mengiar, and Zic, and Gucia, and Gazaria. All these provinces were conquered by king Sain. Before this conquest, they were all Comanians, but they were not under one government, and through their want of union they lost their lands, and were dispersed into different parts of the world, and those who remained were all in a state of serfdom to king Sain. After king Sain reigned king Patu, after him king Berca, next king Mungletemur, then king Totamongur, and lastly Toctai, who now reigns. Having thus given you a list of the kings of the Tartars of the West, we will tell you of a great battle that fell out between Alau, the lord of the East, and Berca, the lord of the West, as well as the cause of the battle, and its result.

CHAPTER 65

Of the war between Alau and Berca, and the battle they fought

In the year 1261 there arose a great quarrel between king Alau, lord of the Tartars of the East, and Berca, king of the Tartars of the West, on account of a province which bordered on each of their territories, which both claimed, and each was too proud to yield it to the other. They mutually defied each other, each declaring that he would go and take it, and he would see who dared hinder him. When things had come to this point, each summoned his followers to his banner, and they exerted themselves to such a degree that within six months each had assembled full three hundred thousand horsemen, very well furnished with all things appertaining to war according to their usage. Alau, lord of the East, now began his march with all his forces, and they rode many days without meeting with any adventure worth mentioning At length they reached an extensive plain, situated between the Iron Gates and the Sea of Sarain, in which they encamped in good order, and there was many a rich pavilion and tent. And there Alau said he would wait to see what course Berca would follow, as this spot was on the borders of the two territories.

CHAPTER 66

How Berca and his host went to meet Alau

Now when king Berca had made all his preparations, and knew that Alau was on his march, he also set out on his way, and in due time reached the same plain where his enemies awaited him, and encamped at about ten miles' distance from him. Berca's camp was quite as richly decked out as that of Alau, and his army was more numerous, for it numbered full three hundred and fifty thousand horsemen. The two armies rested two days, during which Berca called his people together, and addressed them as follows: "Fair sirs," said he, "you know certainly that since I came into possession of the land I have loved you like brothers and sons, and many of you have been in many great battles with me, and you have assisted me to conquer a great part of the lands we hold. You know that I share everything I have with you, and you ought in return to do your best to support my honour, which hitherto you have done. You know what a great and powerful man Alau is, and how in this quarrel he is in the wrong, and we are in the right, and each of you ought to feel assured that we shall conquer him in battle, especially as our number exceeds his; for we know for certain that he has only three hundred thousand horsemen, while we have three hundred and fifty thousand as good men as his and better. For all these reasons, then, you must see clearly that we shall gain the day, but since we have come so great a distance only to fight this battle, it is my will that we give battle three days hence, and we will proceed so prudently and in such good order that we cannot fail of success, and I pray you all to show yourselves on this occasion men of courage, so that all the world shall talk of your deeds. I say no more than that I expect every one of you to be well prepared for the day appointed."

CHAPTER 67

Alau's address to his men

When Alau knew certainly that Berca was come with so great an army, he also assembled his chiefs, and addressed them as follows: "Fair brothers, and sons, and friends," said he, "you know that all my life I have prized you and assisted you, and hitherto you have assisted me to conquer in many battles, nor ever were you in any battle where we failed to obtain the victory, and for that reason are we come here to fight this great man Berca, and I know well that he has more men than we have, but they are not so good, and I doubt not but we shall put them all to flight and discomfiture. We know by our spy that they intend to give us battle three days hence, of which I am very glad, and I pray you all to be ready on that day, and to demean yourselves as you used to do. One thing only I wish to impress upon you, that it is better to die on the field in maintaining our honour, than to suffer discomfiture; so let each of you fight so that our honour may be safe, and our enemies discomfited and slain."

Thus each of the kings encouraged his men, and waited for the day of the battle, and all prepared for it in the best way they could.

CHAPTER 68

Of the great battle between Alau and Berca

When the day fixed for the battle arrived, Alau rose early in the morning, and called his men to arms, and marshalled his army with the utmost skill. He divided it into thirty squadrons, each squadron consisting of ten thousand horsemen; and to each he gave a good leader and a good captain And when all this was duly arranged, he ordered his troops to advance, which they did at a slow pace, until they came half way between the two camps, where they halted and waited for the enemy. On the other side, king Berca had drawn up his army, which was arranged in thirty-five squadrons, exactly in the same manner as that of Alau's, and he also ordered his men to advance, which they did within half-a-mile of the others. There they made a short halt, and then they moved forward again till they came to the distance of about two arbalest shots of each other. It was a fair plain, and wonderfully extensive, as it ought to be, when so many thousands of men were marshalled in hostile array, under the two most powerful warriors in the world, who moreover were near kinsmen, for they were both of the imperial lineage of Jengiz-khan.

After the two armies had remained a short while in face of each other, the nacars at length sounded, upon which both armies let fly such a shower of arrows at each other that you could hardly see the sky, and many were slain, man and horse. When all their arrows were exhausted, they engaged with swords and maces, and then the battle was so fierce that the noise was louder than the thunder of heaven, and the ground was covered with corpses and reddened with blood. Both the kings distinguished themselves by their valour, and their men were not backward in imitating their example The battle continued in this manner till dusk, when Berca began to give way, and fled, and Alau's men pursued furiously, cutting down and slaying without mercy After they had pursued a short distance, Alau recalled them, and they returned to their tents, laid aside their arms, and dressed their wounds; and they were so weary with fighting, that they gladly sought repose Next morning Alau ordered the bodies of the dead to be buried, enemies as well as friends, and the loss was so great on both sides that it would be impossible to describe it. After this was done, Alau returned to his country with all his men who had survived the battle.

CHAPTER 69

How Totamangu was Lord of the Tartars of the West

You must know that in the West there was a king of the Tartars named Mongutemur[1], and the sovereignty descended to Tolobuga, who was a young bachelor, and a very powerful man, named Totamangu[1], slew Tolobuga, with the assistance of another king of the Tartars, named Nogai Thus Totamangu obtained the sovereignty by the aid of Nogai, and, after a short reign, he died, and Toctai, a very able and prudent man, was chosen king. Meanwhile the two sons of Tolobuga had grown to be now capable of bearing arms, and they were wise and prudent The two brothers assembled a very fair company, and went to the court of Toctai, and presented themselves with so much courtesy and humility on their knees that Toctai welcomed them, and told them to stand up. Then the eldest said to the king, "Fair sir Toctai, I will tell you in the best way I can why we are come to court. You know that we are the sons of Tolobuga, who was slain by Totamangu and Nogai Of Totamangu, I have nothing to say, since he is dead; but we claim justice on Nogai for the slaughter of our father, and we pray you as a righteous lord to grant it us. This is the object of our visit to your court."

CHAPTER 70

How Toctai sent for Nogai to Court

When Toctai had heard the youth, he knew that what he said was true, and he replied, "Fair friend, I will willingly yield to your demand of justice upon Nogai,

[1] These are the same names as spelt Mungletemur and Totamongur in Ch. 64, p. 336.

and for that purpose we will summon him to court, and do everything which justice shall require." Then Toctai sends two messengers to Nogai, and ordered him to come to court to answer to the sons of Tolobuga for the death of their father; but Nogai laughed at the message, and told the messengers he would not go. When Toctai heard Nogai's message, he was greatly enraged, and said in the hearing of all who were about him, "With the aid of God, either Nogai shall come before me to do justice to the sons of Tolobuga, or I will go against him with all my men and destroy him." He then sent two other messengers, who rode in all haste to the court of Nogai, and on their arrival they presented themselves before him and saluted him very courteously, and Nogai told them they were welcome. Then one of the messengers said: "Fair sir, Toctai sends you word that if you do not come to his court to render justice to the sons of Tolobuga, he will come against you with all his host, and do you all the hurt he can both to your property and person; therefore resolve what course you will pursue, and return him an answer by us." When Nogai heard Toctai's message, he was very angry, and replied to the messenger as follows: "Sir messenger," said he, "now return to your lord and tell him from me, that I have small fear of his hostility; and tell him further, that if he should come against me, I will wait for him at the entrance of my territory, for I will meet him half way. This is the message you shall carry back to your lord" The messenger hastened back, and when Toctai received this answer, he immediately sent his messengers to all parts which were under his rule, and summoned his people to be ready to go with him against king Nogai, and he had soon collected a great army. When Nogai knew certainly that Toctai was preparing to come against him with so large a host, he also made great preparation, but not so great as Toctai, because, though a great and powerful king, he was not so great or powerful as the other.

CHAPTER 71

How Toctai proceeded against Nogai

When Toctai's army was ready, he commenced his march at the head of two hundred thousand horsemen, and in due time reached the fine and extensive plain of Nerghi, where he encamped to wait for his opponent. With him were the two sons of Tolobuga, who had come with a fair company of horsemen to avenge the death of their father. Nogai also was on his march, with a hundred and fifty thousand horsemen, all young and brave men, and much better soldiers than those of Toctai He arrived in the plain where Toctai was encamped two days after him, and established his camp at a distance of ten miles from him. Then king Toctai assembled his chiefs, and said to them: "Sirs, we are come here to fight king Nogai and his men, and we have great reason to do so, for you know that all this hatred and rancour has arisen from Nogai's refusal to do justice to the sons of Tolobuga; and since our cause is just, we have every reason to hope for victory. Be therefore of good hope; but at all events I know that you are all brave men, and that you will do your best to destroy our enemies." Nogai also addressed his men in the following terms. "Fair brothers and friends," said he, "you know that we have gained many great and hard fought battles, and that we have overcome better men than these Therefore be of good cheer. We have right on our side; for you know well that Toctai was not my superior to summon me to his court to do justice to others. I will only further urge you to demean yourselves so in this battle that we shall be talked of everywhere, and that ourselves and our heirs will be the more respected for it." Next day they prepared for battle. Toctai drew up his army in twenty squadrons, each with a good leader and captain; and Nogai's army was formed in fifteen squadrons. After a long and desperate battle, in which the two kings, as well as the two sons of Tolobuga, distinguished themselves by their reckless valour, the army of Toctai was entirely defeated, and pursued from the field with great slaughter by Nogai's men, who, though less numerous, were much better soldiers than their opponents. Full sixty thousand men were slain in this battle, but king Toctai, as well as the two sons of Tolobuga, escaped.

Conclusion

AND now ye have heard all that we can tell you about the Tartars and the Saracens and their customs, and likewise about the other countries of the world as far as our researches and information extend Only we have said nothing whatever about the GREATER SEA and the provinces that lie round it, although we know it thoroughly. But it seems to me a needless and useless task to speak about places which are visited by people every day. For there are so many who sail all about that sea constantly, Venetians, and Genoese, and Pisans, and many others, that everybody knows all about it, and that is the reason that I pass it over and say nothing of it.

Of the manner in which we took our departure from the Court of the Great Khan you have heard at the beginning of the Book, in that chapter where we told you of all the vexation and trouble that Messer Maffeo and Messer Nicolo and Messer Marco had about getting the Great Khan's leave to go; and in the same chapter is related the lucky chance that led to our departure. And you may be sure that but for that lucky chance, we should never have got away in spite of all our trouble, and never have got back to our country again But I believe it was God's pleasure that we should get back in order that people might learn about the things that the world contains. For according to what has been said in the introduction at the beginning of the Book, there never was a man, be he Christian or Saracen or Tartar or Heathen, who ever travelled over so much of the world as did that noble and illustrious citizen of the City of Venice, Messer Marco the son of Messer Nicolo Polo.

Thanks be to God! Amen! Amen!

INDEX

The place-names are indexed in most cases under their modern forms

Ingram Content Group UK Ltd.
Milton Keynes UK
UKHW021117180423
420361UK00006B/574